D1327273

Molecular pathogenesis of virus infections

Virus and prion diseases remain a major public health threat, in both developed and developing countries. The worldwide HIV pandemic is but one example of a newly emerged virus disease; other potential threats come from exotic viruses such as SARS, Ebola and Hantaan viruses. Older human viruses such as influenza, papilloma, herpes and the hepatitis viruses still cause major health problems. Furthermore, as well as causing acute infections, some viruses may also establish persistent infections which can lead to the development of chronic diseases, including cancer. This symposium book covers central factors that influence the pathogenicity of virus and prion infections. Topics range from innate and adaptive immune responses and virus evasion of host defences to details of selected virus–host interactions, including those involving dengue virus, HIV, influenza viruses, coronaviruses, hepatitis C virus, herpesviruses, papillomaviruses, African swine fever virus and poxviruses.

Paul Digard is a Lecturer in Virology in the Department of Pathology at the University of Cambridge, UK.

Anthony A. Nash is Professor of Veterinary Pathology in the Division of Veterinary Biomedical Sciences at the University of Edinburgh, UK.

R. E. Randall is Professor of Molecular Virology in the School of Biology at the University of St Andrews, UK.

Symposia of the Society for General Microbiology

Managing Editor: Dr Melanie Scourfield, SGM, Reading, UK
Volumes currently available:

SIXTY-FOURTH SYMPOSIUM OF THE
SOCIETY FOR GENERAL MICROBIOLOGY
HELD AT HERIOT-WATT UNIVERSITY APRIL 2005

Edited by
P. Digard, A. A. Nash & R. E. Randall

molecular pathogenesis of virus infections

Published for the Society for General Microbiology

CAMBRIDGE UNIVERSITY PRESS
Cambridge, New York, Melbourne, Madrid, Cape Town,
Singapore, São Paulo

Cambridge University Press
The Edinburgh Building, Cambridge CB2 2RU, UK

Published in the United States of America by
Cambridge University Press, New York

www.cambridge.org
Information on this title: www.cambridge.org/9780521832489

First published 2005

Printed in the United Kingdom at the University Press, Cambridge

A catalogue record for this book is available from the British Library

ISBN 13   978 0 521 83248 9 hardback
ISBN 10   0 521 83248 9 hardback

Typeface   Sabon (Adobe) 10·5/13·5 pt   *System*   QuarkXPress™   [SGM]

Front cover illustration: Coloured scanning electron micrograph of a cluster of
coronavirus particles. Eye of Science / Science Photo Library.

CONTENTS

CONTRIBUTORS

Alcami, A.
Department of Medicine, University of Cambridge, Addenbrooke's Hospital, Cambridge, UK, and Department of Molecular and Cellular Biology, Centro Nacional de Biotecnología (CSIC), Campus Universidad Autónoma, Cantoblanco 28049 Madrid, Spain

Alejo, A.
Department of Medicine, University of Cambridge, Addenbrooke's Hospital, Cambridge, UK, and Department of Molecular and Cellular Biology, Centro Nacional de Biotecnología (CSIC), Campus Universidad Autónoma, Cantoblanco 28049 Madrid, Spain

Andino, R.
Department of Microbiology and Immunology, University of California, San Francisco, CA 94143-2280, USA

Barron, R. M.
Institute for Animal Health, Neuropathogenesis Unit, Ogston Building, West Mains Road, Edinburgh EH9 3JF, UK

Barry, G.
Centre for Infectious Diseases, College of Medicine and Veterinary Medicine, University of Edinburgh, Edinburgh EH9 1QH, UK

Borrow, P.
Viral Immunology Group, The Edward Jenner Institute for Vaccine Research, Compton, Newbury RG20 7NN, UK

Breakwell, L.
Centre for Infectious Diseases, College of Medicine and Veterinary Medicine, University of Edinburgh, Edinburgh EH9 1QH, UK

Decman, V.
Department of Ophthalmology and Graduate Program in Immunology, University of Pittsburgh School of Medicine, Pittsburgh, PA 15213, USA

Dixon, C. M.
Centre for Infectious Diseases, College of Medicine and Veterinary Medicine, University of Edinburgh, Edinburgh EH9 1QH, UK

Dixon, L. K.
Institute for Animal Health, Pirbright Laboratory, Ash Road, Pirbright, Woking GU24 0NF, UK

Dye, C.
Department of Molecular and Cellular Medicine, University of Bristol, Bristol BS8 1TD, UK

Ebrahimi, B.
Centre for Comparative Infectious Diseases, University of Liverpool, Duncan Building,
Daulby Street, Liverpool L69 3GA, UK

Fazakerley, J. K.
Centre for Infectious Diseases, College of Medicine and Veterinary Medicine, University of
Edinburgh, Edinburgh EH9 1QH, UK

Freeman, M. L.
Department of Ophthalmology and Graduate Program in Molecular Virology and
Microbiology, University of Pittsburgh School of Medicine, Pittsburgh, PA 15213, USA

Gray, L.
Bute Medical School, University of St Andrews, Bute Medical Buildings, Westburn Lane,
St Andrews, Fife KY16 9TS, UK

Haller, O.
Abteilung Virologie, Institut für Medizinische Mikrobiologie und Hygiene, Universität
Freiburg, D-79008 Freiburg, Germany

Hartman, A. L.
Special Pathogens Branch, Division of Viral and Rickettsial Diseases, Centers for Disease
Control and Prevention, Atlanta, GA 30306, USA

Hendricks, R. L.
Departments of Ophthalmology, Immunology and Molecular Genetics and Biochemistry,
University of Pittsburgh School of Medicine, Pittsburgh, PA 15213, USA

Herrington, C. S.
Bute Medical School, University of St Andrews, Bute Medical Buildings, Westburn Lane,
St Andrews, Fife KY16 9TS, UK

Hoffmann, E.
Division of Virology, Department of Infectious Diseases, St. Jude Children's Research
Hospital, 332 North Lauderdale St, Memphis, TN 38105, USA

Hughes, D.
Centre for Comparative Infectious Diseases, University of Liverpool, Duncan Building,
Daulby Street, Liverpool L69 3GA, UK

Jolly, C.
Bute Medical School, University of St Andrews, Bute Medical Buildings, Westburn Lane,
St Andrews, Fife KY16 9TS, UK

Kochs, G.
Abteilung Virologie, Institut für Medizinische Mikrobiologie und Hygiene, Universität
Freiburg, D-79008 Freiburg, Germany

Lemon, S. M.
Department of Microbiology & Immunology, Institute for Human Infections & Immunity, University of Texas Medical Branch, Galveston, TX 77555-0428, USA

Li, K.
Department of Microbiology & Immunology, Institute for Human Infections & Immunity, University of Texas Medical Branch, Galveston, TX 77555-0428, USA

Lipatov, A. S.
Division of Virology, Department of Infectious Diseases, St. Jude Children's Research Hospital, 332 North Lauderdale St, Memphis, TN 38105, USA

Manson, J. C.
Institute for Animal Health, Neuropathogenesis Unit, Ogston Building, West Mains Road, Edinburgh EH9 3JF, UK

Mongkolsapaya, J.
Department of Immunology, Hammersmith Hospital, Imperial College, Du Cane Road, London W12 0NN, UK

Nichol, S.
Special Pathogens Branch, Division of Viral and Rickettsial Diseases, Centers for Disease Control and Prevention, Atlanta, GA 30306, USA

Roaden, L.
Centre for Comparative Infectious Diseases, University of Liverpool, Duncan Building, Daulby Street, Liverpool L69 3GA, UK

Ruiz-Argüello, M. B.
Department of Medicine, University of Cambridge, Addenbrooke's Hospital, Cambridge, UK, and Centro de Investigación en Sanidad Animal (INIA), Valdeolmos, Madrid, Spain

Screaton, G.
Department of Immunology, Hammersmith Hospital, Imperial College, Du Cane Road, London W12 0NN, UK

Siddell, S.
Department of Molecular and Cellular Medicine, University of Bristol, Bristol BS8 1TD, UK

Stewart, J. P.
Centre for Comparative Infectious Diseases, University of Liverpool, Duncan Building, Daulby Street, Liverpool L69 3GA, UK

Towner, J. S.
Special Pathogens Branch, Division of Viral and Rickettsial Diseases, Centers for Disease Control and Prevention, Atlanta, GA 30306, USA

Turnbull, E.
Viral Immunology Group, The Edward Jenner Institute for Vaccine Research, Compton, Newbury RG20 7NN, UK

van Rij, R. P.
Department of Microbiology and Immunology, University of California, San Francisco, CA 94143-2280, USA

Weber, F.
Abteilung Virologie, Institut für Medizinische Mikrobiologie und Hygiene, Universität Freiburg, D-79008 Freiburg, Germany

Webster, R. G.
Division of Virology, Department of Infectious Diseases, St. Jude Children's Research Hospital, 332 North Lauderdale St, Memphis, TN 38105, USA

Whitton, J. L.
Department of Neuropharmacology, CVN-9, The Scripps Research Institute, 10550 N Torrey Pines Rd, La Jolla, CA 92037, USA

Adaptive immune responses

J. Lindsay Whitton

Department of Neuropharmacology, CVN-9, The Scripps Research Institute,
10550 N Torrey Pines Rd, La Jolla, CA 92037, USA

INTRODUCTION

Human urbanization, and the associated domestication of wild animals, afforded many viruses the opportunity to colonize, and better adapt to, our species. In the ensuing millennia, they have taken full advantage, killing more than a billion humans, and establish persistent infections with such efficiency that essentially all adult humans carry lifelong virus infections. Widespread immunization has reduced the toll of several acute virus infections: it has permitted the eradication of smallpox virus and the approaching extermination of poliovirus, and many young physicians have never encountered a patient with acute measles, mumps or rubella. Furthermore, as exemplified by the results of hepatitis B virus (HBV) vaccination in Taiwan (Huang & Lin, 2000), protection against acute virus infection can lead to a reduction in the number of persistently infected individuals (and, for HBV, yields an additional bounty – a lower incidence of the related hepatocellular carcinoma). However, despite the success of antiviral vaccination, infectious diseases are responsible for >20 % of deaths worldwide, and remain a leading cause of death even in so-called 'first world' countries.

Public health measures can dramatically reduce exposure to infectious organisms, classically illustrated by Dr John Snow in 1854, when a cholera outbreak in London was interrupted by his removing the handle of the Broad Street water pump (Cameron & Jones, 1983). But, if exposure occurs, what protects the individual from infection and disease? First come the barrier defences (skin and other epithelia, saliva, gastric acid), which constitute a physico-chemical barricade against intrusion. These are vitally important: consider how much more easily one becomes infected when the skin is

SGM symposium 64: Molecular pathogenesis of virus infections.
Editors P. Digard, A. A. Nash & R. E. Randall. Cambridge University Press. ISBN 0 521 83248 9 ©SGM 2005

breached, either by accident (an existing wound is an open invitation to microbes) or by design (penetration by the proboscis of an infected mosquito). Next in line is the multi-faceted innate immune system, whose weaponry includes natural killer (NK) cells, macrophages, polymorphonuclear leukocytes, complement and a plethora of cytokines. The subtlety and complexity of the innate immune response is only now being appreciated, and its role in suppressing virus infection has been highlighted by a recent elegant study of mouse cytomegalovirus (MCMV) infection of SCID mice, which lack an adaptive immune system. It has been known for some time that MCMV infection of SCID mice usually is rapidly lethal, indicating the importance of the adaptive immune response in protecting against this virus. However, occasional mice survive the infection, and appear to clear the virus, indicating the potential effectiveness of innate immunity. Recent work focused on the ultimate fate of these rare survivors, and revealed an ongoing battle; several weeks after their apparent recovery, virus reappeared, and the animals succumbed. DNA sequencing, and other data, indicated that the recrudescent viruses were variants that carried mutations which prevented their recognition by NK cells (French *et al.*, 2004). This is the first study to show that DNA viruses evolve in response to pressure from NK cells, and it further underlines the importance of innate immune functions for holding viruses in check. Nevertheless, the barrier and innate immune systems often provide only incomplete protection, and an additional key component protecting us against microbial onslaught is the adaptive immune system. The importance of adaptive immunity is best revealed by two simple facts. *First*, individuals in whom adaptive immunity is compromised (e.g. SCID mice; or humans with agammaglobulinaemia) are at greatly increased risk of both infection and severe disease. *Second*, the success of vaccination – arguably, the most important single medical advance in history – relies almost entirely on the stimulation of adaptive immune responses. In this chapter, I shall outline the nature of the adaptive immune response, then will describe recent studies from my laboratory, investigating the regulation and maturation of one aspect of the adaptive response over the course of an infection.

THE ADAPTIVE IMMUNE RESPONSE DEPENDS ON LYMPHOCYTES

The adaptive immune response is mediated by lymphocytes, which are driven to expand, and to express their effector functions, by contact with specific antigenic moieties that can be recognized by receptors on the lymphocyte membrane. Usually, on any one lymphocyte, all of the surface receptors are identical; therefore – at least broadly speaking – a lymphocyte can respond to only one antigen (i.e. lymphocytes are *antigen-specific*; and the moiety that a lymphocyte recognizes is termed its *cognate* antigen). However, the receptors vary subtly from one lymphocyte to the next, ensuring that the host can recognize, and mount an immune response to, an enormous diversity of antigens. Any given virus contains several antigens and, therefore, the adaptive immune response to a virus comprises a collection of several distinct antigen-specific responses, each of which is reliant on a lymphocyte expressing appropriate receptor proteins. Upon first encounter with cognate antigen, the host mounts a *primary immune response*, in which antigen-specific effector functions will be somewhat slow to develop, usually becoming detectable only after several days; and this relatively slow response often gives the virus sufficient time to replicate, disseminate, cause illness and spread to the next susceptible host. The somewhat lethargic nature of the primary response results from (i) the host having only a few naïve precursor lymphocytes specific for any one antigen, which means that some time must pass before these antigen-specific lymphocytes can multiply to a biologically meaningful number and (ii) the slow expression of effector functions by the newly activated lymphocytes. In individuals who survive the infection, the primary response wanes (hopefully, after eradication of the microbe), but antigen-specific *memory* lymphocytes remain. These antigen-specific cells are much greater in number than were their naïve progenitors and (as described later for CD8[+] T cells) have optimized their effector functions, so, if the host is again exposed to the same virus, these memory cells allow the host to mount a greatly accelerated *secondary immune response*, which usually rapidly shuts down virus replication. The resulting rapid control of infection has at least two favourable consequences: not only does it protect the infected host against disease, it also diminishes the likelihood of virus transmission to susceptible individuals in the community (this latter benefit is termed herd immunity). Memory lymphocytes are induced not only by infection, but also by immunization, and *these cells serve as the cornerstone of all vaccines.*

Two classes of lymphocyte are involved in the adaptive immune response: B lymphocytes (so named because they were first identified in the avian organ, the bursa of Fabricius) and T lymphocytes (which are derived from the thymus). T lymphocytes can be further subdivided into two groups, characterized by the cell-surface expression of accessory proteins named CD4 and CD8; generally, mature T cells express only one of these proteins. B lymphocytes and their progeny (including plasma cells) are the source of antibodies, soluble effector molecules which act mainly to diminish the infectivity of cell-free virus. In contrast, CD8[+] T lymphocytes eradicate virus-infected cells (the role of CD4[+] T cells is reviewed briefly later). Thus antibodies and CD8[+] T lymphocytes act in a complementary manner (reviewed by Whitton & Oldstone, 2001): antibodies neutralize viruses in the fluid phase (e.g. blood, interstitial spaces), thereby reducing the number of cells that become infected; and CD8[+] T lymphocytes exert their antiviral effects upon infected cells, thereby reducing the production and release of virus into the extracellular milieu. Both antibodies (B cells) and T cells are key components of the antiviral immune response, but in this chapter I shall concentrate on T cells, as they are a major research focus of my laboratory.

T CELLS RECOGNIZE FRAGMENTS OF VIRUS PROTEINS DISPLAYED ON THE CELL SURFACE BY MHC MOLECULES

As noted earlier, each lymphocyte expresses multiple copies of a unique receptor which dictates the antigen specificity of the cell. For both CD4$^+$ and CD8$^+$ T cells, this T-cell receptor (TcR) recognizes short (9–24 amino acid) peptide fragments (*epitopes*) generated inside the 'target' cell by degradation of viral proteins. These epitopes are displayed on the target cell surface, for T-cell perusal, by host glycoproteins encoded in the major histocompatibility complex (MHC); the epitope/MHC complex is, therefore, the antigen for which a TcR (and its lymphocyte) is specific. There are two types of MHC glycoprotein, termed MHC class I and MHC class II. These classes differ in several ways, of which three are particularly important.

(i) Interactions with CD4 or CD8 proteins

MHC molecules on target cells not only present epitopes for TcR recognition, but also are directly recognized by these T-cell accessory proteins. Conserved regions on MHC class I molecules interact directly with CD8 (Salter *et al.*, 1990), and there is a similar interaction between MHC class II proteins and CD4 (Konig *et al.*, 1992). Thus CD8$^+$ T cells can exert their activities only on target cells that express MHC class I; and, similarly, the effects of CD4$^+$ T cells are limited to target cells that express MHC class II.

(ii) Anatomical distribution

MHC class I molecules are expressed on the surface of almost all somatic cells, but expression of MHC class II molecules is restricted to specialized antigen presenting cells (APCs; these cells usually also express class I MHC).

(iii) Source of epitopes

MHC class I molecules present epitopes from proteins *synthesized inside* a target cell, which means that CD8$^+$ T cells can recognize virus-infected cells. In contrast, epitopes presented by MHC class II molecules are derived from proteins *that have been taken up from the extracellular milieu*. This protein uptake and degradation is the purview of APCs, and these cells, therefore, can be recognized by CD4$^+$ T cells even though they are not actively infected.

These three differences in MHC have profound biological implications for their T-cell partners. CD8$^+$ T cells can, in principle, recognize (and exert their antiviral effects

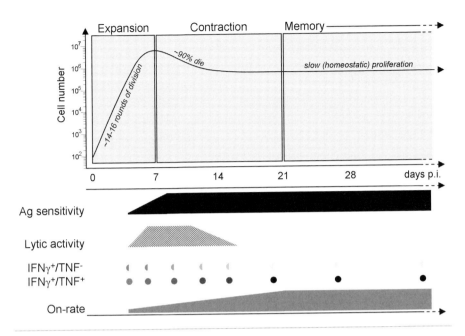

Fig. 1. Quantitative and qualitative summary of the antiviral CD8$^+$ T cell response. The upper part of the figure shows a quantitative summary of a strong CD8$^+$ T cell response against a single viral epitope, assuming a starting number of 100 naïve epitope-specific precursor cells. Note that the y-axis is a $\log_{10}$ scale. Qualitative changes in four aspects of CD8$^+$ T cell effector function, discussed in the text, are shown in the lower part of the figure.

upon) almost any infected cell; thus CD8$^+$ T cells are the foot soldiers of the immune response, engaging in hand-to-hand combat against viruses within cells. In contrast, CD4$^+$ T cells are unable to recognize the majority of infected cells (which do not express MHC class II), and so they seem unlikely to play a direct role in eradicating the microbe; these cells instead orchestrate the antiviral campaign. The majority of studies in my laboratory are aimed at understanding CD8$^+$ T cell responses; these are, therefore, the topic of the remainder of this chapter.

THE THREE PHASES OF THE ANTIVIRAL CD8$^+$ T CELL RESPONSE

The quantitative and qualitative changes that take place in a single population of epitope-specific CD8$^+$ T cells over the course of a virus infection, and beyond, are summarized in Fig. 1. In this figure, the number of precursor cells (i.e. in a naïve mouse) that are specific for a single epitope is assumed to be 10^2, consistent with published data (Blattman *et al.*, 2002). The antiviral CD8$^+$ T cell response is traditionally considered as having three overlapping phases: expansion, contraction and memory.

The expansion phase

Naïve CD8+ T cells express high levels of proteins such as CD62L (Gallatin *et al.*, 1983) and CCR7, which mediate adhesion to lymph node venules, and cause the naïve cells to be retained within lymphoid tissues, which are rich in APCs (Baekkevold *et al.*, 2001). When a naïve antigen-specific CD8+ T cell encounters its cognate antigen presented by an APC, a TcR-dependent signal triggers a programme which leads to the cell's division and differentiation. T-cell activation results in the rapid down-regulation of CD62L and CCR7, and this (along with other factors) allows the newly activated cell to exit the node. Recent studies indicate that a few hours of antigen exposure can trigger a CD8+ T cell to complete its entire developmental programme in the apparent absence of additional antigenic contact (Kaech & Ahmed, 2001; Mercado *et al.*, 2000; van Stipdonk *et al.*, 2001). This rapid loss of antigen dependence implies that a newly triggered CD8+ T cell can very quickly be released from lymphoid tissues (where antigen is most abundant), to take up its duties in peripheral sites, while continuing to divide and differentiate. T cells require only ~6 h between each division, so each activated precursor generates many thousands of progeny cells. The virus-specific T-cell response usually peaks around 7–10 days after infection, after which the cell numbers decline.

The contraction phase (aka the death phase)

This phase – the least well understood of the three – begins as early as 6 days post-infection (and, therefore, overlaps with the end of the expansion phase). Approximately 90 % of T cells die during the contraction phase, a dramatic attrition that is complete by ~21 days post-infection (Badovinac *et al.*, 2002; Badovinac & Harty, 2002; Kaech *et al.*, 2002; Sprent & Surh, 2002). The mechanisms regulating T-cell death remain controversial, but the most widely accepted theory is that death is apoptotic, and results from one of two mechanisms: activation-induced cell death (AICD) or cytokine withdrawal [activated T cell autonomous death (ACAD)]. The probability that one or both of these apoptotic pathways will be activated in any one CD8+ T cell may be determined by, for example, the cell's history (number of prior divisions, antigen contact) and the surrounding microenvironment. Timing appears to play a critical part, as revealed by a recent study in which IL-2 was administered at different times after virus infection. When administered during the expansion phase, this cytokine increased T-cell death (consistent with its inducing AICD), but the same treatment during the contraction phase led to increased T-cell survival (consistent with its preventing ACAD; Blattman *et al.*, 2003).

The memory phase

As noted earlier, CD8+ memory T cells are important in protecting against many viral infections, and experimental vaccines that induce only CD8+ memory T cells (but no antibodies) can confer good protection against viral challenge (del Val *et al.*, 1991; Klavinskis *et al.*, 1989; Whitton *et al.*, 1993). The number of memory cells induced by an infection is related to the degree of CD8[+] T cell expansion during the primary response (Hou *et al.*, 1994; Marshall *et al.*, 2001) and, under normal circumstances, the resting level of memory cells remains relatively stable for months or years after infection or vaccination. This stable level of CD8[+] memory T cells requires ongoing homeostatic division that is regulated by cytokines, in particular by IL-7 and IL-15 (Becker *et al.*, 2002; Schluns *et al.*, 2000, 2002; Tan *et al.*, 2002), and does not require ongoing antigen contact (Lau *et al.*, 1994; Murali-Krishna *et al.*, 1999). Memory T cells are often relatively abundant in non-lymphoid tissues (Mackay *et al.*, 1992; Sprent, 1976). Intuitively, this makes sense: most virus infections begin at mucosal surfaces, and the presence of virus-specific effector cells at or near these locations might improve the host's ability to rapidly eradicate the invaders. However, the ability of memory cells to immediately express certain effector functions is somewhat controversial. For example, some studies suggest that CD8[+] memory T cells in lung and liver might be immediately cytolytic (Masopust *et al.*, 2001), while others have reported that virus-specific CD8[+] memory cells in the lung are not immediately cytolytic (Hogan *et al.*, 2001; Ostler *et al.*, 2001). The factors that regulate the numbers of primary and memory CD8[+] T cells resident in peripheral tissues, and the expression of their effector functions, are important topics; their identification may facilitate the development of better therapeutic and prophylactic approaches to controlling virus infection.

THE ANTIVIRAL EFFECTOR FUNCTIONS OF CD8[+] T CELLS

Perhaps the best-known antiviral effector function of virus-specific CD8[+] T lymphocytes is their cytolytic activity; these cells often can kill infected target cells, usually by secreting the pore-forming protein perforin. Indeed, for this reason, these cells often are termed 'cytotoxic T lymphocytes', and the resulting acronym (CTL) often is used as synonym for all CD8[+] T cells. Perforin-mediated cytotoxicity is required for the clearance of several virus infections [e.g. lymphocytic choriomeningitis virus (LCMV): Kagi *et al.*, 1994; Walsh *et al.*, 1994]. As shown in Fig. 1, CTL activity usually is first detected at ~4–5 days post-infection, peaks at around 7–10 days, and declines quite quickly thereafter; we have shown that, by ~15 days p.i., most of the remaining virus-specific CD8[+] T cells have lost their capacity to rapidly lyse infected target cells (Rodriguez *et al.*, 2001). However, cytolytic activity is not always the primary means by which CD8[+] T cells control virus infections, and it plays little part in eradicating vaccinia, Semliki Forest and vesicular stomatitis viruses (Kagi *et al.*, 1995), rotaviruses (Franco *et al.*, 1997), coxsackieviruses (Gebhard *et al.*, 1998) and HBV (Guidotti *et al.*, 1996, 1999). In these, and other, cases, CD8[+] T cells can control virus infection by a non-lytic mechanism; they can secrete cytokines, such as interferon-gamma (IFN-γ) and tumour necrosis factor (TNF), which can directly reduce virus replication, and may even 'cure' infected cells by inactivating virus replication in the absence of cell death (Estcourt et al., 1998; Guidotti & Chisari, 1996; Levy et al., 1996; reviewed by Slifka & Whitton, 2000a; Walker et al., 1991). Thus, in terms of the effector functions (lytic activity/cytokine synthesis) that are induced by antigen contact, CD8[+] T cells fall into four broad classes: (i) non-lytic, and unable to immediately make abundant cytokines (typical of naïve precursor cells); (ii) lytic, but unable to make cytokines (such cells are unusual); (iii) lytic and cytokine[competent] (fully fledged antiviral cells, found between ~7 and 15 days post-infection); and (iv) non-lytic, but cytokine[competent] (as shown diagrammatically in Fig. 1, after day 15 most virus-specific CD8[+] T cells appear to be in this category). The existence of class (iv) appears to be overlooked by some investigators, as shown by two misnomers that are rampant in the literature. *First*, many investigators reserve the term 'effector cell' for a CD8[+] T cell that is lytic, thereby implying that non-lytic cells are 'non-effectors'; however, cells in class (iv) produce antiviral cytokines and, even in the absence of immediate lytic activity, can be profoundly antiviral, and most certainly deserve the 'effector' title. *Second*, the existence of these cells indicates that the term CTL is an inappropriate synonym for virus-specific CD8[+] T cells; in my view, 'CTL' should be used only to describe cells for which lytic activity has been demonstrated.

CD8[+] T CELL EFFECTOR FUNCTIONS MATURE OVER THE COURSE OF INFECTION

Antibody responses mature over the course of infection; there is a change both in the type of antibody produced ('class switching') and in the affinity of the antigen-binding region, which can increase dramatically following re-encounter with the antigen. The mechanisms underlying these antibody maturation processes are reviewed elsewhere, and are mentioned here mainly as a counterpoint to CD8[+] T cells, which, until recently, were not known to undergo a marked maturation over the course of infection. However, my laboratory has identified three distinct ways in which CD8[+] T cell responses improve over time.

CD8[+] T cells become more sensitive to low levels of epitope/MHC on target cells

We investigated the antigen-responsiveness of virus-specific T cells at various times post-infection, and found that, between ~4 and ~8 days post-infection, the quantity of epitope/MHC needed to trigger cytokine production or cytolytic activity by virus-specific CD8[+] T cells diminished by ~70-fold, and remained stable thereafter, essentially for the lifetime of the animal (Slifka & Whitton, 2001). How might this benefit the host? We proposed that, by optimizing their ability to be triggered by very low levels of epitope/MHC, CD8[+] T cells ensure that they can recognize cells very soon (minutes/hours) after they have become infected, at a time when epitope levels on the infected cell surface are still rising. Recent studies have shown that as few as 10 peptide/MHC complexes may be sufficient to stimulate highly activated T cells (Irvine *et al.*, 2002), and this extraordinary sensitivity should allow the T cells' effector functions (cytokine production, cytolytic activity) to be exerted prior to virus maturation, thereby preventing the production and release of infectious particles. However, if a T cell can see as few as 10 epitope/MHC complexes, this is close to the lowest possible limit of antigen concentration on the cell surface, and it is difficult to conceive of any significant further enhancement of T-cell sensitivity. How else might T cells improve their biological functions?

CD8$^+$ T cells accelerate the initiation of IFN-γ production

We reasoned that, having maximized their ability to be triggered by low levels of antigen, T cells might improve their biological efficacy by improving their cytokine production (for example, by producing more cytokine, and/or by increasing the speed with which they begin cytokine production). We have evaluated the speed with which various epitope-specific populations of CD8$^+$ T cells can initiate IFN-γ synthesis in response to antigen contact by measuring their 'on-rate' (Liu *et al.*, 2004). For some epitope-specific populations, the initiation of cytokine synthesis became progressively faster over the course of infection, reaching optimal performance by ~21 days p.i., and retaining it thereafter. Typically, the on-rate of IFN-γ synthesis at day 21 was ~two- to fourfold faster than at day 8, an improvement of ~1–3 h. This change may seem small, but we consider it likely to be biologically significant, because the life cycle of most viruses is very short (a few hours); consequently, even a small increase in the speed of cytokine production will increase the probability that virus replication in the target cell can be interrupted, thereby substantially benefiting the host.

Different patterns of cytokine production by CD8$^+$ primary and memory T cells

Some time ago, we showed that IFN-γ and TNF production by CD8$^+$ T cells takes place only when the cell is in contact with cognate antigen (Slifka *et al.*, 1999). However, the pattern of cytokines produced by virus-specific CD8$^+$ T cells changes over the course of virus infection (Slifka & Whitton, 2000b). During the primary immune response to LCMV infection, two broad populations of CD8$^+$ T cells can be distinguished: one produces only IFN-γ, while the other produces both IFN-γ and TNF. As the primary response declines, the ratio of these two populations changes, and double-positive cells outnumber single-positive cells by ~5 : 1; and this process continues into the memory phase, at which time almost all cells respond to antigen contact by immediately producing IFN-γ and TNF. These observations have been confirmed by others, using models of influenza virus infection (Belz *et al.*, 2001) and murine gamma herpesvirus infection (Liu *et al.*, 2002). If this difference in cytokine responses between primary and memory cells is, indeed, a characteristic of most infections, then it seems likely to be biologically significant; however, this remains to be proven.

In conclusion, CD8$^+$ T cell responses to virus infection are important. Many laboratories have focused on quantitative analyses of cell numbers in primary or secondary lymphoid tissues and, although groundbreaking, these studies are only now being complemented by thorough investigations of CD8$^+$ cell numbers in peripheral sites. Furthermore, our understanding of the subtleties of T-cell effector functions, in both lymphoid and non-lymphoid tissues, remains rudimentary, and such qualitative evaluations will, doubtless, yield many fascinating results in the years to come.

ACKNOWLEDGEMENTS

This work, supported by NIH R-01 grants AI 27028 and AI 52351, could not have been done without the invaluable contributions of many past and present members of my laboratory, who are appropriately named on the related primary research papers. I thank all of them for their outstanding efforts. I am grateful to Annette Lord for excellent secretarial support. This is manuscript number 16905-NP from the Scripps Research Institute.

REFERENCES

Badovinac, V. P. & Harty, J. T. (2002). CD8$^+$ T-cell homeostasis after infection: setting the 'curve'. *Microbes Infect* **4**, 441–447.

Badovinac, V. P., Porter, B. B. & Harty, J. T. (2002). Programmed contraction of CD8$^+$ T cells after infection. *Nat Immunol* **3**, 619–626.

Baekkevold, E. S., Yamanaka, T., Palframan, R. T., Carlsen, H. S., Reinholt, F. P., von Andrian, U. H., Brandtzaeg, P. & Haraldsen, G. (2001). The CCR7 ligand elc (CCL19) is transcytosed in high endothelial venules and mediates T cell recruitment. *J Exp Med* **193**, 1105–1112.

Becker, T. C., Wherry, E. J., Boone, D., Murali-Krishna, K., Antia, R., Ma, A. & Ahmed, R. (2002). Interleukin 15 is required for proliferative renewal of virus-specific memory CD8 T cells. *J Exp Med* **195**, 1541–1548.

Belz, G. T., Xie, W. & Doherty, P. C. (2001). Diversity of epitope and cytokine profiles for primary and secondary influenza a virus-specific CD8$^+$ T cell responses. *J Immunol* **166**, 4627–4633.

Blattman, J. N., Antia, R., Sourdive, D. J., Wang, X., Kaech, S. M., Murali-Krishna, K., Altman, J. D. & Ahmed, R. (2002). Estimating the precursor frequency of naive antigen-specific CD8 T cells. *J Exp Med* **195**, 657–664.

Blattman, J. N., Grayson, J. M., Wherry, E. J., Kaech, S. M., Smith, K. A. & Ahmed, R. (2003). Therapeutic use of IL-2 to enhance antiviral T-cell responses *in vivo*. *Nat Med* **9**, 540–547.

Cameron, D. & Jones, I. G. (1983). John Snow, the broad street pump and modern epidemiology. *Int J Epidemiol* **12**, 393–396.

del Val, M., Schlicht, H. J., Volkmer, H., Messerle, M., Reddehase, M. J. & Koszinowski, U. H. (1991). Protection against lethal cytomegalovirus infection by a recombinant vaccine containing a single nonameric T-cell epitope. *J Virol* **65**, 3641–3646.

Estcourt, M. J., Ramshaw, I. A. & Ramsay, A. J. (1998). Cytokine responses in virus infections: effects on pathogenesis, recovery and persistence. *Curr Opin Microbiol* **1**, 411–418.

Franco, M. A., Tin, C., Rott, L. S., van Cott, J. L., McGhee, J. R. & Greenberg, H. B. (1997). Evidence for CD8[+] T-cell immunity to murine rotavirus in the absence of perforin, fas, and γ interferon. *J Virol* **71**, 479–486.

French, A. R., Pingel, J. T., Wagner, M., Bubic, I., Yang, L., Kim, S., Koszinowski, U., Jonjic, S. & Yokoyama, W. M. (2004). Escape of mutant double-stranded DNA virus from innate immune control. *Immunity* **20**, 747–756.

Gallatin, W. M., Weissman, I. L. & Butcher, E. C. (1983). A cell-surface molecule involved in organ-specific homing of lymphocytes. *Nature* **304**, 30–34.

Gebhard, J. R., Perry, C. M., Harkins, S., Lane, T., Mena, I., Asensio, V. C., Campbell, I. L. & Whitton, J. L. (1998). Coxsackievirus B3-induced myocarditis: perforin exacerbates disease, but plays no detectable role in virus clearance. *Am J Pathol* **153**, 417–428.

Guidotti, L. G. & Chisari, F. V. (1996). To kill or to cure: options in host defense against viral infection. *Curr Opin Immunol* **8**, 478–483.

Guidotti, L. G., Ishikawa, T., Hobbs, M. V., Matzke, B., Schreiber, R. & Chisari, F. V. (1996). Intracellular inactivation of the hepatitis B virus by cytotoxic T lymphocytes. *Immunity* **4**, 25–36.

Guidotti, L. G., Rochford, R., Chung, J., Shapiro, M., Purcell, R. & Chisari, F. V. (1999). Viral clearance without destruction of infected cells during acute HBV infection. *Science* **284**, 825–829.

Hogan, R. J., Usherwood, E. J., Zhong, W., Roberts, A. A., Dutton, R. W., Harmsen, A. G. & Woodland, D. L. (2001). Activated antigen-specific CD8[+] T cells persist in the lungs following recovery from respiratory virus infections. *J Immunol* **166**, 1813–1822.

Hou, S., Hyland, L., Ryan, K. W., Portner, A. & Doherty, P. C. (1994). Virus-specific CD8[+] T-cell memory determined by clonal burst size. *Nature* **369**, 652–654.

Huang, K. & Lin, S. (2000). Nationwide vaccination: a success story in Taiwan. *Vaccine* **18** (Suppl. 1), S35–S38.

Irvine, D. J., Purbhoo, M. A., Krogsgaard, M. & Davis, M. M. (2002). Direct observation of ligand recognition by T cells. *Nature* **419**, 845–849.

Kaech, S. M. & Ahmed, R. (2001). Memory CD8[+] T cell differentiation: initial antigen encounter triggers a developmental program in naïve cells. *Nat Immunol* **2**, 415–422.

Kaech, S. M., Wherry, E. J. & Ahmed, R. (2002). Effector and memory T-cell differentiation: implications for vaccine development. *Nat Rev Immunol* **2**, 251–262.

Kagi, D., Ledermann, B., Burki, K., Seiler, P., Odermatt, B., Olsen, K. J., Podack, E. R., Zinkernagel, R. M. & Hengartner, H. (1994). Cytotoxicity mediated by T cells and natural killer cells is greatly impaired in perforin-deficient mice. *Nature* **369**, 31–37.

Kagi, D., Seiler, P., Pavlovic, J., Ledermann, B., Burki, K., Zinkernagel, R. M. & Hengartner, H. (1995). The roles of perforin- and Fas-dependent cytotoxicity in protection against cytopathic and noncytopathic viruses. *Eur J Immunol* **25**, 3256–3262.

Klavinskis, L. S., Oldstone, M. B. A. & Whitton, J. L. (1989). Designing vaccines to induce cytotoxic T lymphocytes: protection from lethal viral infection. In *Vaccines 89. Modern Approaches to New Vaccines Including Prevention of AIDS*, pp. 485–489.

Edited by F. Brown, R. Chanock, H. Ginsberg & R. Lerner. Cold Spring Harbor, NY: Cold Spring Harbor Laboratory.

Konig, R., Huang, L. Y. & Germain, R. N. (1992). MHC class II interaction with CD4 mediated by a region analogous to the MHC class I binding site for CD8. *Nature* **356**, 796–798.

Lau, L. L., Jamieson, B. D., Somasundaram, T. & Ahmed, R. (1994). Cytotoxic T-cell memory without antigen. *Nature* **369**, 648–652.

Levy, J. A., Mackewicz, C. E. & Barker, E. (1996). Controlling HIV pathogenesis: the role of the noncytotoxic anti-HIV response of CD8$^+$ T cells. *Immunol Today* **17**, 217–224.

Liu, H., Andreansky, S., Diaz, G., Hogg, T. & Doherty, P. C. (2002). Reduced functional capacity of CD8$^+$ T cells expanded by post-exposure vaccination of gamma-herpesvirus-infected CD4-deficient mice. *J Immunol* **168**, 3477–3483.

Liu, F., Whitton, J. L. & Slifka, M. K. (2004). The rapidity with which virus-specific CD8$^+$ T cells initiate IFNγ synthesis increases markedly over the course of infection, and correlates with immunodominance. *J Immunol* **173**, 456–462.

Mackay, C. R., Marston, W. L., Dudler, L., Spertini, O., Tedder, T. F. & Hein, W. R. (1992). Tissue-specific migration pathways by phenotypically distinct subpopulations of memory T cells. *Eur J Immunol* **22**, 887–895.

Marshall, D. R., Turner, S. J., Belz, G. T. & 7 other authors (2001). Measuring the diaspora for virus-specific CD8$^+$ T cells. *Proc Natl Acad Sci U S A* **98**, 6313–6318.

Masopust, D., Vezys, V., Marzo, A. L. & Lefrancois, L. (2001). Preferential localization of effector memory cells in nonlymphoid tissue. *Science* **291**, 2413–2417.

Mercado, R., Vijh, S., Allen, S. E., Kerksiek, K., Pilip, I. M. & Pamer, E. G. (2000). Early programming of T cell populations responding to bacterial infection. *J Immunol* **165**, 6833–6839.

Murali-Krishna, K., Lau, L. L., Sambhara, S., Lemonnier, F., Altman, J. D. & Ahmed, R. (1999). Persistence of memory CD8 T cells in MHC class I-deficient mice. *Science* **286**, 1377–1381.

Ostler, T., Hussell, T., Surh, C. D., Openshaw, P. & Ehl, S. (2001). Long-term persistence and reactivation of T cell memory in the lung of mice infected with respiratory syncytial virus. *Eur J Immunol* **31**, 2574–2582.

Rodriguez, F., Slifka, M. K., Harkins, S. & Whitton, J. L. (2001). Two overlapping subdominant epitopes identified by DNA immunization induce protective CD8$^+$ T-cell populations with differing cytolytic activities. *J Virol* **75**, 7399–7409.

Salter, R. D., Benjamin, R. J., Wesley, P. K. & 7 other authors (1990). A binding site for the T cell co-receptor CD8 on the α3 domain of HLA-A2. *Nature* **345**, 41–46.

Schluns, K. S., Kieper, W. C., Jameson, S. C. & Lefrancois, L. (2000). Interleukin-7 mediates the homeostasis of naive and memory CD8 T cells in vivo. *Nat Immunol* **1**, 426–432.

Schluns, K. S., Williams, K., Ma, A., Zheng, X. X. & Lefrancois, L. (2002). Cutting edge: requirement for IL-15 in the generation of primary and memory antigen-specific CD8 T cells. *J Immunol* **168**, 4827–4831.

Slifka, M. K. & Whitton, J. L. (2000a). Antigen-specific regulation of T cell-mediated cytokine production. *Immunity* **12**, 451–457.

Slifka, M. K. & Whitton, J. L. (2000b). Activated and memory CD8$^+$ T cells can be distinguished by their cytokine profiles and phenotypic markers. *J Immunol* **164**, 208–216.

Slifka, M. K. & Whitton, J. L. (2001). Functional avidity maturation of CD8$^+$ T cells without selection of higher affinity TCR. *Nat Immunol* **2**, 711–717.

Slifka, M. K., Rodriguez, F. & Whitton, J. L. (1999). Rapid on/off cycling of cytokine production by virus-specific CD8$^+$ T cells. *Nature* **401**, 76–79.

Sprent, J. (1976). Fate of H2-activated T lymphocytes in syngeneic hosts. I. Fate in lymphoid tissues and intestines traced with 3H-thymidine, 125I-deoxyuridine and 51chromium. *Cell Immunol* **21**, 278–302.

Sprent, J. & Surh, C. D. (2002). T cell memory. *Annu Rev Immunol* **20**, 551–579.

Tan, J. T., Ernst, B., Kieper, W. C., LeRoy, E., Sprent, J. & Surh, C. D. (2002). Interleukin (IL)-15 and IL-7 jointly regulate homeostatic proliferation of memory phenotype CD8$^+$ cells but are not required for memory phenotype CD4+ cells. *J Exp Med* **195**, 1523–1532.

van Stipdonk, M. J., Lemmens, E. E. & Schoenberger, S. P. (2001). Naive CTLs require a single brief period of antigenic stimulation for clonal expansion and differentiation. *Nat Immunol* **2**, 423–429.

Walker, C. M., Erickson, A. L., Hsueh, F. C. & Levy, J. A. (1991). Inhibition of human immunodeficiency virus replication in acutely infected CD4$^+$ cells by CD8$^+$ cells involves a noncytotoxic mechanism. *J Virol* **65**, 5921–5927.

Walsh, C. M., Matloubian, M., Liu, C. C. & 7 other authors (1994). Immune function in mice lacking the perforin gene. *Proc Natl Acad Sci U S A* **91**, 10854–10858.

Whitton, J. L. & Oldstone, M. B. A. (2001). The immune response to viruses. In *Fields Virology*, 4th edn, pp. 285–320. Edited by D. M. Knipe & P. M. Howley. Philadelphia: Lippincott Williams & Wilkins.

Whitton, J. L., Sheng, N., Oldstone, M. B. A. & McKee, T. A. (1993). A 'string-of-beads' vaccine, comprising linked minigenes, confers protection from lethal-dose virus challenge. *J Virol* **67**, 348–352.

T-cell responses and dengue haemorrhagic fever

Gavin Screaton and Juthathip Mongkolsapaya

Department of Immunology, Hammersmith Hospital, Imperial College, Du Cane Road, London W12 0NN, UK

Infection by dengue virus has become a major public health threat in tropical and subtropical countries. The virus belongs to the family *Flaviviridae* and circulates as four major serotypes. The virus is still evolving and the serotypes differ in sequence by around 30 %. Dengue is transmitted to humans following a bite from an infected mosquito, usually *Aedes aegypti*.

Following the bite, there is an incubation period of 3–8 days, which is followed by a symptomatic phase, although school-based serological surveys suggest that 50 % or more of infections are asymptomatic (Endy *et al.*, 2002). Symptoms range from a mild undifferentiated fever to more severe fever, headache, muscle, joint and bone pains accompanied later on in the illness by a maculopapular rash. This more severe clinical syndrome is classified as dengue fever and may be accompanied by thrombocytopenia (low platelet count), leucopenia (low white cell count) and petechial haemorrhage (minor bleeding into the skin) (World Health Organization, 2002).

More severe manifestations of the disease are classified as dengue haemorrhagic fever (DHF). This is a potentially life-threatening condition with up to 20 % mortality without expert medical care. DHF is characterized by plasma leakage and thrombocytopenia and can be divided into four clinical grades with increasing severity of bleeding, vascular leakage, hypovolaemia and shock.

One highly characteristic and interesting feature of DHF is that the severe symptoms of bleeding and circulatory collapse seem to occur coincidentally with the time of

SGM symposium 64: Molecular pathogenesis of virus infections.
Editors P. Digard, A. A. Nash & R. E. Randall. Cambridge University Press. ISBN 0 521 83248 9 ©SGM 2005

defervescence, the point at which the fever subsides. Before this and for the 2–7 days of fever these patients will resemble cases of dengue fever.

The mainstay of treatment for DHF is careful management of fluid status with replacement with isotonic saline, colloidal plasma expanders or blood where required. With careful monitoring and attention to fluid balance the mortality can be reduced to substantially below 1 %.

Despite this, dengue infection still remains a major public health issue in a number of tropical and subtropical countries. About 2·5 billion people are at risk of dengue infection and there are estimated to be 50–100 million infections annually. The disease occurs in an epidemic fashion and therefore puts a huge strain on health care services; in Thailand for instance there are around 100 000 cases of DHF annually.

The incidence of dengue infection has increased rapidly since the Second World War. DHF, which was rare or absent in many parts of the world, has also become much more frequent. Many explanations of this huge expansion of disease have been put forward, including a drive toward urbanization and travel allowing rapid dissemination and co-circulation of viral strains (Mackenzie *et al.*, 2004).

Careful epidemiological analyses often in island populations have yielded very valuable information about the pathogenesis of severe dengue infection or DHF. Some of the best such evidence has come from Cuba (Guzman *et al.*, 2000). In 1977, there was an epidemic of dengue fever caused by the DEN-1 serotype of virus. The majority of the population was dengue naïve and, interestingly, during this outbreak there were few cases of severe disease. Infection with the DEN-1 serotype reduced in subsequent years until in 1981 there was a second epidemic caused by the DEN-2 serotype. On this occasion, the spectrum of disease was markedly different; there were nearly 350 000 cases of dengue fever, over 10 000 cases of DHF and around 150 deaths.

This and a number of similar studies have shown that the risk of developing DHF is much higher in people who have been previously exposed to one serotype of dengue and who are subsequently exposed to a secondary or sequential infection with a virus of different serotype (Sangkawibha *et al.*, 1984). In this case, it therefore appears that pre-existing immunity or memory for the previous infection can actually potentiate disease. A further epidemic of DEN-2 in Cuba in 1997 demonstrates that the risk from previous infection (DEN-1 1977) can last for 20 years or more.

Antibody-dependent enhancement has been suggested as a mechanism of disease potentiation (Halstead & O'Rourke, 1977). This was proposed by Scott Halstead in 1977 and was based on similar observations on the related flavivirus tick-borne encephalitis virus (Phillpotts *et al.*, 1985). It is proposed that following a primary dengue infection an antibody response is mounted to envelope proteins on the infecting virus. Since different viral serotypes will differ by around 30 % in sequence, antibodies directed to one serotype may not confer complete neutralizing protection to another serotype. Following primary infection, as the titre of these antibodies falls, it is proposed that there becomes a point where the antibodies may bind to virus and rather than neutralize infection completely, actually target uptake of opsonized virus by Fc-receptor-bearing cells, principally macrophages, thereby driving higher virus replication.

This phenomenon can be demonstrated *in vitro* and there is some evidence that it can operate *in vivo* (Halstead, 1979). Indeed, antibody-dependent enhancement has been invoked to explain a small peak in severe dengue infection in children below the age of 1 year which can occur during a primary exposure and is proposed to be driven by the fall in titre of maternally acquired anti-dengue antibodies which occurs with age (Halstead *et al.*, 2002).

Recently, viral loads have been measured throughout the course of infection using either virus culture or reverse transcriptase PCR. There appears to be a correlation between higher viral titres and more severe disease (Vaughn *et al.*, 2000). Interestingly, however, the virus load falls precipitously at the time that the fever remits. At this point when symptoms peak virus titre is low or undetectable. The lack of correlation between virus load and severe symptoms has led some to speculate that the sequelae of severe dengue infection may be driven more by the immune response to the virus than virus load per se.

Several groups including our own have measured the levels of inflammatory cytokines during dengue infection. Many of these are raised, and in some the peak in their levels coincides with the onset of severe symptoms. High levels of TNF-α, IFN-γ, IL-10, IL-6, etc., have been recorded and many of these can be the products of activated T cells (Rothman & Ennis, 1999). These findings have led some to suggest that it is T cells that cause the damage and vascular leak characteristic of dengue, either by causing direct cytotoxic tissue damage or by producing a variety of inflammatory cytokines.

Like B cells, T cells belong to the acquired immune system in that the antigen receptor is not germ line encoded as with molecules characteristic of the innate immune system but instead constructed following a series of genomic recombinations which occur in the thymus. The T-cell receptor is a non-covalently linked heterodimer, the major form consisting of a pairing of alpha and beta chains whilst a minor form consists of a

Table 1. List of GTS variants obtained from dengue sequences published in GenBank

Modified version of data previously published in Mongkolsapaya *et al.* (2003). Printed with permission from the Nature Publishing Group.

Variant	Sequence	Number of sequences found/total
DEN 1.1	GTSGSPIV**NRE**	5/5
DEN 2.1	GTSGSPII**DKK**	29/39
DEN 2.2	GTSGSPIV**DKK**	6/39
DEN 2.3	GTSGSPIV**DRK**	3/39
DEN 2.4	GTSGSPIA**DKK**	1/39
DEN 3.1	GTSGSPII**NRE**	2/2
DEN 4.1	GTSGSPII**NRK**	2/2

pairing of gamma and delta chains. T-cell receptors recognize short antigenic peptides which are bound in the antigen-binding groove of MHC molecules. MHC-I presents 8–11-mer peptides which are derived predominantly from intracellularly produced proteins to cytotoxic T cells bearing the CD8 co-receptor. MHC-II presents slightly longer, predominantly endocytosed exogenous antigen to helper T cells bearing the CD4 co-receptor.

To study T-cell responses in detail, it is necessary to define the sequence of the antigenic peptides from dengue presented to T cells. Since MHC is highly polymorphic there will likely be a large number of dengue peptides which can bind to different MHC molecules. Several T-cell epitopes for dengue have been previously described but these are mostly restricted by Western MHC types, which makes their utility to examine T-cell responses in endemic areas such as SE Asia limited.

To overcome this problem, we undertook a systematic search for dengue T-cell epitopes using an overlapping peptide approach (Altfeld *et al.*, 2000). MHC molecules bind linear peptides between 8 and 11 amino acids in length and the peptide binding groove can accommodate, although with less efficiency, short peptide extensions at either end. Therefore, a nested set of peptides 15–20 amino acids in length and overlapping by 10–11 amino acids will contain all possible 10–11 linear amino acid sequences, in effect covering all possible antigens, and can be used for antigen discovery.

The presence of responding T cells can then be assessed by culturing peripheral blood mononuclear cells from dengue-immune individuals with pools of these overlapping peptides. If responding T cells are present then they will be activated by peptide and this activation can then be read by assaying cytokine production from the activated T cells.

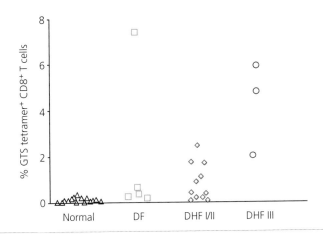

Fig. 1. Frequency of dengue-specific CD8+ T cells at convalescent day 14 from patients with different disease severities, dengue fever (DF), dengue haemorrhagic fever grades I and II (DHF I/II) and DHF grade III, compared with normal healthy dengue-immune individuals (normal). Modified version of data previously published in Mongkolsapaya *et al.* (2003). Printed with permission from the Nature Publishing Group.

We used the gamma interferon ELISPOT assay for this purpose, where a sandwich ELISA for interferon gamma allows the enumeration of the responding T cells.

Using this assay, we were able to discover a number of T-cell epitopes for both CD8- and CD4-positive T cells. We were particularly interested in those epitopes which were restricted by commonly expressed MHC-I alleles in SE Asia such as A11, A24 and A33. An immunodominant epitope from the NS3 protein GTSGSPIIDKK which was restricted by HLA-A11 was found and studied in detail (Mongkolsapaya *et al.*, 2003). T-cell responses to this epitope were examined in a cohort of children admitted to hospital in Khon Kaen in North East Thailand.

Since dengue virus serotypes show about 30 % amino acid difference we searched sequence databases to ascertain whether differences were found in this A11 epitope, and found six variants of the epitope (Table 1) which were expressed by the variant viruses. Responses to these variant epitopes were analysed in the patients' samples using both ELISPOT and MHC tetramer analysis using FACS (McMichael & O'Callaghan, 1998).

All of the patients in this study were suffering from secondary dengue infections and they showed diverse T-cell responses to the variant peptide epitopes derived from the different dengue serotypes. During the acute illness, dengue-specific T cells were highly activated and almost all also showed signs of cell proliferation and apoptosis. There was also a correlation between the magnitude of the T-cell response and the severity of the dengue illness (Fig. 1).

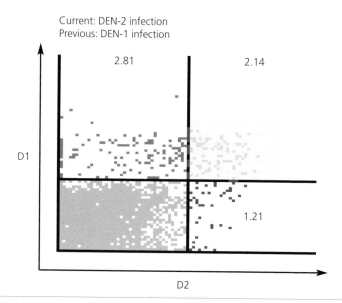

Fig. 2. Secondary dengue infection showing the original antigenic sin phenomenon. PBMCs from DEN-2-infected patients were simultaneously stained with the GTS tetramers from DEN-1 (D1) and DEN-2 (D2) and analysed by flow cytometry. A substantial number of T cells are of higher affinity for the previously encountered DEN-1 as opposed to the currently infecting virus DEN-2. Modified version of data previously published in Mongkolsapaya *et al.* (2003).Printed with permission from the Nature Publishing Group.

When we looked at the fine specificity of the T-cell response we found paradoxically that many of the T cells had a relatively low affinity for the currently infecting viral serotype and showed higher affinity for serotypes which we presume had been encountered before (Fig. 2).

This phenomenon whereby a response to a variant of a previously encountered epitope is constructed mainly from memory T or B cells rather than being generated *de novo* by fresh priming is termed original antigenic sin (Fazekas de St Groth & Webster, 1966a, b; McMichael, 1998). Original antigenic sin has the advantage that a cross-reactive response can be rapidly recalled from memory but has the disadvantage that the response may not be optimal and may contain some lower affinity T cells. In some experimental circumstances, such as murine infection with variants of the lymphocytic choriomeningitis virus, original antigenic sin has been shown to be detrimental, leading to slower clearance of virus in a secondary as opposed to primary infection (Klenerman & Zinkernagel, 1998).

In summary, it seems likely that the increased severity of a secondary or sequential dengue infection is immunologically driven and multifactorial. Antibody-dependent enhancement may drive virus internalization into macrophages and enhance virus replication. At the same time, original antigenic sin may initially delay an effective high-affinity T-cell response to the virus, allowing further virus replication whilst a high-affinity T-cell response is generated. This will lead to a dangerous situation where there is finally the collision of a large number of T cells with a high antigen load. This will lead to massive T-cell activation, which will cause both direct T-cell-mediated cytotoxicity and the release of a variety of inflammatory cytokines. These in turn will lead to tissue damage and the syndrome of vascular leak which occurs coincidentally with virus clearance in DHF.

REFERENCES

Altfeld, M. A., Trocha, A., Eldridge, R. L. & 9 other authors (2000). Identification of dominant optimal HLA-B60- and HLA-B61-restricted cytotoxic T-lymphocyte (CTL) epitopes: rapid characterization of CTL responses by enzyme-linked immunospot assay. *J Virol* **74**, 8541–8549.

Endy, T. P., Chunsuttiwat, S., Nisalak, A., Libraty, D. H., Green, S., Rothman, A. L., Vaughn, D. W. & Ennis, F. A. (2002). Epidemiology of inapparent and symptomatic acute dengue virus infection: a prospective study of primary school children in Kamphaeng Phet, Thailand. *Am J Epidemiol* **156**, 40–51.

Fazekas de St Groth, S. & Webster, R. G. (1966a). Disquisitions on Original Antigenic Sin. I. Evidence in man. *J Exp Med* **124**, 331–345.

Fazekas de St Groth, S. & Webster, R. G. (1966b). Disquisitions on Original Antigenic Sin. II. Proof in lower creatures. *J Exp Med* **124**, 347–361.

Guzman, M. G., Kouri, G., Valdes, L., Bravo, J., Alvarez, M., Vazques, S., Delgado, I. & Halstead, S. B. (2000). Epidemiologic studies on Dengue in Santiago de Cuba, 1997. *Am J Epidemiol* **152**, 793–799; discussion 804.

Halstead, S. B. (1979). In vivo enhancement of dengue virus infection in rhesus monkeys by passively transferred antibody. *J Infect Dis* **140**, 527–533.

Halstead, S. B. & O'Rourke, E. J. (1977). Antibody-enhanced dengue virus infection in primate leukocytes. *Nature* **265**, 739–741.

Halstead, S. B., Lan, N. T., Myint, T. T. & 7 other authors (2002). Dengue hemorrhagic fever in infants: research opportunities ignored. *Emerg Infect Dis* **8**, 1474–1479.

Klenerman, P. & Zinkernagel, R. M. (1998). Original antigenic sin impairs cytotoxic T lymphocyte responses to viruses bearing variant epitopes. *Nature* **394**, 482–485.

Mackenzie, J. S., Gubler, D. J. & Petersen, L. R. (2004). Emerging flaviviruses: the spread and resurgence of Japanese encephalitis, West Nile and dengue viruses. *Nat Med* **10** (12 Suppl.), S98–S109.

McMichael, A. J. (1998). The original sin of killer T cells. *Nature* **394**, 421–422.

McMichael, A. J. & O'Callaghan, C. A. (1998). A new look at T cells. *J Exp Med* **187**, 1367–1371.

Mongkolsapaya, J., Dejnirattisai, W., Xu, X. N. & 11 other authors (2003). Original antigenic sin and apoptosis in the pathogenesis of dengue hemorrhagic fever. *Nat Med* **9**, 921–927.

Phillpotts, R. J., Stephenson, J. R. & Porterfield, J. S. (1985). Antibody-dependent enhancement of tick-borne encephalitis virus infectivity. *J Gen Virol* **66**, 1831–1837.

Rothman, A. L. & Ennis, F. A. (1999). Immunopathogenesis of Dengue hemorrhagic fever. *Virology* **257**, 1–6.

Sangkawibha, N., Rojanasuphot, S., Ahandrik, S., Viriyapongse, S., Jatanasen, S., Salitul, V., Phanthumachinda, B. & Halstead, S. B. (1984). Risk factors in dengue shock syndrome: a prospective epidemiologic study in Rayong, Thailand. I. The 1980 outbreak. *Am J Epidemiol* **120**, 653–669.

Vaughn, D. W., Green, S., Kalayanarooj, S. & 8 other authors (2000). Dengue viremia titer, antibody response pattern, and virus serotype correlate with disease severity. *J Infect Dis* **181**, 2–9.

World Health Organization (2002). *Dengue and Dengue Haemorrhagic Fever.* Fact sheet no. 117.

The immune response to human immunodeficiency virus type 1 (HIV-1)

Emma Turnbull and Persephone Borrow

Viral Immunology Group, The Edward Jenner Institute for Vaccine Research, Compton, Newbury RG20 7NN, UK

INTRODUCTION

Human immunodeficiency virus type 1 (HIV-1) is a retrovirus in the lentivirus family that establishes a persistent infection in humans, which is ultimately associated with the development of an acquired immunodeficiency syndrome (AIDS). Despite intensive efforts in basic and applied research, HIV-1 infection remains a major global health burden. The severity of the HIV-1 pandemic is reflected in the 2004 statistics provided by UNAIDS (see www.unaids.org/bangkok2004/report.html). Their latest figures show that after more than 20 years and 20 million deaths since the first AIDS diagnosis in 1981, approximately 38 million people are infected with HIV-1 worldwide. Importantly, infection continues to spread at an alarming rate; in sub-Saharan Africa it is estimated that 3 million people became infected in 2003 alone. New epidemics are also advancing unchecked in other places, notably Eastern Europe and Asia. Although antiretroviral therapy has been key in reducing the morbidity and mortality associated with HIV-1 infection, this is not a long-term solution for curbing the pandemic. Therapy does not completely eliminate infection and so must be taken for life; if it is interrupted, the viral load typically becomes highly elevated within a few days. The costs of continuous antiretroviral therapy preclude its use by the majority of infected individuals in under-developed countries, and even in those who can afford therapy there are numerous adverse side effects which contribute to non-compliance.

There is thus an urgent need for the development of effective HIV-1 vaccines, both prophylactic vaccines that may block infection (or at a minimum, allow a 'blunted' infection with a less severe clinical course and reduced chance of virus transmission)

SGM symposium 64: Molecular pathogenesis of virus infections.
Editors P. Digard, A. A. Nash & R. E. Randall. Cambridge University Press. ISBN 0 521 83248 9 ©SGM 2005

and therapeutic vaccines that may eliminate infection or prevent/delay disease progression. The latter could be used in combination with antiretroviral treatment and/or other forms of therapy aimed at improving immune function in HIV-1-infected individuals. Given these clinical needs and the lack of success of attempts to produce effective HIV-1 vaccines using traditional empiric methods, considerable research effort is currently being focused on characterizing the nature of the immune response to HIV-1. Interests lie in defining why the immune response in HIV-1-infected individuals is not more effective in eradicating or containing virus replication and in identifying immune correlates of protection and/or good natural long-term control of virus replication.

Correlates of protection against HIV-1 infection can be studied in animal models where protection against challenge with pathogenic virus can be achieved by prior infection with live attenuated virus, and in individuals who remain HIV-1-uninfected despite repeated exposure to the virus (exposed seronegatives; ESNs). Further, correlates of good natural long-term control of virus replication can be addressed by comparing the immune response in HIV-1-infected individuals who control virus replication with differing efficiency. There is considerable variation in the disease course amongst HIV-1-infected individuals. Some individuals progress to AIDS rapidly within 1–2 years of infection (rapid progressors) whereas others naturally maintain a very low viral load during chronic infection and remain asymptomatic for 10 years or more before progressing to AIDS (long-term non-progressors; LTNPs). Analysis of the immune correlates of slow disease progression may give valuable insights into what type of immune response vaccines should aim to elicit in order to provide a similar protection against disease progression.

A second, related area of research in the field of HIV-1 immunology is focusing on understanding how the immune system is undermined during the course of infection, eventually leading to the characteristic immunodeficiency syndrome and death. This has been the subject of other review articles and will not be covered here (Douek, 2003; McCune, 2001; Stevenson, 2003). Advances in this area may lead to development of therapeutic strategies for preventing or reversing the generalized immunological dysfunction associated with HIV-1 infection.

Here, we will give a brief overview of the course of HIV-1 infection, then will focus on the immune response to the virus. We will review how each arm of the immune response may contribute to the control of virus replication, discuss strategies that the virus employs to evade this control and consider the implications of this knowledge for prophylactic and therapeutic vaccine design. Finally, we will briefly review the current status of the HIV-1 vaccine field and approaches being taken to the challenges ahead.

OVERVIEW OF THE COURSE OF HIV-1 INFECTION

HIV-1 transmission occurs predominantly through sexual contact, although infection by contact with contaminated blood or blood products (for example, via infected needles or transfusion) and transmission from mother to child also contribute to the spread of the virus. Virus replication initially occurs locally (within the mucosal lamina propria following sexual transmission), although spread to systemic sites is rapid. Although activated CD4$^+$ T cells and macrophages constitute the main cellular sites of virus replication, dendritic cells (DCs) and macrophages may play an important role in the spread of infection. DCs have been shown to bind and sequester virions (via DC-SIGN; Geijtenbeek *et al.*, 2000) for transmission to CD4$^+$ T cells, and macrophages, when infected, secrete soluble factors which attract and partially activate CD4$^+$ T cells allowing for their infection by locally released virus (Swingler *et al.*, 2003). Virus spread is reflected in a dramatic burst of virus replication during primary HIV-1 infection, with plasma viral loads typically reaching >10^6 RNA copies ml^{-1} by 4–6 weeks post-infection (Clark *et al.*, 1991; Daar *et al.*, 1991; Piatak *et al.*, 1993). In many individuals, this is associated with an acute, self-resolving (often 'flu-like') illness and a transient drop in peripheral blood CD4$^+$ T cell counts. The acute burst of virus replication is rapidly contained, concomitant with the induction of the virus-specific immune response, but although the plasma viral load is reduced, infection is not eliminated (Fig. 1). A clinically asymptomatic phase of chronic infection then ensues, during which a relatively stable 'set-point' level, typically of between 10^3 and 10^6 HIV-1 RNA copies (ml plasma)$^{-1}$, is achieved. The 'set-point' viral load reached by 6 months post-infection has been shown to be a good indicator of the subsequent rate of disease progression (Lyles *et al.*, 2000; Mellors *et al.*, 1995, 1996). Typically, the higher the persisting viral load established in early infection, the more rapidly an individual will progress to AIDS (Mellors *et al.*, 1996) (illustrated in Fig. 2). The factors determining the set-point viral load remain poorly defined, but may include viral pathogenicity, host genetics and the quality of the immune response generated. There is currently much interest in the field in elucidating the events that occur during early infection that may dictate the subsequent disease course.

Even during the asymptomatic phase of infection, the magnitude and dynamics of virus replication are tremendous. It is estimated at least 10^{10} virions are produced daily, with 140 generational cycles occurring annually (Ho *et al.*, 1995; Wei *et al.*, 1995). At least 99 % of circulating virus is produced by infected, activated CD4$^+$ T cells (having a half-life of less than a day; Markowitz *et al.*, 2003). The peripheral CD4$^+$ T cell pool can normally be replenished by production of new T cells from the thymus and by post-thymic proliferation. However, during the course of HIV-1 infection, there is a gradual undermining of the immune system, including defects in thymic function and an impairment of extra-thymic T cell homeostasis. Over time, the rate of CD4$^+$ T cell

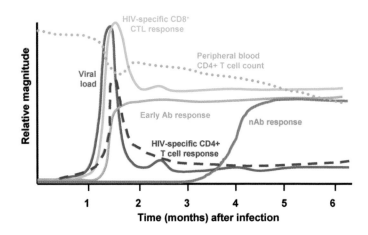

Fig. 1. Immune response in early HIV-1 infection. Primary HIV-1 infection is characterized by a dramatic burst in virus replication which is reflected in the plasma viral RNA load. CD8$^+$ T cells are activated in response to the increasing antigenic stimulus and thereafter their kinetics of expansion and contraction closely follow the viral titre. Activation of HIV-specific CD4$^+$ T cells also occurs during acute HIV-1 infection, but whereas HIV-specific CD8$^+$ T cells are maintained at a high frequency after the acute viral burst, activated CD4$^+$ T cells are rapidly lost and the CD4 response is typically weak throughout chronic infection. Infection causes an initial drop in total peripheral blood CD4$^+$ T cell counts, and over time, CD4$^+$ T cell destruction exceeds homeostatic replacement, leading to a gradual fall in CD4$^+$ T cell frequency. HIV-specific antibodies are detectable as early as a few weeks after infection; however, antibodies capable of neutralizing the virus (nAbs) are not produced until several months after infection.

destruction exceeds that of replenishment and peripheral CD4$^+$ T cell numbers fall, eventually declining below the level required for normal immune function (Douek, 2003; McCune, 2001). As virus replication increases and CD4$^+$ T cell numbers reach very low levels, the individual progresses to AIDS, characterized by susceptibility to otherwise harmless opportunistic infections, development of malignancies, wasting and frequently also neurological impairment.

THE IMMUNE RESPONSE TO HIV-1

Typically, the immune response against a virus infection involves initial containment by innate mechanisms (rapidly induced but lacking in specificity) and subsequent long-term control conferred by the adaptive immune system. Effective integration of the different arms of the immune response is key to efficient control of virus replication, although in a particular infection, one arm of the adaptive immune response (cell-mediated or humoral) may make a more dominant contribution to the control of virus replication. CD8$^+$ cytotoxic T lymphocytes (CTLs) directly destroy cells expressing viral antigens and release soluble factors with antiviral activity that may 'cure' cells of infection. Although CD4$^+$ T cells may also have direct antiviral effector functions, they

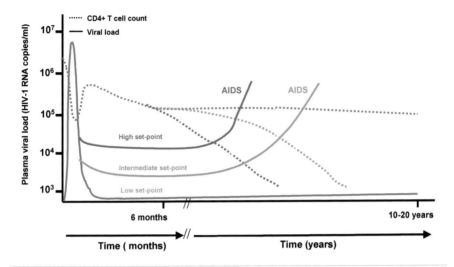

Fig. 2. Set-point viral load after resolution of the acute viral burst is a good indicator of the subsequent rate of progression to AIDS. Primary HIV-1 infection is characterized by a dramatic burst of virus replication (associated with acute symptoms in ~70 % of infected individuals) and an initial, transient drop in the peripheral blood CD4$^+$ T cell count. The acute viraemia is typically resolved within 2–3 months of initial infection; however, the extent to which virus replication is contained varies widely between individuals. The 'set-point' viral load established following resolution of the acute viral burst is a good indicator of the subsequent rate of disease progression. In the presence of a high persisting set-point viral load (typically >40 000 copies ml^{-1}), the rate of peripheral blood CD4$^+$ T cell loss is rapid and the individual may progress to AIDS within as little as 1–3 years in the absence of antiretroviral therapy (rapid progressors; red lines). Individuals who naturally establish very low persisting viral loads (typically <1000 copies ml^{-1}) generally show good preservation of CD4$^+$ T cell counts and may remain asymptomatic for 10–20 years (long-term non-progressors; blue lines). Individuals with intermediate viral loads typically undergo disease progression in around 3–10 years (typical progressors; green lines).

are often more important in providing help for the induction and/or maintenance of other arms of the adaptive response (achieved via cell–cell contact and cytokine secretion). Antiviral antibodies can limit virus spread by preventing infection of new cells and may also contribute to the elimination of virus infection. Importantly, pre-existing antibodies also play a key role in blocking reinfection on re-exposure to a previously encountered virus. Normally these defences are effective in resolving virus infection; however, HIV-1 is an incredibly sophisticated virus which has evolved numerous mechanisms that enable it to evade the full effectiveness of almost every arm of the antiviral immune response and persist long-term. Additionally, the ability of HIV-1 to productively infect and destroy cells of the immune system ultimately results in impairment of immunological functions and inevitable progression to a severe immunodeficiency syndrome. The following sections address the contribution of different arms of the immune response to the suppression of HIV-1 replication and the mechanisms that HIV-1 has evolved to evade immune control.

The innate immune response

Type I interferons (IFNs). The type I IFNs are a family of cytokines (including the IFN-α subtypes and IFN-β) that are rapidly induced following infection with most viruses (Muller *et al.*, 1994). They have important direct antiviral effects and also mediate the activation of cell types within both the innate and adaptive immune systems [e.g. natural killer (NK) cells, DCs and T cells]. Although most cell types can produce type I IFN when infected with virus, the major *in vivo* source in many virus infections is from the plasmacytoid DC population (PDCs) (reviewed by Santini *et al.*, 2002; Tough, 2004), also termed natural IFN-α-producing cells (Siegal *et al.*, 1999). PDCs are rare amongst circulating peripheral blood leukocytes but are found at higher frequency in secondary lymphoid tissues. PDCs are stimulated to produce IFN-α following inter-action of their pattern-recognition receptors with viral components, including viral nucleic acids [e.g. single-stranded viral RNA, which binds Toll-like receptor 7 (TLR7) (Diebold *et al.*, 2004; Heil *et al.*, 2004; Lund *et al.*, 2004), and viral dsDNA, which binds Toll-like receptor 9 (TLR9) (Akira & Hemmi, 2003; Lund *et al.*, 2003)] and virion envelope glycoproteins [likely recognized via lectin receptors (Charley *et al.*, 1991; Ito, 1994)]. Indeed, the HIV-1 gp120 envelope glycoprotein has been implicated in induction of IFN-α (Capobianchi *et al.*, 1992; Francis & Meltzer, 1993; Ito, 1994).

Type I IFN levels are transiently elevated in serum during acute HIV-1 infection, peaking just prior to the peak in primary viraemia (von Sydow *et al.*, 1991). These cytokines have been shown to have potent anti-HIV-1 activity and can inhibit HIV-1 replication *in vitro* and *in vivo* (Hartshorn *et al.*, 1987; Pitha, 1994; Poli *et al.*, 1989). The early burst of type I IFN production may therefore play a role in restricting acute-phase virus replication. In mid-stage infection, the contribution made by type I IFN to control of HIV-1 replication is less clear. Serum type I IFN levels are below detection, although there may be local type I IFN production in tissues, which could help to restrict virus replication. As disease progresses, IFN-α is again detected in the serum of infected individuals (DeStefano *et al.*, 1982; Grunfeld *et al.*, 1991; Krown *et al.*, 1991); however, this has been reported to be an aberrant acid-labile form which may be detrimental through its inhibition of the production or effects of bioactive IFN-α (Hess *et al.*, 1991).

Primary HIV-1 isolates, unlike laboratory-adapted strains, show variable sensitivity to the effects of IFN *in vitro* (Kunzi *et al.*, 1995), suggesting that IFN resistance may be one mechanism contributing to the ability of HIV-1 to persist *in vivo*. HIV-1 persistence may also be promoted by the loss of the principal type I IFN-producing cells following infection. HIV-1-infected individuals show a selective loss of circulating PDCs and markedly reduced production of virus-induced IFN-α on a per cell basis (Feldman *et al.*, 2001). This effect is first evident in primary infection and is enhanced in the advanced stages of the disease. The number and function of circulating IFN-producing PDCs has been negatively correlated with viral load (Donaghy *et al.*, 2001; Feldman *et al.*, 2001; Soumelis *et al.*, 2001) and usually mirrors the loss of CD4$^+$ T cells over time. PDCs are present in higher numbers in LTNPs (Levy *et al.*, 2003; Soumelis *et al.*, 2001) compared with rapid progressors, advanced stage AIDS patients and even uninfected individuals. The mechanisms responsible for loss of PDCs in HIV-1-infected individuals are not clear. One possibility is that their loss may be due to virus infection. PDCs may be predisposed to infection by HIV-1 because they express the proteins used by HIV-1 as entry receptors: CD4 and the chemokine receptors CCR5 and CXCR4. Indeed, there have been reports of HIV-1 infecting PDCs *in vitro* (Fong *et al.*, 2002; Patterson *et al.*, 1999, 2001), and PDCs isolated from HIV-1-infected patients showed evidence of infection and were severely impaired in their ability to stimulate T-lymphocyte proliferation (Donaghy *et al.*, 2003). In addition to potentially facilitating HIV-1 persistence, the loss of PDCs in HIV-1 infection may also be of importance in promoting the opportunistic infections characteristic of AIDS: in HIV-1-infected individuals with low CD4$^+$ T cell counts, the absence of opportunistic infections and cancer has been reported to correlate with normal numbers of PDCs in blood (Soumelis *et al.*, 2001). Importantly, however, the loss of circulating PDCs can be reversed by antiretroviral treatment. Reduction of viral load by antiviral therapy allows for recovery of IFN-α production by PBMCs and can occur several months before full reconstitution of CD4$^+$ T cell numbers (Siegal *et al.*, 2001).

Based on its ability to block different steps in HIV-1 replication *in vitro* and its relative deficiency in HIV-1-infected individuals, IFN-α has been used therapeutically in HIV-1-infected individuals, either alone or in combination with other agents. IFN-α has shown efficacious anti-neoplastic effects against HIV-1-associated Kaposi's sarcoma (de Wit *et al.*, 1988; Francis *et al.*, 1992; Krown, 1998; Krown *et al.*, 1983; Lane *et al.*, 1988) and there is also some evidence that IFN treatment during acute infection may be associated with a more favourable disease prognosis (Lane *et al.*, 1990).

NK cells. NK cells contribute to innate defence against virus infections by performing both effector and immunoregulatory roles. NK cells are typically defined as CD56$^+$ CD3$^-$ lymphocytes, which constitute ~10–15 % of human PBMCs (Cooper *et al.*, 2001), although a rare population of CD56$^-$ CD16$^+$ NK cells has also been defined (Scott-Algara & Paul, 2002). NK cells can lyse infected cells and also produce a range of cytokines and chemokines with direct antiviral activity and/or immunoregulatory properties (Cooper *et al.*, 2001). Activation of NK cell cytotoxicity is controlled via the balance of signalling through inhibitory and activatory NK cell receptors (Lanier, 2001; Leibson, 1997) which have specificity for ligands such as stress-inducible proteins (activating receptors), glycosylation patterns found on viral glycoproteins (activating receptors) and MHC class I molecules (inhibitory receptors) (Biron & Brossay, 2001). Innate cytokines are important for NK cell maturation (particularly IL-15) and activation of cytotoxicity (particularly type I IFNs, IL-12 and IL-18; Biron & Brossay, 2001; Cooper *et al.*, 2001). Indeed, in contrast to resting T cells, NK cells constitutively express receptors for IL-12, IL-15 and IL-18, and in response to stimulation, rapidly produce IFN-γ and other cytokines such as IL-10, IL-13, TNF-α and GM-CSF (Aste-Amezaga *et al.*, 1994; Colucci *et al.*, 2003; Fehniger *et al.*, 1999). Differential expression levels of the CD56 marker delineate phenotypically and functionally distinct NK cell subsets (Lanier *et al.*, 1986). CD56^{dim} NK cells (representing ~90 % of the NK cell pool) are inherently more cytotoxic and express higher levels of perforin and FcγRIII (CD16) than their CD56^{bright} counterparts (Cooper *et al.*, 2001). CD56^{bright} NK cells may play a more immunoregulatory role in the immune response.

NK cells may combat HIV-1 infection via lysis of virus-infected cells and also via non-cytolytic mechanisms such as the secretion of soluble factors, e.g. IFN-γ, which can limit virus replication, and C-C chemokines, which can suppress CCR5-tropic HIV-1 infection (Biron & Brossay, 2001; Fehniger *et al.*, 1998; Oliva *et al.*, 1998). That NK cells could play an important protective role in defence against HIV-1 is evidenced from studies of ESN i.v. drug users. These individuals show increased NK cell lytic activity compared with normal control individuals or drug-using HIV-1-infected individuals before or after seroconversion (Scott-Algara *et al.*, 2003). However, the extent to which NK cells contribute to the long-term containment of HIV-1 replication in infected individuals is questionable. As discussed later, the abnormalities in NK cell numbers and functional capacity induced in the context of HIV-1 infection, together with the existence of virus strategies for evading NK cell recognition and control by NK cell effector functions, may limit their effectiveness in combating virus replication.

Several reports have identified both quantitative and qualitative changes in the NK cell population in HIV-1-infected individuals. Total NK cell numbers are reduced in HIV-1 infection (Sinicco *et al.*, 1993) and this is primarily attributable to a selective loss of the more cytotoxic CD56^{dim} NK cell compartment (Hu *et al.*, 1995; Tarazona *et al.*, 2002). The loss of NK cells, like that of PDCs, can be reversed by antiviral therapy: recovery of normal NK cell frequencies has been demonstrated upon reduction of viral load during therapy (Goodier *et al.*, 2003). Interestingly, the rare CD56$^-$ CD16$^+$ NK cell subset is selectively increased in viraemic, but not aviraemic, HIV-1-infected individuals (Mavilio *et al.*, 2003). NK cell function is also perturbed in HIV-1 infection; decreased NK cell activity is apparent early after infection and becomes more pronounced in individuals with progressive disease (Cai *et al.*, 1990; Ratcliffe *et al.*, 1994). The abnormalities observed in NK cell functions in HIV-1-infected individuals may be due in part to alterations in the subset composition of the NK cell pool seen during HIV-1 infection (a reduction in the proportion of CD56^{dim} cytolytic NK cells is associated with a reduction in the cytolytic capacity of the overall NK population). However, abnormalities in NK cell functions are also observed on a per cell basis, including a reduction in both cytolytic and non-cytolytic effector functions. These abnormalities are associated with aberrant patterns of expression of inhibitory and activatory receptors by NK cells. In patients with low viraemia (either naturally low or suppressed through drug therapy), expression of NK cell natural cytotoxicity receptors is maintained at levels comparable with healthy donors, whereas NK cells from viraemic patients tend to show up-regulation of inhibitory receptors and down-regulation of activatory receptors (Ahmad *et al.*, 2001; Eger & Unutmaz, 2004; Fogli *et al.*, 2004; Kottilil *et al.*, 2004; Mavilio *et al.*, 2003).

The cause of NK cell loss and impaired function during HIV-1 infection is not clear. A subset of CD4$^+$ CCR5$^+$ CXCR4$^+$ NK cells may be productively infected by HIV-1 and die as a result (Valentin *et al.*, 2002); however, this subset is rare within the total NK cell pool (Valentin *et al.*, 2002), hence its loss cannot account for the significant loss of NK cells observed during HIV-1 infection. Most NK cell abnormalities are most marked in patients unable to control viraemia, although it is unclear whether this is cause or effect. NK cell defects may thus be secondary to other immune system defects such as abnormalities in production of cytokines which may be needed for NK cell homeostasis/ maturation.

The reduction in NK cell numbers and their impaired function may contribute to the ability of HIV-1 to escape from this arm of the immune response. In addition, HIV-1 has evolved strategies that enable it to avoid NK cell recognition and resist control by NK effector mechanisms. For example, the HIV-1 Nef protein induces down-regulation of MHC class I expression on infected cells and can impair their recognition by virus-specific CD8$^+$ T cells. Nef is, however, selective in its choice of class I targets to down-regulate. Whereas HLA-A and HLA-B expression is decreased, HLA-C and -E expression is not (Cohen *et al.*, 1999). By ligating inhibitory receptors on NK cells, the latter could shift the balance of signalling away from NK cell activation, thus preventing killing of the infected cell.

Natural T lymphocytes (NT cells/NK-T cells) and $\gamma\delta$ T cells. NT cells (CD56$^+$ CD3$^+$ lymphocytes that express NK cell receptors and T cell receptors) and $\gamma\delta$ T cells play an important bridging role between the innate and adaptive immune systems. The specificity of the T cell receptors expressed by these cells is restricted and they frequently recognize non-peptide antigens (microbial components or other ligands presented on the cell surface) in the context of molecules lacking the classical genetic restriction of polymorphic MHC molecules (Gougeon *et al.*, 2002). Like NK cells, NT and $\gamma\delta$ T cells can recognize target cells deficient in the expression of classical MHC class I molecules (Poccia *et al.*, 2001). Like NK cells, NT cells and $\gamma\delta$ T cells may combat virus infections by (i) direct lysis of virus-infected cells, (ii) production of cytokines and chemokines that block virus transmission and/or replication and (iii) cytokine-dependent activation of other effector cells. *In vitro* stimulation of $V\gamma9V\delta2$ TCR$^+$ T cells with phosphocarbohydrates induces their activation, expansion and secretion of cytokines (IFN-γ and TNF-α; Lang *et al.*, 1995) and chemokines capable of restricting the replication of both R5 and X4 tropic HIV-1 strains (Poccia *et al.*, 1999). Moreover, these effects are observed in co-cultures of these T cells with HIV-1-infected cells (Poccia *et al.*, 1999; Wallace *et al.*, 1996).

The high numbers of $\gamma\delta$ T cells present at epithelial sites, together with their rapid responsiveness, make these cells well suited to combat virus infections such as HIV-1 in the early stages of transmission. Indeed, $\gamma\delta$ TCR$^+$ T cells present in the rectal and vaginal epithelia have been shown to secrete antiviral suppressor factors such as RANTES, MIP-1α and MIP-1β, and their protective potential has been suggested by studies carried out in macaques challenged rectally with simian immunodeficiency virus (SIV), which revealed the presence of increased numbers of $\gamma\delta$ T cells in the rectal mucosa and local iliac lymph nodes in protected compared with infected macaques (Lehner *et al.*, 2000a, b).

In HIV-1-infected individuals, $V\gamma9V\delta2$ TCR$^+$ cells are reduced in number (Gougeon *et al.*, 2002; Sandberg *et al.*, 2002; van der Vliet *et al.*, 2002) and those remaining are anergic, lack IL-2R expression, frequently fail to proliferate in response to ligand stimulation and show impaired secretion of IFN-γ and TNF-α, which cannot be restored by provision of IL-12 and IL-15 (Boullier *et al.*, 1999; Poccia *et al.*, 1996). Interestingly, partial restoration of the immune system under HAART is associated with the recovery of functional $V\gamma9V\delta2$ TCR$^+$ cells (Martini *et al.*, 2000). $V\alpha24^+V\beta11^+$ NT cells have also been found to be dramatically reduced in number in HIV-1$^+$ individuals compared with healthy donors (Motsinger *et al.*, 2002; van der Vliet *et al.*, 2002) and higher numbers of these cells were inversely correlated with viral load (Motsinger *et al.*, 2002). It has, however, been difficult to determine whether the gradual loss of NT/$\gamma\delta$ cells is causally associated with disease progression in these studies (Motsinger *et al.*, 2002; van der Vliet *et al.*, 2002). A proportion of NT/$\gamma\delta$ cell loss may relate to their susceptibility to infection due to expression of CD4 and CCR5 receptors (Unutmaz, 2003) as well as activation-induced cell death during persistent infection (Motsinger *et al.*, 2002; van der Vliet *et al.*, 2002). Dysfunction in the NT/$\gamma\delta$ cell compartment later on in HIV-1 infection may contribute to the increased susceptibility to opportunistic pathogens and tumours (Unutmaz, 2003) associated with progression to AIDS.

The complement system. The complement system is composed of around 30 soluble and cell-bound proteins arranged as a cascade of interacting components, which are normally present in an inactive form but become activated upon detection of conserved microbial patterns. Complement activation targets viruses for destruction in several ways: (i) direct osmotic lysis of virions by formation of a membrane attack complex, (ii) opsonization of virions for subsequent uptake by complement-receptor (CR)-expressing cells and (iii) recruitment and activation of inflammatory cells by activated complement proteins.

HIV-1 can stimulate the three pathways of complement activation: classical, alternative and lectin. The HIV-1 gp41 envelope protein can directly bind and activate C1q protein, leading to C3 deposition on the viral surface (Ebenbichler *et al.*, 1991; Stoiber *et al.*, 1994; Thielens *et al.*, 1993). This antibody-independent mechanism may be important during acute infection. Additional complement activation can be induced by inter-action of the gp120 envelope glycoprotein with components of the classical, alternative and lectin complement pathways (Saifuddin *et al.*, 2000; Spear, 1993). During sero-conversion, the development of HIV-1-specific antibodies further enhances activation of the classical pathway and increases the deposition of C3 cleavage products on HIV-1 virions (Stoiber *et al.*, 2001).

Whether complement is really effective in destroying HIV-1 *in vivo* is debatable. That HIV-1 activates the complement system is evidenced by several findings. Complement breakdown products and virus coated in complement activation products (notably C3) are found in the blood of HIV-1-infected individuals (Spear, 1993). Despite this activation, complement seems to be insufficient for complete virolysis (Stoiber *et al.*, 1997; Sullivan *et al.*, 1996). This implies that HIV-1 may have evolved mechanisms to evade complement-mediated lysis. Among these strategies is the incorporation of host cell complement regulatory proteins such as CD55 (DAF), CD46 (MCP) and CD59 (protectin) into the virion envelope during budding (Dierich *et al.*, 1996; Favoreel *et al.*, 2003; Frank *et al.*, 1996; Speth *et al.*, 2003). Additionally, factor H binding to the viral envelope further enhances the protection of HIV-1 against complement-mediated damage (Stoiber *et al.*, 1996). The action of these regulatory proteins appears to keep the level of complement activation under the threshold needed for lysis of HIV-1 virions (Kacani *et al.*, 2001).

Complement activation may have detrimental consequences in HIV-1 infection. Targeting of HIV-1 via CRs may actually enhance HIV-1 infection by facilitating transmission of bound virus (either directly or indirectly) to susceptible cell types. CR-expressing CD4[+] cells may be directly infected by C3-opsonized HIV-1. Indeed, a subset of CR1[+] CR2[+] CD4[+] T cells were found to be infected more efficiently *in vitro* by HIV-1 in the presence of complement (Delibrias *et al.*, 1993). CR3$^+$ monocytes and macrophages were also found to be more susceptible to HIV-1 infection by C3-coated HIV-1 (Stoiber *et al.*, 1997, 2001). Follicular dendritic cells (FDCs) expressing high levels of CR1, CR2 and CR3 also bind complement-coated HIV-1 very efficiently and can retain infectious virus for extremely long periods (Smith *et al.*, 2001). FDCs can thus act as an important source of infection for T cells migrating through germinal centres (Heath *et al.*, 1995; Hufert *et al.*, 1997). *Ex vivo* experiments have shown that up to 80 % of HIV-1 may be bound extracellularly by FDCs via C3–CR interactions. These cells may thus constitute an extremely important reservoir of HIV-1 in lymphoid tissue. In addition to FDCs, activated germinal centre B cells express CR1 and CR2 and can bind opsonized HIV-1 particles (Doepper *et al.*, 2000). This also facilitates efficient infection of autologous T cells (Doepper *et al.*, 2000; Jakubik *et al.*, 1999).

The adaptive immune response

Humoral immunity – the HIV-1-specific antibody response. Antibodies are rapidly induced in most virus infections. They may help to restrict the spread of infection and also contribute to containment of virus replication, although cell-mediated immunity typically plays a more dominant role in the elimination of established virus infections. Neutralizing antibodies (nAbs) do, however, play a key role in protection against reinfection; they are therefore a critical component of vaccine-induced immunity. Antibodies specific for viral components can bind to and neutralize virus particles via several mechanisms. They may prevent virus attachment to host cell surface receptors, interfere with the conformational changes needed for virion fusion and infection of target cells, or inhibit infection at a post-entry step. Antibodies binding viral proteins can also trigger complement activation, leading to lysis of virions and/or infected cells, and can opsonize cells for antibody-dependent cell cytotoxicity (ADCC) (Casadevall & Pirofski, 2003).

HIV-1-specific antibodies can be detected in the serum as early as 2–6 weeks after initial infection (Aasa-Chapman *et al.*, 2004; Ariyoshi *et al.*, 1992), typically as primary viraemia is being resolved. Antibodies specific for Env gp120, Gag p24 antigen, Env gp41 (Binley *et al.*, 1997) and Gag p17 (Pilgrim *et al.*, 1997) have been identified in acute HIV-1 infection. As seroconversion usually only occurs coincident with the resolution of the acute phase viral burst, it is unlikely that antibodies play a role in initial containment of acute virus replication. This is supported by the normal resolution of acute viraemia in B-cell-depleted, SIV-infected macaques (Schmitz *et al.*, 2003). The contribution made by antibodies to control of HIV-1 replication during the post-acute phase remains controversial. Although seroconversion occurs in acute infection, nAbs are generally not detected until several months after HIV-1 infection (Aasa-Chapman *et al.*, 2004; Ariyoshi *et al.*, 1992; Legrand *et al.*, 1997; Pilgrim *et al.*,

1997; Wei *et al.*, 2003). A similar delay in nAb formation has been reported after SIV infection of macaques (Montefiori *et al.*, 1996; Reimann *et al.*, 1994; Schmitz *et al.*, 2003) and is also seen in certain other infections [e.g. lymphocytic choriomeningitis virus (LCMV) infection of mice]. The reasons for this are not clear, although a recent study in the LCMV infection model suggests that excessive polyclonal B cell activation may play a role (Recher *et al.*, 2004). Although the antibodies generated early after HIV-1 infection are non-neutralizing, they may play an as yet unappreciated role in controlling virus replication during the post-acute phase, through effector functions such as ADCC (Connick *et al.*, 1996; Forthal *et al.*, 2001) and complement-mediated lysis of virions and/or infected cells.

When nAbs do start to be produced in HIV-1-infected individuals, they exert a significant pressure on virus replication. This is evidenced by the rapid acquisition of mutations in the HIV-1 *env* gene that confer resistance to antibody neutralization, which are detectable in the plasma virus population from within a few months after infection (Richman *et al.*, 2003; Wei *et al.*, 2003). Neutralization escape mutants of the animal lentiviruses equine infectious anaemia virus (Montelaro *et al.*, 1984; Salinovich *et al.*, 1986), visna virus (Scott *et al.*, 1979) and SIV (Burns & Desrosiers, 1994; Burns *et al.*, 1993) have also been reported in horses, sheep and monkeys, respectively. Virus escape from antibody neutralization is frequently achieved through changes in the pattern of N-linked glycosylation of gp120, such that exposed sites on the viral glycoprotein against which nAbs are directed are masked by a 'glycan shield' (Wei *et al.*, 2003). These mutations affect antibody neutralization without interfering with receptor binding and virus fusion. In line with this, challenge of macaques with SIV that has been deliberately mutated to remove glycosylation sites around the V1 loop results in an effective antibody response that can control the virus, but only until mutations repair the glycosylation deficits (Koch *et al.*, 2003; Reitter *et al.*, 1998).

Virus evolution to escape from antibody neutralization could be followed by generation of new antibodies capable of combating the mutated virus. However, given that T helper responses are critical for efficient antibody synthesis, and as discussed in the following section, HIV-1-specific CD4$^+$ T cell responses tend to be very weak in the majority of HIV-1-infected individuals, the capacity to elicit and maintain virus-neutralizing antibody responses in the face of continuous emergence of variant viruses during chronic infection may be very limited. Similarly, lack of CD4$^+$ T cell help during persistent LCMV infection of mice leads to a failure of infected mice to elicit effective antibody responses against emerging neutralization escape mutants (Ciurea *et al.*, 2001).

Not only is HIV-1 able to evolve to escape from antibody neutralization during the course of infection, but the virus is also inherently very resistant to antibody attack.

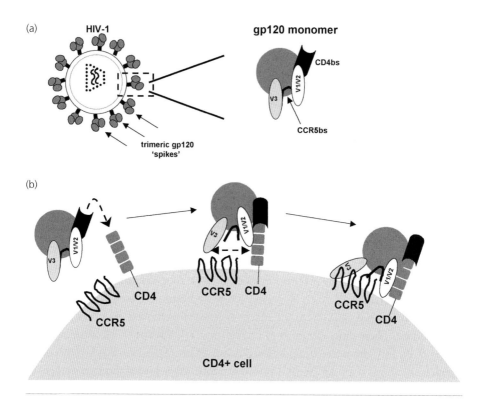

Fig. 3. Conformational changes in HIV-1 gp120 that occur during interaction with CD4 and the chemokine co-receptor on target cells. A pre-requisite for HIV-1 entry into cells is binding of the viral glycoprotein to CD4 and a chemokine co-receptor (e.g. CCR5) on target cells. (a) Schematic of a HIV-1 virion with its trimeric glycoprotein 'spikes' projecting from the virion membrane. In the absence of CD4 binding, the CCR5 binding site is masked by variable loops 1–3 (V1, V2, V3). CD4bs, CD4 binding site; CCR5bs, CCR5 binding site. (b) Interaction between gp120 and CD4 results in a conformational change that exposes the gp120 CCR5 binding site, which is normally masked by variable loops 1–3. Binding of gp120 to CCR5 then triggers further conformational changes in the viral glycoprotein that allow gp41 to insert into the target cell membrane and mediate the fusion event (not shown).

Antibodies that neutralize HIV-1 are mainly directed against the HIV-1 gp160 envelope glycoprotein subunits gp120 and gp41, which mediate virus binding and entry into host cells. The gp120 and gp41 subunit heterodimers are arranged as a trimer, with the gp120 'spikes' projecting from the surface of the virion and membrane-spanning domains of gp41 anchoring the complex into the lipid bilayer. The binding of specific residues in gp120 to the cellular CD4 receptor leads to conformational changes in the viral glycoprotein that expose binding sites on gp120 for binding to cellular CCR5/CXCR4 chemokine co-receptors (Fig. 3). Interaction between gp120 and the chemokine co-receptor then leads to further conformational changes in the glycoprotein, culminating in the insertion of the hydrophobic N-terminus of gp41 into the target cell membrane. Refolding of gp41 then drives the fusion of the viral and cellular membranes, and the viral nucleoprotein is released into the target cell cytosol (Chan & Kim, 1998; Weissenhorn *et al.*, 1999).

Although HIV-1-infected patients and SIV-infected monkeys typically generate high levels of circulating antibodies throughout the course of infection directed against both linear and conformation-dependent epitopes in gp120 and gp41, these antibodies react less well with the fully assembled gp160 complex on virions and on the surface of infected cells, and much of the antibody generated has poor neutralizing activity. This is likely due to a combination of the following factors, which together make HIV-1 extremely resistant to antibody neutralization (Pikora, 2004). (i) Many antibodies are directed against 'decoy' epitopes on gp160 antigen contained within debris from infected cells, or soluble monomeric gp120 shed from the surface of virions/infected cells. These antibodies are not able to bind to the trimeric form of gp160 found on the surface of virions and infected cells, as they recognize epitopes on parts of the glycoprotein that are buried within the trimeric complex (Wyatt *et al.*, 1998). (ii) The viral envelope (particularly gp120) is heavily glycosylated, which leads to masking of a number of epitopes in the conserved core of the gp160 complex to which antibodies can be raised. These glycans, however, are poorly immunogenic in themselves (Reitter *et al.*, 1998). (iii) The exposed highly variable domains within gp120 (V1–V5, four of which form loops on the surface of the glycoprotein) serve as an antigenically variable shield covering the more conserved, functional elements within the core of the complex. When nAbs against the variable loops develop, escape mutants can be selected for without extreme loss to virus fitness (Burton *et al.*, 2004). (iv) Functionally important sites (antibodies to which might have strong neutralizing potential) are only transiently exposed during the virion binding/entry process (Piguet & Trono, 2001), by which time antibody access may be limited by steric constraints. For example, the conserved co-receptor binding site on gp120 is only exposed after gp120 has bound to CD4, by which time antibody access to this site is incredibly restricted due to spatial constraints. The neutralizing activity of many antibodies to this site is thus greater with Fab fragments than with intact antibodies (Labrijn *et al.*, 2003). Likewise, antibodies directed against conserved sites on gp41 that would be able to inhibit the virus–cell fusion process also are impeded from neutralizing viral infectivity by both steric and kinetic constraints. Thus the HIV-1 envelope is inherently difficult to neutralize. SIV shows a similar natural resistance to antibody neutralization. Interestingly, amino acid substitutions, elimination of N-glycan attachment sites or long deletions within variable loops V1 and V2 could dramatically increase the sensitivity of SIVmac239 to antibody neutralization (Johnson *et al.*, 2003). Studies of this nature are invaluable in identifying the regions of the virion envelope that confer resistance to antibody neutralization and in promoting understanding of how we might be able to circumvent this resistance and develop an effective vaccine capable of eliciting broadly cross-reactive nAbs.

Although many nAbs have been identified that efficiently prevent infection of T cell lines *in vitro* by laboratory-adapted strains, most are inefficient in neutralizing primary isolates from HIV-1-infected patients. However, primary HIV-1 isolates from infected individuals of different genetic subtypes can be neutralized by some broadly cross-reactive human monoclonal antibodies such as b12, 2G12, 2F5 and 4E10 (Burton *et al.*, 2004). Furthermore, sera from a minority of HIV-1-infected individuals have been shown to contain antibodies capable of neutralizing a broad range of clade B isolates (Moore *et al.*, 2001). Encouragingly, when passively transferred, such broadly neutralizing antibodies can protect human PBL-reconstituted scid mice against HIV-1 challenge (Gauduin *et al.*, 1997) and protect monkeys against challenge with the SIV/HIV-1 hybrid SHIV-1 (Hofmann-Lehmann *et al.*, 2001a, b; Mascola *et al.*, 2000; Ruprecht *et al.*, 2001). However the antibody titres required are high and might be difficult to achieve in humans through vaccination. Findings made in a recent study suggest that the prospect of vaccine-elicited antibody responses being able to block HIV-1 transmission may not be as bleak as suggested by previous work on HIV-1/SIV neutralization. Analysis of HIV-1 transmission in a cohort of discordant heterosexual couples suggested that there may be preferential transmission and/or enhanced early replication of viruses bearing envelope glycoproteins with shorter variable loops (which may be more neutralization-sensitive; Derdeyn *et al.*, 2004).

In summary, it is apparent that nAb has the potential to control HIV-1 infection and replication *in vivo*, provided that the antibody is directed to an optimal accessible target within the HIV-1 envelope complex. However, as a result of the inherent complexity of the envelope structure and the ability of the virus to escape antibody neutralization, HIV-1 is well equipped to evade the nAb response. This poses great problems for vaccination strategies. Nonetheless, there remains considerable interest in research into the generation of nAbs for prophylactic vaccination and therapy. Some of the approaches currently being pursued are discussed later.

The HIV-1-specific CD4$^+$ T cell response. CD4$^+$ T cells are critical to effective functioning of the adaptive immune response. They provide help (mediated via cell–cell contact and cytokine production) to (i) B cells for efficient antibody synthesis (class switching, affinity maturation and memory cell generation) and (ii) CD8$^+$ T cells for their activation, memory generation and maintenance during persistent virus infection, the latter being evidenced particularly in murine models of chronic virus infection (Battegay *et al.*, 1994; Matloubian *et al.*, 1994; Planz *et al.*, 1997; Thomsen *et al.*, 1996). Additionally, CD4$^+$ T cells may have a direct antiviral effector role via production of cytokines (e.g. IFN-γ) and chemokines (Abdelwahab *et al.*, 2003; Lotti *et al.*, 2002) and via cytotoxic activity (Littaua *et al.*, 1992; Lotti *et al.*, 2002; Norris *et al.*, 2004; Sethi *et al.*, 1988).

CD4[+] T cells are activated through recognition of viral antigens presented by HLA class II molecules on antigen-presenting cells (APCs). HIV-1-specific CD4[+] T cell responses are frequently directed against Gag or other internal virus antigens, whilst responses to the envelope glycoproteins are generally weaker or undetectable (Cohen *et al.*, 2003; Pitcher *et al.*, 1999). HIV-1-specific CD4[+] T cell responses are activated during the acute phase of infection (Gloster *et al.*, 2004; Oxenius *et al.*, 2001); however, the initial increase in activated HIV-1-specific CD4[+] T cells is often very transient (Oxenius *et al.*, 2001) and CD4[+] responses are typically weak in the majority of chronically infected individuals (Musey *et al.*, 1999; Norris *et al.*, 2001). An exception to this is those individuals who naturally control virus replication efficiently over many years (LTNPs), in whom strong virus-specific CD4[+] T cell responses are detected during both acute and chronic infection (Gloster *et al.*, 2004; Rosenberg *et al.*, 1997; Schwartz *et al.*, 1994). Strong HIV-1-specific CD4 responses are also preserved in patients treated with HAART during primary infection (Altfeld *et al.*, 2001; Oxenius *et al.*, 2000; Rosenberg *et al.*, 1997).

The mechanisms responsible for HIV-1-specific CD4[+] T cell responses being so weak in the majority of infected individuals are not completely clear, but it is likely that infection and subsequent loss of virus-specific CD4[+] T cells plays a major role. Activated CD4[+] T cells, particularly HIV-1-specific CD4[+] T cells, are a prime target for infection by HIV-1 (Douek *et al.*, 2002). CD4[+] T cells may encounter virus during cell–cell contact with HIV-1-infected APCs, or non-infected APCs carrying virus bound to the C-type lectin receptor DC-SIGN (Geijtenbeek *et al.*, 2000). Infected CD4[+] T cells are then prone to destruction and loss, which can be induced by (i) direct cytopathic effects of the virus (Fauci, 1988; Levy, 1993), (ii) killing by HIV-1-specific CD8[+] T cells or (iii) ADCC triggered by antibody complexed with gp120 expressed on the infected cell surface (Jewett *et al.*, 1997). There are generally more apoptotic CD4[+] cells than HIV-1-infected cells (Embretson *et al.*, 1993; Finkel *et al.*, 1995) in HIV-1 infection, implying that other mechanisms may also contribute to the loss of CD4[+] T cells. Inactivated HIV-1 virions (Esser *et al.*, 2001) and HIV-1 proteins such as gp120, Tat, Nef and Vpu proteins released into the extracellular environment can induce death in uninfected cells (Alimonti *et al.*, 2003; Azad, 2000). Bystander killing may also be mediated through activation-induced cell death, predominantly through Fas/FasL interactions (Alimonti *et al.*, 2003).

The induction and/or maintenance of HIV-1-specific CD4[+] T cell responses may also be impaired by defects in the activation and/or function of APCs during HIV-1 infection. Incomplete up-regulation of co-stimulatory molecule expression by DCs in secondary lymphoid tissue has been observed in acute HIV-1 infection, but not acute Epstein–Barr virus (EBV) infection (Lore *et al.*, 2002). In addition, the HIV-1 Nef protein is able to down-regulate expression of HLA class II/peptide complexes on the cell surface (Schindler *et al.*, 2003; Stumptner-Cuvelette *et al.*, 2001, 2003), which may impact on both CD4$^+$ T cell activation and the effector functions of HIV-1-specific CD4$^+$ T cells. The interaction between CD4$^+$ T cells and APCs or infected cells may also be impaired by gp120, which has been suggested to interfere with CD4 binding to HLA class II molecules (Fidler *et al.*, 1996). It has also been shown that antibodies directed against the CD4-binding domain of gp120 can prevent the activation of T helper responses to gp120 (Hioe *et al.*, 2001). The mechanism underlying this phenomenon has only recently been elucidated. Gp120, when complexed to anti-CD4 binding antibodies, becomes more resistant to lysosomal degradation within APCs, such that peptide epitopes are not released and presented efficiently to gp120-specific CD4$^+$ T cells. Antibodies to other gp120 regions do not confer this effect (Chien *et al.*, 2004).

Another mechanism by which HIV-1 can evade the virus-specific CD4$^+$ T cell response is via the evolution of virus variants capable of avoiding recognition by this arm of the immune response. Although the role of antigenic variation in HLA class I-restricted epitopes has been widely studied, work on the influence of HIV-1 antigenic variation on HLA class II-restricted CD4$^+$ T cell responses has been limited. This is probably due to difficulties in detecting these responses in HIV-1-infected individuals at any stage of infection. *In vitro* studies of HIV-1-specific class II-restricted CD4$^+$ T helper cell lines derived from SHIV-1-immunized rhesus monkeys and HIV-1-infected individuals have shown that certain synthetic peptide variants of known CD4$^+$ T cell epitopes fail to drive CD4$^+$ T cell proliferation or cytokine secretion, despite being able to bind to the relevant class II molecule (Lekutis & Letvin, 1998; Siliciano *et al.*, 1988). There is also evidence that such variant epitopes evolve *in vivo*. Viral sequence variation occurring within known CD4$^+$ T cell epitopes during the natural course of infection has been demonstrated by analysis of viral RNA or proviral sequences derived from HIV-1-infected patients' PBMCs (Harcourt *et al.*, 1998; Meddows-Taylor *et al.*, 2004), and some of the variant sequences detected were poorly recognized by epitope-specific CD4$^+$ T cells. These findings suggest that in the context of HIV-1 infection, CD4$^+$ T cell activity may exert a pressure on the virus to mutate and escape immune recognition.

Given that CD4$^+$ T helper responses are known to make a critical contribution to the control of many other human and animal virus infections, it follows that impairment of CD4$^+$ responses in chronically infected patients may be central to the eventual breakdown in control of HIV-1 replication. The importance of CD4$^+$ T cells in HIV-1 infection is suggested by associations between the magnitude of HIV-1-specific T cell responses and the efficiency of control of virus replication *in vivo*, either naturally or after drug therapy. LTNPs who naturally maintain very low viral loads over many years show strong CD4$^+$ T cell proliferative responses to HIV-1 antigens (Norris *et al.*, 2004;

Rosenberg *et al.*, 1997; Schwartz *et al.*, 1994), and other studies have shown an inverse correlation between viral load and proliferative responses to Gag p24 antigen (Kalams *et al.*, 1999b; Rosenberg *et al.*, 1997). Likewise, cytokine production by HIV-1-specific CD4[+] T cells has also been documented in patients who control virus replication well (Pitcher *et al.*, 1999) and loss of CD4[+] T cell function *in vitro* predicts progression to disease and a decrease in survival (Dolan *et al.*, 1995; Roos *et al.*, 1995). Furthermore, strong virus-specific CD4[+] T cell responses are typically seen in HIV-2-infected compared with HIV-1-infected individuals (Berry *et al.*, 1998, 2002), the former typically showing a much less severe disease course compared with the latter (Poulsen *et al.*, 1997). As it is thought that impaired virus-specific CD4[+] T cell responses may be the result of infection of CD4[+] T cells by HIV-1 and their subsequent loss, it is difficult to determine cause and effect where associations are observed between strong virus-specific CD4[+] responses and efficient containment of virus replication. However, the further association between strong CD4[+] T cell responses and strong CD8[+] T cell responses in HIV-1 infection (Kalams *et al.*, 1999b) is suggestive of an impact of the former on the latter, and may help to explain the eventual loss of overall immune control of virus replication in patients with very poor HIV-1-specific CD4[+] T cell responses.

The HIV-1-specific CD8[+] T cell response. CD8[+] T cells play a key role in the control of established virus infections. They act to combat virus replication in two ways: (i) by recognition and lysis of infected cells expressing virus-derived epitopes in conjunction with HLA class I molecules and (ii) via production of soluble factors with antiviral activity (Fig. 4).

Activated HIV-1-specific CD8[+] T cells exert cytotoxic effector functions mediated via the release of perforin and granzymes (Gulzar & Copeland, 2004; Smyth *et al.*, 2001) and may also limit HIV-1 replication by their secretion of several types of soluble factor. These include cytokines such as IFN-γ (Emilie *et al.*, 1992; Meylan *et al.*, 1993; Wells *et al.*, 1991) that trigger intracellular pathways for inhibition of virus replication; chemokines such as RANTES, MIP-1α and MIP-1β (Cocchi *et al.*, 1995; Price *et al.*, 1998; Wagner *et al.*, 1998) which help to prevent infection of new host cells by competing for binding to, or inducing down-regulation of, the CCR5 chemokine receptor required for HIV-1 entry; and an as-yet-unidentified soluble factor with antiviral activity. This factor, termed CAF (CD8[+] cell-derived Anti-HIV-1 inhibitory Factor), suppresses HIV-1 replication at the transcriptional level and in a non-cytotoxic fashion (Levy *et al.*, 1996; Mackewicz & Levy, 1992). Interestingly, CAF is active against many different strains of HIV-1, in addition to HIV-2 and SIV (Barker, 1999).

During primary HIV-1 infection, virus-specific CD8[+] T cells are activated (Borrow *et al.*, 1994; Koup *et al.*, 1994) and clonally expand in response to the increasing

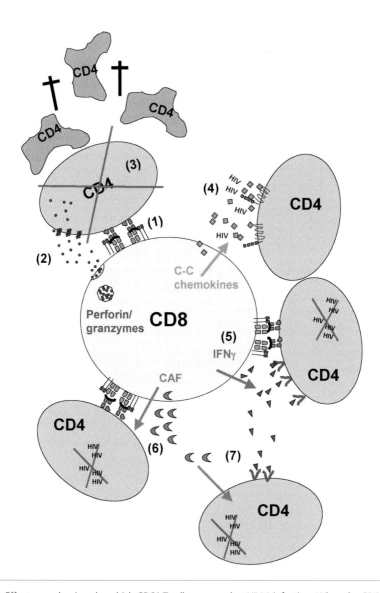

Fig. 4. Effector mechanisms by which CD8+ T cells may combat HIV-1 infection. When the CD8+ T cell receptor engages virus-derived peptides presented by HLA class I molecules on the surface of infected cells (e.g. CD4+ T cells) (1), the CD8+ cell is activated and may initiate cytotoxic mechanisms and/or secrete soluble factors that inhibit HIV-1 replication. Via the exocytosis of pre-formed granules containing cytotoxic effector molecules such as perforin and granzymes (2), HIV-1-infected cells are induced to undergo apoptosis (3). Secretion of CCR5 co-receptor-binding C-C chemokines such as RANTES and MIP-1α/β blocks the interaction between HIV-1 gp120 and CCR5, thus preventing virus entry (4). IFN-γ is able to trigger a variety of intracellular pathways for inhibition of virus replication upon binding its cognate receptor on infected cells (5). The mode of action of CD8+ cell-derived Anti-HIV inhibitory Factor (CAF) is poorly defined, but it is thought to suppress HIV-1 replication at the transcriptional level (6). IFN-γ and CAF may also inhibit virus replication in neighbouring (bystander) cells (7).

antigenic load (Pantaleo et al., 1994; Wilson et al., 2000). The expansion of CD8[+] T cells in acute infection can be huge, with up to around 50 % of total CD8[+] T cells being HIV-1-specific at the peak of the response (McMichael & Phillips, 1997). As primary viraemia resolves, the expanded anti-HIV-1 CD8[+] T cell population partially contracts (Wilson et al., 2000) as a result of apoptosis (Roos et al., 1994; Selin & Welsh, 1994); however, HIV-1-specific CD8[+] T cell frequencies remain elevated throughout the chronic asymptomatic phase of infection, with T cells responsive to more dominant HIV-1 epitopes typically constituting around 1–3 % of total CD8[+] T cells at this time (Altman et al., 1996; Ogg et al., 1998). Maintenance of HIV-1-specific CD8[+] T cells at high frequencies is likely due to continued antigenic stimulation; this is supported by the finding that treatment with HAART during chronic HIV-1 infection results in a decay of HIV-1-specific CD8[+] T cell numbers (Dalod et al., 1999; Gray et al., 1999; Kalams et al., 1999a; Ogg et al., 1999). In the absence of antiretroviral therapy, high numbers of HIV-1-specific CD8[+] T cells may persist into late infection, and in some patients, may even transiently increase as the viral load elevates during progression to AIDS. However, HIV-1-specific CD8[+] T cell frequencies typically decline in the end stage of infection as control of virus replication is ultimately lost.

Multiple lines of evidence support a crucial role for CD8[+] T cells in control of HIV-1 replication in infected individuals. CD8[+] T cells can limit HIV-1 replication *in vitro* (Baier et al., 1995; Buseyne & Riviere, 1993; Buseyne et al., 1996; Cocchi et al., 1995; Vella & Daniels, 2003; Walker et al., 1987) and lyse HIV-1-infected CD4[+] T cells *in vitro* (Yang et al., 1996). *In vivo*, HIV-1-specific CTLs are induced in primary infection prior to the production of nAbs, their expansion coinciding temporally with the containment of the acute burst of virus replication (Borrow et al., 1994; Connick et al., 2001; Koup et al., 1994; Wilson et al., 2000). Good HIV-1-specific CD8[+] T cell responses are also sustained over time in LTNPs (Greenough et al., 1994; Harrer et al., 1994; Klein et al., 1995; Pantaleo et al., 1995; Rinaldo et al., 1995). Importantly, *in vivo* depletion of CD8[+] T cells leads to abrogated control of SIV replication in macaques (Jin et al., 1999; Schmitz et al., 1999). Further direct evidence that CD8[+] T cells exert strong antiviral effects *in vivo* is provided by the emergence of virus variants bearing mutations that confer escape from epitope-specific CD8[+] T cell responses (Borrow et al., 1997; Goulder et al., 1997a; Price et al., 1997). In addition, the association of certain HLA class I alleles with slower or more rapid progression to AIDS also implicates class I-restricted CD8[+] T cells in the control of HIV-1 infection (Carrington et al., 1999; Kaslow et al., 1996). That HIV-1-specific CD8[+] T cell responses have been observed in ESN individuals (Fowke et al., 2000; Kaul et al., 2000, 2001; Rowland-Jones et al., 1998) suggests that the CD8[+] response may also be effective in preventing full-blown infection that would otherwise culminate in seroconversion.

(i) Mechanisms by which HIV-1 may evade control by the virus-specific CD8[+] T cell response. Although the virus-specific CD8[+] T cell response plays an important role in the control of viraemia during HIV-1 infection, virus replication is not completely contained after the acute viral burst but continues at moderate to high levels in the majority of infected individuals. A variety of mechanisms are thought to contribute to HIV-1's ability to sustain high levels of replication in the face of the host CD8[+] T cell response. These include virus strategies for resisting control by CD8[+] T cell effector mechanisms and for avoiding recognition by CD8[+] T cells (summarized in Table 1).

As discussed earlier, one of the mechanisms by which CD8[+] T cells combat virus infections is by inducing destruction of infected cells via the apoptotic pathway. Some of the cell types infected by HIV-1, e.g. macrophages, may show an inherent resistance to CTL lysis (Kedzierska & Crowe, 2002; Lum & Badley, 2003). HIV-1 has also developed mechanisms to prevent or delay apoptosis of the cells it infects (Alimonti *et al.*, 2003; Benedict *et al.*, 2002). The HIV-1 Nef protein has been shown to inhibit Fas and TNF-R-mediated apoptosis of infected cells by binding to and inhibiting apoptosis signal regulating kinase-1, ASK-1 (Geleziunas *et al.*, 2001; Yoon *et al.*, 2001). Nef also phosphorylates and inactivates the pro-apoptotic protein Bad (Wolf *et al.*, 2001) and blocks the mitochondrial-induced apoptotic pathway in infected cells. In addition, Vpr has been shown to stimulate an increase in anti-apoptotic Bcl-2 while simultaneously decreasing pro-apoptotic Bax (Conti *et al.*, 1998), although the involvement of Vpr in inhibiting CTL-mediated apoptosis of HIV-1-infected cells is not clear (Lewinsohn *et al.*, 2002). HIV-1 may also provide infected cells with a means of counter-attack against CTLs. The HIV-1 Nef protein can up-regulate FasL expression on the surface of infected cells, which, through interaction with Fas expressed on CTLs, can trigger their death (Xu *et al.*, 1999). Apoptosis of CD8[+] T cells in HIV-1-infected patients might also be mediated by HIV-1-infected macrophages. HIV-1 gp120 binding to the CXCR4 chemokine receptor can up-regulate the expression of membrane-bound TNF-α and TNFRII on macrophages and CD8[+] T cells, respectively, and promote CD8[+] T cell apoptosis. This may represent another mechanism by which HIV-1-infected cells may avoid CTL killing *in vivo* (Herbein *et al.*, 1998).

HIV-1 also employs several strategies for avoiding recognition by the host CD8[+] T cell response. Latently infected cells, in which no viral proteins are expressed, are essentially 'invisible' to virus-specific T cells. Such cells constitute an important reservoir of infection, from which virus can subsequently be reactivated. HIV-1 also replicates within immune privileged sites in the body, such as the brain, to which access of T cells is limited, and where MHC expression levels are low and an immunoregulatory environment typically prevails (infection of immune privileged sites and establishment of latency are reviewed in Hamer, 2004; Persaud *et al.*, 2003; Saksena & Potter, 2003;

Table 1. Strategies used by HIV-1 to evade CD8$^+$ T cell control

Evasion strategy	Reference(s)
Resistance to CD8$^+$ cytotoxic T cell (CTL)-mediated cell death	
HIV-1 may infect cell types with an inherent resistance to CTL lysis	Lum & Badley (2003); Kedzierska & Crowe (2002)
HIV-1 Nef can inhibit Fas/FasL-mediated and TNF-R-mediated apoptosis of infected cells	Geleziunas *et al.* (2001); Yoon *et al.* (2001)
Nef-mediated inactivation of the pro-apoptotic protein Bad can lead to inhibition of cellular apoptosis induced by CD8$^+$ T cells	Wolf *et al.* (2001)
Counter-attack of HIV-infected cells against CTL	
Nef-mediated up-regulation of FasL on the surface of infected cells can allow for engagement of Fas on CD8$^+$ T cells and subsequent CD8$^+$ T cell apoptosis	Xu *et al.* (1999)
Gp120/CXCR4-induced up-regulation of membrane-bound TNF-α on macrophages can stimulate apoptosis in TNFRII-expressing CD8$^+$ T cells	Herbein *et al.* (1998)
Avoidance of recognition	
Establishment of latency in selected cell types (where no viral protein is expressed) allows the virus to 'hide' away from the immune response	Reviewed by Hamer (2004); Saksena & Potter (2003); Wassef *et al.* (2003); Persaud *et al.* (2003)
Infection of cells within immune privileged sites prevents CD8$^+$ T cell recognition and killing	As above
Nef-mediated down-regulation of HLA class I expression on infected cells can reduce/abrogate recognition by CD8$^+$ T cells	Swann *et al.* (2001); Le Gall *et al.* (2000); Collins *et al.* (1998)
The evolution of appropriate mutations within the viral genome may allow for escape from the epitope-specific CD8$^+$ T cell response. Mutations within epitope flanking regions may affect processing of the epitope and its subsequent presentation to T cells. Mutations within the epitope itself may reduce/abrogate binding to MHC class I or alter TCR contact residues, leading to impaired recognition of peptide : MHC complexes by CD8$^+$ T cells.	Draenert *et al.* (2004b); Allen *et al.* (2004); Yokomaku *et al.* (2004); Borrow *et al.* (1997); Goulder *et al.* (1997a); Price *et al.* (1997); Couillin *et al.* (1994); Rammensee & Monaco (1994); McMichael & Phillips (1997); McAdam *et al.* (1995); Klenerman *et al.* (1994)

Shearer, 1998; and Wassef *et al.*, 2003). However, active HIV-1 replication also occurs in multiple systemic sites, including within lymphoid tissues. Here, CD8$^+$ T cell recognition of infected cells might be impaired by down-regulation of surface MHC class I (Kerkau *et al.*, 1989; Scheppler *et al.*, 1989), resulting in a reduction in the density of antigenic complexes available for CD8$^+$ T cell activation. Three HIV-1 proteins have been shown to affect MHC class I expression in infected cells *in vitro* and these may well act in a similar fashion *in vivo* (reviewed by Kamp *et al.*, 2000). (i) In the presence of Nef, class I molecules are synthesized normally and transported towards the cell surface; however, HLA-A and -B molecules are ultimately redirected to the *trans*-Golgi network for degradation (Le Gall *et al.*, 1998; Schwartz *et al.*, 1996; Swann *et al.*, 2001). Nef-mediated class I down-regulation has been shown to protect HIV-1-infected cells from killing by CD8⁺ CTLs (Collins *et al.*, 1998). (ii) An alternatively spliced 'two-exon' form of HIV-1 Tat has been shown to significantly decrease the promoter activity of the MHC class I gene, thus representing a second mechanism by which class I expression on the surface of HIV-1-infected cells is reduced (Howcroft *et al.*, 1993; Weissman *et al.*, 1998). (iii) HIV-1 Vpu has been shown to down-regulate surface MHC class I molecule expression by inducing rapid loss of newly synthesized class I alpha chains (Kerkau *et al.*, 1997).

In spite of these mechanisms, the CD8⁺ T cell response nonetheless exerts a considerable pressure on virus replication in HIV-1-infected individuals. In response to this selection pressure, the virus may evolve 'escape' variants bearing amino acid changes that result in reduced or ablated recognition/control by key epitope-specific components of the virus-specific CD8 response. It is increasingly clear that CTL escape represents a major immune evasion mechanism for HIV-1 (Goulder & Watkins, 2004).

HIV-1's rapid replication kinetics (10^9–10^{10} new virions are produced each day; Ho *et al.*, 1995; Wei *et al.*, 1995) and the lack of proofreading activity by the error-prone viral reverse transcriptase enzyme (the viral mutation rate is ~1 nucleotide change per 10^5 bases; Mansky & Temin, 1995) result in the generation of virus variants at an incredible rate (every possible mutation at each site in the genome is typically generated each day; Coffin, 1995). Only mutations that alter the encoded amino acid sequence (non-synonymous mutations) can have detrimental consequences for the continued surveillance of HIV-1 by the immune system. The rate at which mutations are generated in the HIV-1 genome might suggest that the CD8⁺ T cell response would be escaped so rapidly that it would have little impact on control of virus replication. In fact, however, the vast majority of mutations in the HIV-1 genome are highly deleterious to the virus (Nietfield *et al.*, 1995), resulting in a complete lack of viability or conferring a severe replicative disadvantage. For a virus variant to outgrow wild-type virus and establish itself as a dominant quasispecies *in vivo*, the benefit gained by escaping the CD8⁺ T cell response must outweigh the cost to intrinsic virus fitness. Hence the capacity of the virus to escape, particularly from responses directed against functionally important regions of the HIV-1 genome (Kelleher *et al.*, 2001; Wagner *et al.*, 1999), is limited.

Amino acid alterations may affect epitope recognition by CD8⁺ T cells in a number of ways (Koup, 1994). (i) Variations in the regions flanking CD8⁺ T cell epitopes (or within epitopes themselves) can influence antigen-processing mechanisms and impair/prevent generation of the epitope. This phenomenon has been appreciated for some time (Del Val *et al.*, 1991), although selection during the natural course of HIV-1 infection for extra-epitopic mutations that confer escape by affecting the processing of known CD8⁺

T cell epitopes has only recently been documented (Allen *et al.*, 2004; Draenert *et al.*, 2004b; Jones *et al.*, 2004; Yokomaku *et al.*, 2004). (ii) Amino acid changes within the epitope sequence can affect the interaction of the epitope with class I molecules. In some cases, they may completely abrogate binding, whilst in other cases, the affinity of binding is reduced, leading to an increased 'off-rate' (Borrow *et al.*, 1997; Couillin *et al.*, 1994; Goulder *et al.*, 1997a; Price *et al.*, 1997; Rammensee & Monaco, 1994). (iii) Amino acid changes within CD8[+] T cell epitopes can also impact on the interaction of the MHC class I : peptide complex with the T cell receptor (TCR). The TCR contact surface may be altered by amino acid changes at positions within the epitope that are involved in the interaction with the TCR and by changes elsewhere in the epitope that affect the conformation of the peptide : MHC complex. Altered interaction of the peptide : MHC complex with the TCR can result in loss of TCR recognition or in abnormal T cell signalling, such that proliferation but not effector functions are triggered and T cells unable to control the mutant virus might be expanded (McAdam *et al.*, 1995; Rammensee & Monaco, 1994). Alternatively, there may be an induction of partial anergy, such that the response to the wild-type virus is also impaired (CTL antagonism; Klenerman *et al.*, 1994).

It is now appreciated that CTL escape occurs frequently in HIV-1 (and SIV) infection. Evidence for CTL escape being extremely common in HIV-1 infection was provided by analysis of HIV-1 reverse transcriptase sequences in a large cohort of chronically infected individuals, which revealed associations between viral amino acid poly- morphisms and particular HLA class I alleles, a number of the former lying within epitopes known to be presented by the allele concerned (Moore *et al.*, 2002). In line with this, analysis of the extent and kinetics of escape over the course of SIV infection in the macaque model has shown that rapid CTL escape is a typical feature of acute SIV infection (Allen *et al.*, 2000b; O'Connor *et al.*, 2002b, 2003), and that escape also occurs during chronic infection, including during progression to AIDS (D. T. Evans *et al.*, 1999). Similarly comprehensive studies of escape throughout the course of HIV-1 infection have not been carried out, but the picture emerging here appears very similar. Again, rapid escape from epitope-specific CTL responses has been documented in primary HIV-1 infection (Borrow *et al.*, 1997; Price *et al.*, 1997), with a recent study showing that there can be sequential selection for mutations in multiple CTL epitopes within the viral quasispecies in a single individual during the first 6 months of infection (Jones *et al.*, 2004). During acute HIV-1 infection, escape may be promoted by the very high levels of virus replication and the associated huge expansions of CD8[+] T cells, particularly those directed against immunodominant viral epitopes (Borrow *et al.*, 1997; Pantaleo *et al.*, 1994). The fact that nAbs are typically not detectable at this time – and hence the CD8[+] T cell response may be the major controlling force on virus replication – may also add to the incentive for virus escape.

During mid-stage HIV-1 infection, escape virus variants continue to evolve (Brander *et al.*, 1998; Goulder *et al.*, 1997b), but frequently are not selected for to completion in the viral quasispecies. A complex picture of shifting hierarchies of immunodominance among epitope-specific CD8[+] T cell responses and corresponding oscillations in the composition of the viral quasispecies has been drawn (Nowak *et al.*, 1995). Escape may be less common at this time because (i) virus-specific CD8[+] T cell frequencies are not as high as in acute infection; (ii) the CD8[+] T cell response is broader than in primary infection (Draenert *et al.*, 2004a); (iii) any epitope-specific component of the response making a dominant contribution to control of virus replication is likely directed against epitopes where escape would be associated with high costs to intrinsic virus fitness (or it would probably otherwise have been escaped during acute infection); and (iv) control of virus replication is now being mediated by both humoral and cell-mediated immunity (hence the advantage gained by evading CTL control is lower).

Emergence of virus escape variants has also been documented during late infection and may contribute to loss of control of virus replication at this time (Goulder *et al.*, 1997a). It is hypothesized that responses escaped in late infection may be directed against epitopes where escape is associated with high fitness costs to virus replication, but where compensatory mutations that reduce the fitness costs associated with the primary escape mutation can ultimately be evolved, allowing eventual escape. The best documented example of such a response is that to the HLA-B2705-restricted epitope KRWIILGLNK (KK10). This epitope is located within a structurally important region of Gag p24 which is involved in the conformational multimerization of p24 during capsid formation (Zhang *et al.*, 1996). Few amino acid substitutions within or near this epitope can thus be tolerated. HLA-B2705[+] individuals typically make a strong response to the KK10 epitope, which is associated with good control of virus replication. Ultimately, in HIV-1[+] HLA-B27[+] individuals infected for many years, substitution of the arginine residue at position 2 of the epitope with lysine or glycine can occur, resulting in impaired HLA-B27 binding and subsequent escape from the epitope-specific CD8[+] T cell response (Feeney *et al.*, 2004; Goulder *et al.*, 1997a; Kelleher *et al.*, 2001). However, the acquisition of this mutation is always preceded by a mutation at position 6 of the epitope which is thought to compensate for the reduction in virus fitness caused by the initial position 2 mutation (although virus reversion to wild-type after loss of the HLA-B27-restricted CTL response suggests that the fitness compensation is only partial; Kelleher *et al.*, 2001).

The potential for CD8[+] T cells to achieve efficient long-term control of HIV-1 replication, and the key role played by evolution of escape virus variants in evasion of CTL control are suggested by associations between immunodominant CD8[+] T cell responses that are not escaped (or are escaped only at high fitness costs to the virus) and good control of virus replication. Examples include the HLA-B27-restricted response to the Gag KK10 epitope discussed earlier and also the response to the HLA-B57-restricted Gag TSTLQEQIGW (TW10) epitope, escape from which is again associated with high costs to virus fitness and frequently involves a combination of amino acid changes (Altfeld *et al.*, 2003). Interestingly, lack of escape from the response to this epitope is associated with very effective long-term control of HIV-1 replication (Migueles *et al.*, 2003). By contrast, in typical HIV-1-infected individuals, CD8+ T cell responses that make a dominant contribution to control of virus replication may be escaped in primary infection, leaving less efficacious responses to contain virus replication over time. The extent to which the virus-specific CD8 response is evaded by CTL escape in the acute phase of chronic infection may thus be one of the factors determining the subsequent efficiency of control of virus replication (Jones *et al.*, 2004).

CTL escape variant selection may be important not only in determining the course of infection in a given infected individual, but also at a population level. Transmission of virus variants bearing escape-conferring mutations to individuals sharing the same HLA type may impact on the immune response and efficiency of containment of virus replication in the newly infected individual. This was evidenced by analysis of mother-to-child transmission of virus bearing the HLA-B27 KK10 epitope escape mutation; this mutation was found to be maintained following transmission, and HLA-B27+ children infected with the escape virus mounted a response to otherwise subdominant epitopes and controlled virus replication poorly (Goulder *et al.*, 2001). However, other studies have shown reversion of escape mutations to the wild-type sequence following transmission (Friedrich *et al.*, 2004; Leslie *et al.*, 2004), probably because these mutations were associated with higher costs to virus fitness, hence were not maintained in the absence of the selection pressure that originally drove them. It might therefore be predicted that over time, although some CD8+ T cell epitopes will be lost from the transmitted HIV-1 population, those where escape involves high costs to virus fitness will be retained – perhaps ultimately resulting in greater focusing of CD8+ T cell responses against these less-readily escaped (and hence more valuable) epitopes.

We have described in detail the mechanisms by which HIV-1 can reduce its chances of being recognized and eradicated by the CD8+ T cell response. In addition, there is evidence to suggest that the functional capacity of CD8+ T cells may not be optimal; this may also contribute to HIV-1 persistence *in vivo*. A high proportion of HIV-1-specific CD8+ T cells in most HIV-1-infected individuals differ phenotypically from antigen-specific CD8+ T cells in other persistent virus infections such as EBV and cytomegalovirus infection. Although phenotypically activated (Altman *et al.*, 1996; Appay *et al.*, 2000; Shankar *et al.*, 2000), HIV-1-specific CD8+ T cells fail to show a fully mature effector phenotype (Appay *et al.*, 2000, 2002; Petrovas *et al.*, 2004), have a low perforin content (indicative of poor cytolytic capacity; Appay *et al.*, 2000, 2002; Champagne *et al.*, 2001) and have a limited ability to proliferate *ex vivo* (Migueles *et al.*, 2002). In addition, there may be abnormalities in their ability to produce the effector cytokine IFN-γ, although this remains controversial (Goepfert *et al.*, 2000; Goulder *et al.*, 2000; Kostense *et al.*, 2001; Shankar *et al.*, 2000). It has been suggested that HIV-1-specific CD8$^+$ T cells may be anergic, as their secretion of cytokines (Shankar *et al.*, 2000) and lytic activity can be dramatically up-regulated after short exposure to IL-2 *in vitro* (Shankar *et al.*, 2000; Trimble & Lieberman, 1998). The underlying cause of inappropriate CD8$^+$ T cell maturation/T cell anergy is unclear; however, it has been proposed that a lack of CD4$^+$ T cell help (perhaps in particular CD4$^+$ T cell-derived IL-2) may be involved. As discussed earlier, T helper cell function is impaired early after HIV-1 infection; this may impact on the induction and/or maintenance of effective HIV-specific CD8$^+$ T cell responses. Indeed, the frequency of Gag-specific CD4$^+$ T cells producing both IL-2 and IFN-γ has been shown to inversely correlate with viral load and positively correlate with IFN-γ^+ CD8$^+$ T cell frequency in HIV-1 infection (Boaz *et al.*, 2002). Viraemic individuals show a reduced frequency of circulating IL-2-producing CD4$^+$ T cells (Iyasere *et al.*, 2003; Younes *et al.*, 2003), and it was found that exogenously added IL-2 could restore proliferation of these cells *in vitro* (Iyasere *et al.*, 2003).

(ii) Definition of features of the HIV-1-specific CD8$^+$ T cell response associated with efficient control of virus replication. There has been much interest in defining features of the CD8$^+$ T cell response that may be associated with good control of virus replication and favourable disease prognosis in HIV-1-infected individuals. Early research into the role of CD8$^+$ T cells in controlling HIV-1 replication *in vivo* questioned whether restriction of virus replication and subsequent disease prognosis were associated with the magnitude of the HIV-1-specific CD8$^+$ T cell response. Several studies have demonstrated an inverse correlation between HIV-1-specific CD8$^+$ T cell frequency and viral load and/or the rate of disease progression (Greenough *et al.*, 1997; Ogg *et al.*, 1998; Rinaldo *et al.*, 1995; van Baalen *et al.*, 1997), suggesting that a higher frequency CD8$^+$ T cell response is associated with better HIV-1 containment and a more favourable clinical course. However, results from other studies do not support this conclusion (Addo *et al.*, 2003; Cao *et al.*, 2003; Dalod *et al.*, 1999; Draenert *et al.*, 2004b; Edwards *et al.*, 2002; Gea-Banacloche *et al.*, 2000; Masemola *et al.*, 2004). Different conclusions may have been reached depending on the method used to quantitate HIV-1-specific CD8$^+$ T cells (e.g. tetramer, IFN-γ secretion, CTL activity) and whether responses to selected or total HIV-1 epitopes/proteins were studied. There is also increasing evidence to suggest that qualitative aspects of the HIV-1-specific CD8$^+$ T cell response may be as or more important than its magnitude in determining the efficiency of control of virus replication. Among these qualitative factors are the functional capacity of the HIV-1-specific CD8$^+$ T cells and a series of repertoire-related aspects of the HIV-1-specific CD8 response.

As discussed earlier, there is some evidence to suggest that there may be defects in the maturation state of HIV-1-specific CD8$^+$ T cells in the majority of infected individuals, with these cells expressing an 'immature' phenotype associated with expression of low levels of perforin, and low *ex vivo* IL-2 production and proliferative potential (Appay *et al.*, 2000, 2002; Champagne *et al.*, 2001; Migueles *et al.*, 2002; van Baarle *et al.*, 2002). Comparative analysis of HIV-1-specific CD8$^+$ T cells in LTNPs and typical progressors revealed that although high frequencies of specific CD8$^+$ T cells were present in both groups, only cells from the former maintained a high proliferative capacity, and that this proliferation was coupled to increases in perforin expression (Migueles *et al.*, 2002). A more recent study also reported an association between the proportion of HIV-1-specific CD8$^+$ T cells with a CD45RA$^+$ CCR7$^-$ phenotype (considered to be 'fully differentiated effectors') and the efficiency of control of virus replication following withdrawal of therapy in patients who had been treated with antiretroviral agents during primary infection (Hess *et al.*, 2004). It is thus hypothesized that the maturation state of HIV-1-specific CD8$^+$ T cells may be one of the factors that impacts on the efficiency of long-term control of virus replication. It is, however, possible that this may be a reflection (rather than the primary cause) of high levels of virus replication and associated antigenic stimulation (Appay & Rowland-Jones, 2002).

Turning to repertoire-related aspects of the HIV-1-specific CD8$^+$ T cell response, the specific epitopes against which the antiviral CD8 response is directed may be one of the factors that determines the efficiency with which virus replication is contained. It has been convincingly demonstrated that CD8$^+$ T cell responses directed against epitopes in proteins that are expressed early during the viral life cycle (such as Nef, Tat and Rev) control virus replication *in vitro* much more efficiently than responses directed against epitopes in proteins that are expressed later (Ali *et al.*, 2004; van Baalen *et al.*, 2002). This may be because the former have a longer 'window' of opportunity in which to recognize and combat infected cells prior to the release of infectious virus particles (reviewed by Gruters *et al.*, 2002). In addition, epitope display by infected cells may be more efficient in the early stages of infection before Nef-induced down-regulation of surface MHC class I expression occurs. There is also evidence to suggest that CD8$^+$ T cell responses directed at epitopes in early viral proteins may have a high *in vivo* efficacy too. This includes observations of slower disease progression in HIV-1-infected individuals and better control of viraemia in SIV-infected macaques when strong CTL responses were mounted to early proteins (Geretti *et al.*, 1999; van Baalen *et al.*, 1997), reports of very rapid escape from Nef, Tat and Rev-specific CTL responses during acute HIV-1 and SIV infection (Allen *et al.*, 2000b; D. T. Evans *et al.*, 1999; Gruters *et al.*,

2002; Jones *et al.*, 2004; O'Connor *et al.*, 2001, 2003; Price *et al.*, 1997), and efficient control of viraemia associated with vaccine-induced Tat-specific CD8$^+$ T cell responses in SIV/SHIV challenge models (Cafaro *et al.*, 1999; Maggiorella *et al.*, 2004; Stittelaar *et al.*, 2002). The epitope specificity of the HIV-1-specific CD8$^+$ T cell response may also be of importance in determining the efficiency of control of virus replication because, as discussed earlier, CD8$^+$ T cell responses directed against epitopes in regions of viral proteins where there are strong structural/functional constraints on amino acid sequence may not be readily escaped because of the fitness costs associated with acquisition of escape-conferring mutations. Responses directed against such epitopes may thus be able to maintain good control of virus replication over time.

The importance of CTL escape as a means of HIV-1 evasion of control by the CD8 response means that other aspects of the repertoire of CD8$^+$ T cells involved in the anti-viral response that may impact on the extent/kinetics of viral escape may also be key determinants of the efficiency of control of virus replication. Studies in the murine LCMV infection model have illustrated that CTL escape is promoted when CD8$^+$ T cell pressure is focused on a highly immunodominant epitope, and is restricted by broader distribution of the response across a larger number of epitopes (Pircher *et al.*, 1990; Weidt *et al.*, 1995). Likewise, in SIV/SHIV macaque models, rapid escape from narrowly directed vaccine-induced CD8$^+$ T cell responses has been observed (Barouch *et al.*, 2002; Mortara *et al.*, 1998; O'Connor *et al.*, 2002a). The epitope breadth and/or extent of biasing of the HIV-1-specific CD8$^+$ T cell response towards immunodominant (or particularly efficacious) epitopes may likewise impact on the control of virus replication.

The observations that maximum HLA heterozygosity at class I A, B and C loci in HIV-1-infected individuals is associated with slower progression to AIDS, whereas individuals who are homozygous for one or more loci progress rapidly to AIDS (Carrington *et al.*, 1999), and that the number of HLA allele-specific peptide motifs in HIV-1 proteins correlates with the association of HLA alleles with relative rates of disease progression (Nelson *et al.*, 1997), are both consistent with the epitope breadth of the HIV-1-specific CD8$^+$ T cell response being related to the efficiency of contain-ment of virus replication. However, whereas some studies have found an inverse correlation between viral load and the number of different epitopes targeted (Altfeld *et al.*, 2001), others found no relationship between these variables (Addo *et al.*, 2003; Masemola *et al.*, 2004). The relationship between the number of epitopes targeted by the HIV-1-specific CD8$^+$ T cell response and efficiency of control of virus replication might not be expected to be clear-cut, because other factors such as the extent of biasing of CD8$^+$ T cell pressure against dominant epitopes, and the number of (parti-cularly dominant) responses that can be escaped by the virus early in infection might be of importance. We recently documented extensive acute-phase CTL escape and establishment of a high persisting viral load in two patients whose primary HIV-1-specific CD8[+] T cell response was highly biased against a smaller number of dominant epitopes, and much more limited acute-phase escape in a third patient whose dominant CD8[+] T cell response was more co-dominantly directed against a broader range of epitopes and who established a low persisting viral load (Jones *et al.*, 2004).

Not only the epitopes targeted by the HIV-1-specific CD8[+] T cell response, but also the T cell receptors used to recognize them, may impact on how well virus replication is controlled. Several studies have documented associations between the clonal breadth of the HIV-1-specific CD8 responses and control of viraemia. HIV-1-infected individuals showing mono/oligoclonal T cell expansions during primary HIV-1 infection were found to establish high persisting viral loads (Pantaleo *et al.*, 1994, 1997). Likewise, analysis of chronically infected patients suggested that epitope-specific responses that are dominated by limited numbers of T cell clones were associated with poorer control of virus replication (Kalams *et al.*, 1994). In addition, antigen-specific CD8[+] T cells from HIV-2-infected individuals (who typically show a much less severe disease course than HIV-1-infected individuals) showed a broader TCR usage than those from HIV-1-infected individuals, which was associated with an enhanced potential for T cell expansion and IFN-γ production on cross-recognition of variant epitopes (Lopes *et al.*, 2003). As suggested by the latter study, greater heterogeneity of TCR usage in an epitope-specific CD8 response may be beneficial by allowing recognition of variant versions of the epitope peptide, and hence reducing the potential for virus evasion of the response via the acquisition of mutations that affect TCR recognition of the MHC:peptide complex. However, it could equally by envisaged that biasing of TCR usage towards a single (or closely related set of) receptor(s) with a high capacity to recognize potential epitope variants would also be beneficial, and, indeed, an example of this has recently been reported (Dong *et al.*, 2004). TCR use may also affect the avidity of interaction between CD8[+] T cells and the APC/target cells. There is some evidence from both *in vitro* (Yang *et al.*, 2003) and *in vivo* (O'Connor *et al.*, 2002b) studies to suggest that TCR affinity may impact on the efficiency with which CD8[+] T cells mediate control of virus replication, although some of the other variables discussed earlier may be more important (Yang *et al.*, 2003).

What determines the breadth of the HIV-1-specific CD8[+] T cell response is currently unclear. It may be related to several factors including previous pathogen experience (Selin & Welsh, 1994), the efficiency of epitope presentation (Siliciano & Soloski, 1995), and/or the availability of CD4[+] T cell help at the time the response is primed. In addition, given that HIV-1-specific CD8[+] T cells recognize peptides presented in the context of HLA class I, the HLA type of an individual may play a role. There is now strong evidence for associations between certain HLA class I alleles and rate of progression to AIDS (Carrington & Bontrop, 2002; Carrington & O'Brien, 2003; O'Brien *et al.*, 2001; Roger, 1998). Alleles including HLA-B57 (Gillespie *et al.*, 2002; Goulder *et al.*, 1996; Migueles *et al.*, 2000; Tang *et al.*, 2002) and HLA-B27 (den Uyl *et al.*, 2004; Goulder *et al.*, 1997a, 2001; McNeil *et al.*, 1996; O'Brien *et al.*, 2001) are over-represented in LTNPs and are associated with more efficient restriction of virus replication *in vivo* and a longer disease-free survival. Alleles including HLA-B8 (Candore *et al.*, 1998; Kaslow *et al.*, 1990; McNeil *et al.*, 1996), HLA-A29 (Hendel *et al.*, 1999) and HLA-B3502/03/5301 (Gao *et al.*, 2001; Jin *et al.*, 2002) are associated with a poorer restriction of virus replication and more rapid progression to AIDS. Similar associations have been reported between HLA type and progression in HIV-2 infection (Diouf *et al.*, 2002) and between MHC class I type and disease progression in SIV infection of rhesus monkeys (Carrington & Bontrop, 2002; O'Connor *et al.*, 2003). From the previous discussion, it is evident that HLA associations with the outcome of HIV-1 infection may be explained in a number of ways. (i) The dominant HIV-1-specific response(s) typically induced in the context of 'good' alleles may be directed against epitopes where escape is restricted by high costs to virus fitness, as has been suggested in the case of HLA-B57 and HLA-B27 (den Uyl *et al.*, 2004; Goulder & Watkins, 2004). (ii) The breadth of HIV-1 epitopes that can be presented by a given HLA allele may play a role (Nelson *et al.*, 1997). The HLA-B35 alleles provide an excellent example of how subtle differences in peptide binding may influence clinical outcome in virus infection. The HLA-B3501 (Py) and HLA-B3503 (Px) alleles differ by one amino acid at position 116 (which forms the floor of the peptide-binding pocket). This residue dictates the size of the C-terminal amino acid residue that can be bound and directly interacts with residue 9 of the bound peptide. HLA-B3501 peptide binding is more promiscuous in that it will tolerate several amino acids (aside from tyrosine) at position 9 whereas HLA-B3502/03 alleles will only bind peptides with a tyrosine residue at position 9 (Carrington & Bontrop, 2002). Although the mechanism is not clear, this difference probably underlies the marked difference in rate of progression to AIDS observed between HLA-B3501[+] and HLA-B3502/03/5301[+] individuals (Gao *et al.*, 2001). (iii) The breadth and/or functional flexibility of TCRs used to recognize the dominant epitopes presented by a given HLA allele may also be of importance.

In summary, our understanding of what may constitute the most efficacious type of HIV-1-specific CD8[+] T cell response is now increasing, providing opportunities, as discussed next, to exploit this knowledge in HIV-1 vaccine design.

PROSPECTS FOR HIV-1 VACCINE DEVELOPMENT

As reviewed in the Introduction, there is an urgent need to develop safe, effective prophylactic and therapeutic vaccines to combat HIV-1 infection – but it is proving extremely difficult to do so.

One of the reasons for this is that we still have much to learn about the immune correlates of protection against HIV-1 infection and disease progression. Clearance of infection by the host immune system and development of natural immunity to reinfection are not seen in HIV-1 infection, which raises doubts about the feasibility of inducing a host immune response capable of eliminating this infection. There are individuals who, despite repeated exposure to HIV-1, remain persistently seronegative. Efforts to determine the basis of the apparent resistance of ESN individuals to infection have revealed that they frequently exhibit HIV-1-specific CD4$^+$ and CD8$^+$ T cell responses (Clerici *et al.*, 1994; Kaul *et al.*, 2000; Pinto *et al.*, 1995; Rowland-Jones *et al.*, 1995, 1998), and that HIV-1-specific antibodies can be detected at mucosal sites (Belec *et al.*, 1989; Broliden *et al.*, 2001; Devito *et al.*, 2000, 2002; Mazzoli *et al.*, 1997), although whether these responses merely reflect exposure to virus or constitute the basis of protection against widespread systemic infection remains unclear. Interestingly, in one cohort of repeatedly exposed HIV-1-resistant individuals, an increased rate of infection was observed following a decline in their frequency of high-risk exposures (Kaul *et al.*, 2001). Such an effect might be expected if protection against infection was mediated by innate defences and/or required maintenance of virus-specific T cells in an activated, effector state; either way, the finding does not bode well for the success of vaccines that would elicit only transient innate activation and antigenic stimulation.

Another approach to the identification of immune correlates of protection against infection has been to study the basis of vaccine-induced protection in animal models. Infection of macaques with very low doses of SIV or with attenuated SIV can afford considerable protection against subsequent challenge with fully virulent SIV (Almond & Stott, 1999; Daniel *et al.*, 1992; Johnson & Desrosiers, 1998; Johnson *et al.*, 1999). Safety concerns preclude the use of similar approaches in humans (in the latter case because there are concerns about the possibility of attenuated mutants reverting to wild-type and causing disease; Baba *et al.*, 1995, 1999; Greenough *et al.*, 1999; Learmont *et al.*, 1999; Whatmore *et al.*, 1995; Wyand *et al.*, 1997), but it was hoped that definition of the immunologic basis of protection would give insight into approaches that could be pursued to elicit similar protection by safer means. Unfortunately, identification of the immune correlates of protection in such models has not been definitively achieved, possibly suggesting that innate defences or a non-immunologic mechanism of super-infection resistance may be involved. However, other studies in animal models have supported the potential for virus-specific cell-mediated and humoral responses to mediate protection against challenge infection. Passive transfer of human nAbs to non-human primates before or early after infection with wild-type virus has demonstrated protection from infection or an attenuated disease course (Baba *et al.*, 2000; Emini *et al.*, 1992; Haigwood *et al.*, 1996; Mascola *et al.*, 1999, 2000; Prince *et al.*, 1991; Putkonen *et al.*, 1991; Shibata *et al.*, 1999) and illustrates that nAbs could potentially confer

Table 2. Approaches pursued in prophylactic HIV-1 vaccine development

Approach	Reference(s)
Inactivated virus vaccines	
Formalin-inactivated SIV was shown to confer protection in macaques challenged with virulent SIV, but protection was subsequently found to be dependent on the cells in which the virus was grown	Murphey-Corb et al. (1989); Stott (1991)
Live attenuated virus vaccines	
Attenuated SIV was shown to protect macaques against challenge with wild-type SIV, but use of this approach in humans has been precluded by safety concerns	Daniel et al. (1992); Reviewed by Whitney & Ruprecht (2004)
Recombinant monomeric envelope immunogens	
Shown to protect non-human primates against challenge with homologous virus, but failed to elicit antibodies able to neutralize primary HIV-1 isolates in Phase I/II human trials and have failed to show efficacy in Phase III trials	Berman et al. (1990); Reviewed by Letvin et al. (2002); Cohen (2003)
Vaccines designed to elicit virus-specific T cell responses	
Multiple strategies aimed at eliciting HIV-1-specific T cell responses are being developed: these include plasmid DNA vaccines and recombinant virus vectors [e.g. modified vaccinia virus Ankara (MVA), fowlpox virus (FPV), canarypox virus (CPV), replication-deficient adenovirus, Semliki Forest virus (SFV) and Venezuelan equine encephalitis virus (VEEV)]. Combinations of these immunogens are also being employed in prime–boost regimes.	Reviewed by McMichael & Hanke (2003)
CD8[+] T cell-inducing vaccines have been shown to protect against SHIV challenge in macaque models	Barouch et al. (2000); Rose et al. (2001); Shiver et al. (2002); Amara et al. (2001)
Many T cell-inducing vaccines are currently in Phase I/II human trials	Reviewed by Garber et al. (2004) and Estcourt et al. (2004)
Novel immunogens designed to elicit broadly neutralizing antibodies	
Human mAbs capable of neutralizing a broad range of primary HIV-1 isolates have been shown to confer passive protection against SHIV challenge	Mascola et al. (2000)
Many strategies for eliciting broadly neutralizing antibodies are currently being developed (e.g. envelope trimers, deglycosylated/variable loop deleted envelopes, envelope-receptor intermediates and gp41 fusion intermediates)	Reviewed by Estcourt et al. (2004) and Letvin et al. (2002)

sterilizing immunity. Further, vaccination strategies designed to induce virus-specific CD8[+] (+/– CD4[+]) T cell responses have also been shown to be capable of attenuating acute-phase virus replication and/or conferring protection against subsequent development of disease in SIV/SHIV challenge models (Barouch et al., 2000; Rose et al., 2001; Shiver et al., 2002; Amara et al., 2001). One potential caveat to such findings is that it is unclear how the challenge infections used in such model systems relate to human infection. Paradoxically, protective effects may be easier to demonstrate using some of the most pathogenic challenge strains (e.g. SHIV89.6P; Amara *et al.*, 2001; Barouch *et al.*, 2000; Rose *et al.*, 2001; Shiver *et al.*, 2002). Nonetheless, findings from many vaccine studies in animal models lend support to the efforts to design HIV-1 vaccine strategies aimed at eliciting virus-specific nAb and cell-mediated immune responses.

This is corroborated by our current understanding of the immune evasion strategies used by HIV-1 and immune correlates of natural control of HIV-1 replication, as exhibited by LTNPs. A wealth of evidence (reviewed in the previous sections) suggests that optimal long-term containment of HIV-1 infection is associated with efficacious virus-specific $CD8^+$ T cell responses and maintenance of strong HIV-1-specific $CD4^+$ T cell responses. Characteristics of an efficacious HIV-1-specific $CD8^+$ T cell response appear to include determinants of the antiviral capacity of component T cells (e.g. recognition of epitopes in proteins expressed early in the virus life cycle, high avidity and sustained functionality) and, importantly, factors that determine the difficulty encountered by the virus in escaping from the immune response [which can be maximized by directing the response against epitopes where acquisition of escape mutations is associated with high costs to intrinsic virus fitness; broad direction of CTL pressure against multiple sites in the virus; and use of T cell receptor(s) with a high capacity to recognize variant versions of their cognate epitope peptide]. In the context of adequate $CD4^+$ T cell help, the virus-specific nAb response may be able to evolve to contain newly emerging neutralization-resistant virus variants, and virus-specific $CD8^+$ T cells may be maintained in an optimally functional state.

A second important problem faced in HIV-1 vaccine development is the difficulty associated with eliciting the type(s) of immune response that we currently think may be best able to protect against establishment of persistent infection and/or provide efficient long-term control of virus replication. A summary of the approaches taken to prophylactic HIV-1 vaccine development is shown in Table 2; approaches to HIV-1 vaccine design are more extensively reviewed by Estcourt *et al.* (2004); Lehner *et al.* (1999); Letvin *et al.* (2002); McMichael & Hanke (2003); Nathanson & Mathieson (2000); Ruprecht (1999); Ruprecht *et al.* (2000); Shearer *et al.* (1999); Garber *et al.* (2004).

One focus of efforts in the HIV-1 vaccine field has been on development of a vaccine that will protect via induction of an HIV-1-specific antibody response. The efficacy of other commercially licensed vaccines typically correlates with their ability to induce a protective antibody response, but HIV-1's resistance to antibody neutralization (discussed earlier) has so far thwarted efforts to develop a vaccine that will elicit nAbs capable of conferring protection against this infection. Early HIV-1 vaccine development focused on testing whether recombinant envelope subunit vaccines (such as gp120 and gp160) could stimulate nAb responses *in vivo* and afford protection from challenge with wild-type virus. Monomeric recombinant HIV-1 envelope subunits were evaluated as immunogens in non-human primate disease models (where they were shown to confer protection in chimpanzees challenged with homologous HIV-1; Berman *et al.*, 1990; Lubeck *et al.*, 1997) and in human volunteers. Unfortunately, most anti-HIV-1 vaccines of this nature tested to date in humans have only been able to induce antibody that neutralizes laboratory-adapted HIV-1 strains or virus identical to the vaccine strain *in vitro*. Phase I and II trials using recombinant envelope subunit vaccines demonstrated considerable differences between the antibody responses induced by envelope subunit vaccination and natural HIV-1 infection, with the former showing very limited neutralization of primary HIV-1 isolates (Mascola *et al.*, 1996; Montefiori & Evans, 1999). Furthermore, in Phase II trials, subjects vaccinated with recombinant gp120 or a placebo control showed no difference in virological or clinical criteria when naturally infected with HIV-1 during the course of the study (Connor *et al.*, 1998; Graham *et al.*, 1998). Recently, the first Phase III trials assessing the efficacy of a recombinant gp120 vaccine (AIDSVAX) in high-risk human volunteers failed to show any protective effect against naturally acquired HIV-1 infection (Cohen, 2003) (see www.vaxgen.com).

Although some further trials involving similar immunogens are (contentiously) going ahead, it is widely believed that more sophisticated envelope immunogens will be required to elicit broadly neutralizing antibodies. Attempts are currently being made to generate envelope molecules that more closely mimic native envelope trimers (rather than monomers); to engineer gp120 so that nAb epitopes are exposed (e.g. by deletion of glycosylation sites); and to produce immunogens that will elicit antibodies to envelope structures transiently generated during receptor interaction or fusion, although to date such approaches have met with only limited success (Barnett *et al.*, 2001; Binley *et al.*, 1998, 2000; Earl *et al.*, 1994, 1997, 2001; Jeffs *et al.*, 1996; Kang *et al.*, 1999; Paul *et al.*, 1993; Quinones-Kochs *et al.*, 2002; Reitter *et al.*, 1998; Rossio *et al.*, 1998; Wyatt *et al.*, 1993; Yang *et al.*, 2002).

Other efforts in the HIV-1 vaccine field focus on eliciting HIV-1-specific CD8[+] and/ or CD4[+] T cell responses to control the level of virus replication *in vivo*. Whilst, as reviewed earlier, there is strong evidence to support the potential for CD8[+] T cells to have a substantial impact on the *in vivo* viral load, the utility of vaccine-elicited CD8[+] T cell responses in effecting protection in human virus infections is currently unproven, and remains questionable (McMichael & Hanke, 2003; Zinkernagel, 2003). Immunization strategies for eliciting strong, efficacious CD8[+] T cell responses to vaccine antigens in humans also need to be optimized, and techniques for modulating epitope and clonal breadth developed. Current approaches are predominantly focused on the use of DNA plasmids and live (typically virus) vectors for delivery of HIV-1 antigens.

DNA vaccination uses eukaryotic expression vectors to deliver genes encoding whole protein fragments and/or specific T cell epitopes to the host for induction of an immune response (Estcourt *et al.*, 2004). When taken up by host cells, the DNA is transcribed within the nucleus and mRNA is subsequently translated into protein within the cytosol, which can then be directly presented or cross-presented for priming of a CD8[+] T cell response. Although this approach has proved very effective for eliciting T cell responses in animal models, it appears less effective in humans. This may relate in part to the differential ability of CpG motifs in the DNA plasmids used in most studies to activate innate responses in different species. Different CpG motifs have differential effects on immune cells from different species (Mutwiri *et al.*, 2003; Rankin *et al.*, 2001). A variety of approaches are being attempted to try and enhance the efficacy of plasmid DNA as a vector for immunization. These include incorporation of 'humanized' CpG motifs into the DNA, repeated deliveries of DNA vaccines to boost immunity, improving the efficiency of delivering the DNA into cells, and the use of genetically encoded adjuvants such as cytokines, chemokines and co-stimulatory molecules to boost immune responses (Estcourt *et al.*, 2004).

A variety of live vectors are also being used to deliver encoded HIV-1 proteins, protein subunits or T cell epitopes to the host for the induction of humoral and cell-mediated immunity. Early studies showed that immunization with recombinant vaccinia virus was able to elicit HIV-1-specific cellular and humoral immunity in non-human primates (Hu *et al.*, 1987; Shen *et al.*, 1991). Modified vaccinia virus Ankara (MVA) (Meyer *et al.*, 1991) has also been shown to be highly immunogenic in monkeys and constructs bearing HIV-1 genes have protected macaques from SIV- and SHIV-1-induced disease (Barouch *et al.*, 2001; Hirsch *et al.*, 1996; Ourmanov *et al.*, 2000; Seth *et al.*, 2000). Several other viral and bacterial vectors have been used to deliver HIV-1 antigens, including replication-deficient adenovirus (T. G. Evans *et al.*, 1999; Ferrari *et al.*, 1997; Shiver *et al.*, 2002), canarypox (Egan *et al.*, 1995; Pialoux *et al.*, 1995), single-stranded RNA viruses (Davis *et al.*, 2000; Mossman *et al.*, 1996), BCG (Yasutomi *et al.*, 1993) and enteric bacteria (Fouts *et al.*, 1995; Hone *et al.*, 1996). One drawback to the use of some of these vectors in humans is that pre-existing host immunity to the (virus) vector may limit their effectiveness. Another drawback is that immune responses are induced to both the vector and the encoded HIV-1 antigen(s). There may thus be problems with antigenic competition between vector and vaccine antigens, and efforts to boost responses by repeated immunizations may be complicated by vector-specific antibody production.

To overcome the latter problems, mixed modality or 'prime/boost' vaccines, typically involving an initial immunization with plasmid DNA encoding HIV-1 gene(s), followed by a boost with adjuvanted protein or a virus vector encoding the same HIV-1 gene(s), are increasingly being employed to improve the potency and durability of HIV-1-specific immune responses. Prime–boost regimes (particularly DNA immunization followed by a boosting with poxvirus vectors) have been shown to confer significant protective effects in rhesus macaques challenged with SHIV (Allen *et al.*, 2000a; Amara *et al.*, 2001; Cherpelis *et al.*, 2001; Hanke *et al.*, 1999; Letvin *et al.*, 1997; Robinson *et al.*, 1999). A number of ongoing Phase I and II trials are focusing on similar prime–boost regimes. Unfortunately, recent immunogenicity results from one such trial were not encouraging (Cohen, 2004), although this type of approach warrants further evaluation.

Two further problems which both nAb and T cell-based HIV-1 vaccines confront are the possibility that high levels of immunity may need to be induced at mucosal sites to achieve protection (Lehner *et al.*, 1999), and the challenge posed by the tremendous diversity of HIV-1 variants worldwide. HIV-1 clades can differ at the amino acid level by as much as 30 % in Env, and 10–15 % in more conserved viral proteins such as Gag; and even within a clade, amino acid sequence diversity in some proteins may be as high as 10 %. It seems very unlikely that a single immunogen, even if designed to target multiple conserved viral epitope(s), would be able to elicit a response capable of cross-recognizing and protecting against such a pattern of viral strains. A recent study emphasized how even broadly directed CD8$^+$ T cell responses may fail to provide cross-protection against a closely related HIV-1 strain (Altfeld *et al.*, 2002). Potential solutions may be to design vaccines that include a combination of immunogens (possibly representing the consensus sequence, or the sequence of a 'representative' member of different clades), or to develop different vaccines for use in geographic locations where particular HIV-1 subtypes are prevalent.

Development of therapeutic HIV-1 vaccines constitutes an even greater challenge than development of a prophylactic vaccine. Here, it will be necessary to induce protective immunity in the face of a previously primed response (where expansion of T and B cell responses of new specificities may be limited due to the phenomenon of 'original antigenic sin'; reviewed by Muller, 2004), and in the context of the many immunological defects induced during the course of infection. The latter may include not only defects induced in the adaptive response (e.g. destruction of HIV-1-specific CD4$^+$ T cells), but also infection-associated defects in the innate immune system which may impact on vaccine efficiency. Notably, one of the most successful therapeutic vaccination studies in the SIV macaque model involved immunization with DCs pulsed with inactivated HIV-1 (Lu *et al.*, 2003). The approach currently being pursued is to try and use therapeutic immunization in the context of antiretroviral therapy to boost virus-specific immunity so that therapy can be reduced or withdrawn (Autran *et al.*, 2003). It is hoped that the (at least partial) restoration of immune function that is known to occur when virus replication is suppressed by antiretroviral treatment (Autran *et al.*, 1997, 1999) may allow vaccination strategies a better chance of success in augmenting antiviral immunity. However, there are many hurdles to be overcome before this goal can be realized.

In summary, more than 20 years after the discovery of HIV-1, development of both prophylactic and therapeutic vaccines to combat this infection still remains a tremendous challenge. Nonetheless, advances in our understanding of the interaction between this virus and the host immune response, coupled with international efforts to develop and test novel antiviral vaccine strategies, constitute positive steps towards eventual success. Much of the experience gained in HIV-1 vaccine development will also be of value in the design of prophylactic and therapeutic vaccines for other chronic infections and for tumours.

ACKNOWLEDGEMENTS

The authors are supported by core funding from The Edward Jenner Institute for Vaccine Research, and by NIH grant AI 41530. This is publication number 99 from The Edward Jenner Institute for Vaccine Research.

REFERENCES

Aasa-Chapman, M. M., Hayman, A., Newton, P., Cornforth, D., Williams, I., Borrow, P., Balfe, P. & McKnight, A. (2004). Development of the antibody response in acute HIV-1 infection. *AIDS* **18**, 371–381.

Abdelwahab, S. F., Cocchi, F., Bagley, K. C., Kamin-Lewis, R., Gallo, R. C., DeVico, A. & Lewis, G. K. (2003). HIV-1-suppressive factors are secreted by CD4+ T cells during primary immune responses. *Proc Natl Acad Sci U S A* **100**, 15006–15010.

Addo, M. M., Yu, X. G., Rathod, A. & 17 other authors (2003). Comprehensive epitope analysis of human immunodeficiency virus type 1 (HIV-1)-specific T-cell responses directed against the entire expressed HIV-1 genome demonstrate broadly directed responses, but no correlation to viral load. *J Virol* **77**, 2081–2092.

Ahmad, R., Sindhu, S. T., Tran, P., Toma, E., Morisset, R., Menezes, J. & Ahmad, A. (2001). Modulation of expression of the MHC class I-binding natural killer cell receptors, and NK activity in relation to viral load in HIV-infected/AIDS patients. *J Med Virol* **65**, 431–440.

Akira, S. & Hemmi, H. (2003). Recognition of pathogen-associated molecular patterns by TLR family. *Immunol Lett* **85**, 85–95.

Ali, A., Lubong, R., Ng, H., Brooks, D. G., Zack, J. A. & Yang, O. O. (2004). Impacts of epitope expression kinetics and class I downregulation on the antiviral activity of human immunodeficiency virus type 1-specific cytotoxic T lymphocytes. *J Virol* **78**, 561–567.

Alimonti, J. B., Ball, T. B. & Fowke, K. R. (2003). Mechanisms of CD4+ T lymphocyte cell death in human immunodeficiency virus infection and AIDS. *J Gen Virol* **84**, 1649–1661.

Allen, T. M., Vogel, T. U., Fuller, D. H. & 11 other authors (2000a). Induction of AIDS virus-specific CTL activity in fresh, unstimulated peripheral blood lymphocytes from rhesus macaques vaccinated with a DNA prime/modified vaccinia virus Ankara boost regimen. *J Immunol* **164**, 4968–4978.

Allen, T. M., O'Connor, D. H., Jing, P. & 16 other authors (2000b). Tat-specific cytotoxic T lymphocytes select for SIV escape variants during resolution of primary viraemia. *Nature* **407**, 386–390.

Allen, T. M., Altfeld, M., Yu, X. G. & 17 other authors (2004). Selection, transmission, and reversion of an antigen-processing cytotoxic T-lymphocyte escape mutation in human immunodeficiency virus type 1 infection. *J Virol* **78**, 7069–7078.

Almond, N. & Stott, J. (1999). Live attenuated SIV – a model of a vaccine for AIDS. *Immunol Lett* **66**, 167–170.

Altfeld, M., Rosenberg, E. S., Shankarappa, R. & 15 other authors (2001). Cellular immune responses and viral diversity in individuals treated during acute and early HIV-1 infection. *J Exp Med* **193**, 169–180.

Altfeld, M., Allen, T. M., Yu, X. G. & 13 other authors (2002). HIV-1 superinfection despite broad CD8+ T-cell responses containing replication of the primary virus. *Nature* **420**, 434–439.

Altfeld, M., Addo, M. M., Shankarappa, R. & 13 other authors (2003). Enhanced detection of human immunodeficiency virus type 1-specific T-cell responses to highly variable regions by using peptides based on autologous virus sequences. *J Virol* **77**, 7330–7340.

Altman, J. D., Moss, P. A. H., Goulder, P. J. R., Barouch, D. H., McHeyzer-Williams, M. G., Bell, J. I., McMicheal, A. J. & Davis, M. M. (1996). Phenotypic analysis of antigen-specific T lymphocytes. *Science* **274**, 94–96.

Amara, R. R., Villinger, F., Altman, J. D. & 19 other authors (2001). Control of a mucosal challenge and prevention of AIDS by a multiprotein DNA/MVA vaccine. *Science* **292**, 69–74.

Appay, V. & Rowland-Jones, S. L. (2002). Premature ageing of the immune system: the cause of AIDS? *Trends Immunol* **23**, 580–585.

Appay, V., Nixon, D. F., Donahoe, S. M. & 13 other authors (2000). HIV-specific CD8(+) T cells produce antiviral cytokines but are impaired in cytolytic function. *J Exp Med* **192**, 63–75.

Appay, V., Dunbar, P. R., Callan, M. & 18 other authors (2002). Memory CD8+ T cells vary in differentiation phenotype in different persistent virus infections. *Nat Med* **8**, 379–385.

Ariyoshi, K., Harwood, E., Chiengsong-Popov, R. & Weber, J. (1992). Is clearance of HIV-1 viraemia at seroconversion mediated by neutralising antibodies? *Lancet* **340**, 1257–1258.

Aste-Amezaga, M., D'Andrea, A., Kubin, M. & Trinchieri, G. (1994). Cooperation of natural killer cell stimulatory factor/interleukin-12 with other stimuli in the induction of cytokines and cytotoxic cell-associated molecules in human T and NK cells. *Cell Immunol* **156**, 480–492.

Autran, B., Carcelain, G., Li, T. S., Blanc, C., Mathez, D., Tubiana, R., Katlama, C., Debre, P. & Leibowitch, J. (1997). Positive effects of combined antiretroviral therapy on CD4+ T cell homeostasis and function in advanced HIV disease. *Science* **277**, 112–116.

Autran, B., Carcelaint, G., Li, T. S. & 9 other authors (1999). Restoration of the immune system with anti-retroviral therapy. *Immunol Lett* **66**, 207–211.

Autran, B., Debre, P., Walker, B. & Katlama, C. (2003). Therapeutic vaccines against HIV need international partnerships. *Nat Rev Immunol* **3**, 503–508.

Azad, A. A. (2000). Could Nef and Vpr proteins contribute to disease progression by promoting depletion of bystander cells and prolonged survival of HIV-infected cells? *Biochem Biophys Res Commun* **267**, 677–685.

Baba, T. W., Jeong, Y. S., Pennick, D., Bronson, R., Greene, M. F. & Ruprecht, R. M. (1995). Pathogenicity of live, attenuated SIV after mucosal infection of neonatal macaques. *Science* **267**, 1820–1825.

Baba, T. W., Liska, V., Khimani, A. H. & 8 other authors (1999). Live attenuated, multiply deleted simian immunodeficiency virus causes AIDS in infant and adult macaques. *Nat Med* **5**, 194–203.

Baba, T. W., Liska, V., Hofmann-Lehmann, R. & 16 other authors (2000). Human neutralizing monoclonal antibodies of the IgG1 subtype protect against mucosal simian-human immunodeficiency virus infection. *Nat Med* **6**, 200–206.

Baier, M., Werner, A., Bannert, N., Metzner, K. & Kurth, R. (1995). HIV suppression by interleukin-16. *Nature* **378**, 563.

Barker, E. (1999). CD8+ cell-derived anti-human immunodeficiency virus inhibitory factor. *J Infect Dis* **179** (Suppl. 3), S485–488.

Barnett, S. W., Lu, S., Srivastava, I. & 15 other authors (2001). The ability of an oligomeric human immunodeficiency virus type 1 (HIV-1) envelope antigen to elicit neutralizing antibodies against primary HIV-1 isolates is improved following partial deletion of the second hypervariable region. *J Virol* **75**, 5526–5540.

Barouch, D. H., Santra, S., Schmitz, J. E. & 26 other authors (2000). Control of viremia and prevention of clinical AIDS in rhesus monkeys by cytokine-augmented DNA vaccination. *Science* **290**, 486–492.

Barouch, D. H., Santra, S., Kuroda, M. J. & 13 other authors (2001). Reduction of simian-human immunodeficiency virus 89.6p viremia in rhesus monkeys by recombinant modified vaccinia virus Ankara vaccination. *J Virol* **75**, 5151–5158.

Barouch, D. H., Kunstman, J., Kuroda, M. J. & 11 other authors (2002). Eventual AIDS vaccine failure in a rhesus monkey by viral escape from cytotoxic T lymphocytes. *Nature* **415**, 335–339.

Battegay, M., Moskophidis, D., Rahemtulla, A., Hengartner, H., Mak, T. W. & Zinkernagel, R. M. (1994). Enhanced establishment of a virus carrier state in adult CD4+ T-cell-deficient mice. *J Virol* **68**, 4700–4704.

Belec, L., Georges, A. J., Steenman, G. & Martin, P. M. (1989). Antibodies to human immunodeficiency virus in vaginal secretions of heterosexual women. *J Infect Dis* **160**, 385–391.

Benedict, C. A., Norris, P. S. & Ware, C. F. (2002). To kill or be killed: viral evasion of apoptosis. *Nat Immunol* **3**, 1013–1018.

Berman, P. W., Gregory, T. J., Riddle, L. & 7 other authors (1990). Protection of chimpanzees from infection by HIV-1 after vaccination with recombinant glycoprotein gp120 but not gp160. *Nature* **345**, 622–625.

Berry, N., Ariyoshi, K., Jaffar, S., Sabally, S., Corrah, T., Tedder, R. & Whittle, H. (1998). Low peripheral blood viral HIV-2 RNA in individuals with high CD4 percentage differentiates HIV-2 from HIV-1 infection. *J Hum Virol* **1**, 457–468.

Berry, N., Jaffar, S., Van Der Loeff, M. S. & 9 other authors (2002). Low level viremia and high CD4% predict normal survival in a cohort of HIV type-2-infected villagers. *AIDS Res Hum Retrovir* **18**, 1167–1173.

Binley, J. M., Klasse, P. J., Cao, Y., Jones, I., Markowitz, M., Ho, D. D. & Moore, J. P.

(1997). Differential regulation of the antibody responses to Gag and Env proteins of human immunodeficiency virus type 1. *J Virol* **71**, 2799–2809.

Binley, J. M., Wyatt, R., Desjardins, E., Kwong, P. D., Hendrickson, W., Moore, J. P. & Sodroski, J. (1998). Analysis of the interaction of antibodies with a conserved enzymatically deglycosylated core of the HIV type 1 envelope glycoprotein 120. *AIDS Res Hum Retrovir* **14**, 191–198.

Binley, J. M., Sanders, R. W., Clas, B. & 8 other authors (2000). A recombinant human immunodeficiency virus type 1 envelope glycoprotein complex stabilized by an intermolecular disulfide bond between the gp120 and gp41 subunits is an antigenic mimic of the trimeric virion-associated structure. *J Virol* **74**, 627–643.

Biron, C. A. & Brossay, L. (2001). NK cells and NKT cells in innate defense against viral infections. *Curr Opin Immunol* **13**, 458–464.

Boaz, M. J., Waters, A., Murad, S., Easterbrook, P. J. & Vyakarnam, A. (2002). Presence of HIV-1 Gag-specific IFN-gamma+IL-2+ and CD28+IL-2+ CD4 T cell responses is associated with nonprogression in HIV-1 infection. *J Immunol* **169**, 6376–6385.

Borrow, P., Lewicki, H., Hahn, B. H., Shaw, G. M. & Oldstone, M. B. (1994). Virus-specific CD8$^+$ cytotoxic T-lymphocyte activity associated with control of viremia in primary human immunodeficiency virus type 1 infection. *J Virol* **68**, 6103–6110.

Borrow, P., Lewicki, H., Wei, X. & 8 other authors (1997). Antiviral pressure exerted by HIV-1-specific cytotoxic T lymphocytes (CTLs) during primary infection demonstrated by rapid selection of CTL escape virus. *Nat Med* **3**, 205–211.

Boullier, S., Poquet, Y., Debord, T., Fournie, J. J. & Gougeon, M. L. (1999). Regulation by cytokines (IL-12, IL-15, IL-4 and IL-10) of the Vgamma9Vdelta2 T cell response to mycobacterial phosphoantigens in responder and anergic HIV-infected persons. *Eur J Immunol* **29**, 90–99.

Brander, C., Hartman, K. E., Trocha, A. K. & 8 other authors (1998). Lack of strong immune selection pressure by the immunodominant, HLA-A*0201-restricted cytotoxic T lymphocyte response in chronic human immunodeficiency virus-1 infection. *J Clin Invest* **101**, 2559–2566.

Broliden, K., Hinkula, J., Devito, C. & 7 other authors (2001). Functional HIV-1 specific IgA antibodies in HIV-1 exposed, persistently IgG seronegative female sex workers. *Immunol Lett* **79**, 29–36.

Burns, D. P. & Desrosiers, R. C. (1994). Envelope sequence variation, neutralizing antibodies, and primate lentivirus persistence. *Curr Top Microbiol Immunol* **188**, 185–219.

Burns, D. P., Collignon, C. & Desrosiers, R. C. (1993). Simian immunodeficiency virus mutants resistant to serum neutralization arise during persistent infection of rhesus monkeys. *J Virol* **67**, 4104–4113.

Burton, D. R., Desrosiers, R. C., Doms, R. W. & 7 other authors (2004). HIV vaccine design and the neutralizing antibody problem. *Nat Immunol* **5**, 233–236.

Buseyne, F. & Riviere, Y. (1993). HIV-specific CD8+ T-cell immune responses and viral replication. *AIDS* **7** (Suppl. 2), S81–85.

Buseyne, F., Fevrier, M., Garcia, S., Gougeon, M. L. & Riviere, Y. (1996). Dual function of a human immunodeficiency virus (HIV)-specific cytotoxic T-lymphocyte clone: inhibition of HIV replication by noncytolytic mechanisms and lysis of HIV-infected CD4+ cells. *Virology* **225**, 248–253.

Cafaro, A., Caputo, A., Fracasso, C. & 16 other authors (1999). Control of SHIV-89. 6P-infection of cynomolgus monkeys by HIV-1 Tat protein vaccine. *Nat Med* **5**, 643–650.

Cai, Q., Huang, X. L., Rappocciolo, G. & Rinaldo, C. R., Jr (1990). Natural killer cell responses in homosexual men with early HIV infection. *J Acquir Immune Defic Syndr* **3**, 669–676.

Candore, G., Romano, G. C., D'Anna, C., Di Lorenzo, G., Gervasi, F., Lio, D., Modica, M. A., Potestio, M. & Caruso, C. (1998). Biological basis of the HLA-B8,DR3-associated progression of acquired immune deficiency syndrome. *Pathobiology* **66**, 33–37.

Cao, J., McNevin, J., Holte, S., Fink, L., Corey, L. & McElrath, M. J. (2003). Comprehensive analysis of human immunodeficiency virus type 1 (HIV-1)-specific gamma interferon-secreting CD8+ T cells in primary HIV-1 infection. *J Virol* **77**, 6867–6878.

Capobianchi, M. R., Ankel, H., Ameglio, F., Paganelli, R., Pizzoli, P. M. & Dianzani, F. (1992). Recombinant glycoprotein 120 of human immunodeficiency virus is a potent interferon inducer. *AIDS Res Hum Retrovir* **8**, 575–579.

Carrington, M. & Bontrop, R. E. (2002). Effects of MHC class I on HIV/SIV disease in primates. *AIDS* **16** (Suppl. 4), S105–114.

Carrington, M. & O'Brien, S. J. (2003). The influence of HLA genotype on AIDS. *Annu Rev Med* **54**, 535–551.

Carrington, M., Nelson, G. W., Martin, M. P. & 7 other authors (1999). HLA and HIV-1: heterozygote advantage and B*35-Cw*04 disadvantage. *Science* **283**, 1748–1752.

Casadevall, A. & Pirofski, L. A. (2003). Antibody-mediated regulation of cellular immunity and the inflammatory response. *Trends Immunol* **24**, 474–478.

Champagne, P., Ogg, G. S., King, A. S. & 12 other authors (2001). Skewed maturation of memory HIV-specific CD8 T lymphocytes. *Nature* **410**, 106–111.

Chan, D. C. & Kim, P. S. (1998). HIV entry and its inhibition. *Cell* **93**, 681–684.

Charley, B., Lavenant, L. & Delmas, B. (1991). Glycosylation is required for coronavirus TGEV to induce an efficient production of IFN alpha by blood mononuclear cells. *Scand J Immunol* **33**, 435–440.

Cherpelis, S., Shrivastava, I., Gettie, A., Jin, X., Ho, D. D., Barnett, S. W. & Stamatatos, L. (2001). DNA vaccination with the human immunodeficiency virus type 1 SF162-DeltaV2 envelope elicits immune responses that offer partial protection from simian/human immunodeficiency virus infection to CD8(+) T-cell-depleted rhesus macaques. *J Virol* **75**, 1547–1550.

Chien, P. C., Jr, Cohen, S., Tuen, M., Arthos, J., Chen, P. D., Patel, S. & Hioe, C. E. (2004). Human immunodeficiency virus type 1 evades T-helper responses by exploiting antibodies that suppress antigen processing. *J Virol* **78**, 7645–7652.

Ciurea, A., Hunziker, L., Klenerman, P., Hengartner, H. & Zinkernagel, R. M. (2001). Impairment of CD4(+) T cell responses during chronic virus infection prevents neutralizing antibody responses against virus escape mutants. *J Exp Med* **193**, 297–305.

Clark, S. J., Saag, M. S., Decker, W. D., Campbell-Hill, S., Roberson, J. L., Veldkamp, P. J., Kappes, J. C., Hahn, B. H. & Shaw, G. M. (1991). High titers of cytopathic virus in plasma of patients with symptomatic primary HIV-1 infection. *N Engl J Med* **324**, 954–960.

Clerici, M., Levin, J. M., Kessler, H. A., Harris, A., Berzofsky, J. A., Landay, A. L. & Shearer, G. M. (1994). HIV-specific T-helper activity in seronegative health care workers exposed to contaminated blood. *JAMA (J Am Med Assoc)* **271**, 42–46.

Cocchi, F., DeVico, A. L., Garzino-Demo, A., Arya, S. K., Gallo, R. C. & Lusso, P. (1995). Identification of RANTES, MIP-1 alpha, and MIP-1 beta as the major HIV-suppressive factors produced by CD8+ T cells. *Science* **270**, 1811–1815.

Coffin, J. M. (1995). HIV population dynamics in vivo: implications for genetic variation, pathogenesis, and therapy. *Science* **267**, 483–489.

Cohen, G. B., Gandhi, R. T., Davis, D. M., Mandelboim, O., Chen, B. K., Strominger, J. L. & Baltimore, D. (1999). The selective downregulation of class I major histocompatibility complex proteins by HIV-1 protects HIV-infected cells from NK cells. *Immunity* **10**, 661–671.

Cohen, J. (2003). HIV/AIDS. Vaccine results lose significance under scrutiny. *Science* **299**, 1495.

Cohen, J. (2004). AIDS vaccines. HIV dodges one-two punch. *Science* **305**, 1545–1547.

Cohen, S., Tuen, M. & Hioe, C. E. (2003). Propagation of CD4+ T cells specific for HIV type 1 envelope gp120 from chronically HIV type 1-infected subjects. *AIDS Res Hum Retrovir* **19**, 793–806.

Collins, K. L., Chen, B. K., Kalams, S. A., Walker, B. D. & Baltimore, D. (1998). HIV-1 Nef protein protects infected primary cells against killing by cytotoxic T lymphocytes. *Nature* **391**, 397–401.

Colucci, F., Caligiuri, M. A. & Di Santo, J. P. (2003). What does it take to make a natural killer? *Nat Rev Immunol* **3**, 413–425.

Connick, E., Marr, D. G., Zhang, X. Q., Clark, S. J., Saag, M. S., Schooley, R. T. & Curiel, T. J. (1996). HIV-specific cellular and humoral immune responses in primary HIV infection. *AIDS Res Hum Retrovir* **12**, 1129–1140.

Connick, E., Schlichtemeier, R. L., Purner, M. B. & 8 other authors (2001). Relationship between human immunodeficiency virus type 1 (HIV-1)-specific memory cytotoxic T lymphocytes and virus load after recent HIV-1 seroconversion. *J Infect Dis* **184**, 1465–1469.

Connor, R. I., Korber, B. T., Graham, B. S. & 17 other authors (1998). Immunological and virological analyses of persons infected by human immunodeficiency virus type 1 while participating in trials of recombinant gp120 subunit vaccines. *J Virol* **72**, 1552–1576.

Conti, L., Rainaldi, G., Matarrese, P. & 7 other authors (1998). The HIV-1 vpr protein acts as a negative regulator of apoptosis in a human lymphoblastoid T cell line: possible implications for the pathogenesis of AIDS. *J Exp Med* **187**, 403–413.

Cooper, M. A., Fehniger, T. A. & Caligiuri, M. A. (2001). The biology of human natural killer-cell subsets. *Trends Immunol* **22**, 633–640.

Couillin, I., Culmann-Penciolelli, B., Gomard, E., Choppin, J., Levy, J.-P., Guillet, J.-G. & Saragosti, S. (1994). Impaired cytotoxic T lymphocyte recognition due to genetic variations in the main immunogenic region of the human immunodeficiency virus 1 NEF protein. *J Exp Med* **180**, 1129–1134.

Daar, E. S., Moudgil, T., Meyer, R. D. & Ho, D. D. (1991). Transient high levels of viremia in patients with primary human immunodeficiency virus type 1 infection. *N Engl J Med* **324**, 961–964.

Dalod, M., Dupuis, M., Deschemin, J. C., Sicard, D., Salmon, D., Delfraissy, J. F., Venet, A., Sinet, M. & Guillet, J. G. (1999). Broad, intense anti-human immunodeficiency virus (HIV) ex vivo CD8(+) responses in HIV type 1-infected patients: comparison with anti-Epstein-Barr virus responses and changes during antiretroviral therapy. *J Virol* **73**, 7108–7116.

Daniel, M. D., Kirchhoff, F., Czajak, S. C., Seghal, P. K. & Desrosiers, R. C. (1992). Protective effects of a live attenuated SIV vaccine with a deletion in the nef gene. *Science* **258**, 1938–1941.

Davis, N. L., Caley, I. J., Brown, K. W. & 9 other authors (2000). Vaccination of macaques against pathogenic simian immunodeficiency virus with Venezuelan equine encephalitis virus replicon particles. *J Virol* **74**, 371–378.

Delibrias, C. C., Kazatchkine, M. D. & Fischer, E. (1993). Evidence for the role of CR1 (CD35), in addition to CR2 (CD21), in facilitating infection of human T cells with opsonized HIV. *Scand J Immunol* **38**, 183–189.

Del Val, M., Schlicht, H. J., Ruppert, T., Reddehase, M. J. & Koszinowski, U. H. (1991). Efficient processing of an antigenic sequence for presentation by MHC class I molecules depends on its neighboring residues in the protein. *Cell* **66**, 1145–1153.

den Uyl, D., van der Horst-Bruinsma, I. E. & van Agtael, M. (2004). Progression of HIV to AIDS: a protective role for HLA-B27? *AIDS Rev* **6**, 89–96.

Derdeyn, C. A., Decker, J. M., Bibollet-Ruche, F. & 11 other authors (2004). Envelope-constrained neutralization-sensitive HIV-1 after heterosexual transmission. *Science* **303**, 2019–2022.

DeStefano, E., Friedman, R. M., Friedman-Kien, A. E., Goedert, J. J., Henriksen, D., Preble, O. T., Sonnabend, J. A. & Vilcek, J. (1982). Acid-labile human leukocyte interferon in homosexual men with Kaposi's sarcoma and lymphadenopathy. *J Infect Dis* **146**, 451–459.

Devito, C., Broliden, K., Kaul, R. & 10 other authors (2000). Mucosal and plasma IgA from HIV-1-exposed uninfected individuals inhibit HIV-1 transcytosis across human epithelial cells. *J Immunol* **165**, 5170–5176.

Devito, C., Hinkula, J., Kaul, R. & 10 other authors (2002). Cross-clade HIV-1-specific neutralizing IgA in mucosal and systemic compartments of HIV-1-exposed, persistently seronegative subjects. *J Acquir Immune Defic Syndr* **30**, 413–420.

de Wit, R., Schattenkerk, J. K., Boucher, C. A., Bakker, P. J., Veenhof, K. H. & Danner, S. A. (1988). Clinical and virological effects of high-dose recombinant interferon-alpha in disseminated AIDS-related Kaposi's sarcoma. *Lancet* **2**, 1214–1217.

Diebold, S. S., Kaisho, T., Hemmi, H., Akira, S. & Reis e Sousa, C. (2004). Innate antiviral responses by means of TLR7-mediated recognition of single-stranded RNA. *Science* **303**, 1529–1531.

Dierich, M. P., Stoiber, H. & Clivio, A. (1996). A "complement-ary" AIDS vaccine. *Nat Med* **2**, 153–155.

Diouf, K., Sarr, A. D., Eisen, G., Popper, S., Mboup, S. & Kanki, P. (2002). Associations between MHC class I and susceptibility to HIV-2 disease progression. *J Hum Virol* **5**, 1–7.

Doepper, S., Stoiber, H., Kacani, L., Sprinzl, G., Steindl, F., Prodinger, W. M. & Dierich, M. P. (2000). B cell-mediated infection of stimulated and unstimulated autologous T lymphocytes with HIV-1: role of complement. *Immunobiology* **202**, 293–305.

Dolan, M. J., Clerici, M., Blatt, S. P. & 7 other authors (1995). In vitro T cell function, delayed-type hypersensitivity skin testing, and CD4+ T cell subset phenotyping independently predict survival time in patients infected with human immuno-deficiency virus. *J Infect Dis* **172**, 79–87.

Donaghy, H., Pozniak, A., Gazzard, B., Qazi, N., Gilmour, J., Gotch, F. & Patterson, S. (2001). Loss of blood CD11c(+) myeloid and CD11c(-) plasmacytoid dendritic cells in patients with HIV-1 infection correlates with HIV-1 RNA virus load. *Blood* **98**, 2574–2576.

Donaghy, H., Gazzard, B., Gotch, F. & Patterson, S. (2003). Dysfunction and infection of freshly isolated blood myeloid and plasmacytoid dendritic cells in patients infected with HIV-1. *Blood* **101**, 4505–4511.

Dong, T., Stewart-Jones, G., Chen, N. & 14 other authors (2004). HIV-specific cytotoxic T cells from long-term survivors select a unique T cell receptor. *J Exp Med* **200**, 1547–1557.

Douek, D. C. (2003). Disrupting T-cell homeostasis: how HIV-1 infection causes disease. *AIDS Rev* **5**, 172–177.

Douek, D. C., Brenchley, J. M., Betts, M. R. & 13 other authors (2002). HIV preferentially infects HIV-specific CD4+ T cells. *Nature* **417**, 95–98.

Draenert, R., Verrill, C. L., Tang, Y. & 9 other authors (2004a). Persistent recognition of autologous virus by high-avidity CD8 T cells in chronic, progressive human immunodeficiency virus type 1 infection. *J Virol* **78**, 630–641.

Draenert, R., Le Gall, S., Pfafferott, K. J. & 20 other authors (2004b). Immune selection for altered antigen processing leads to cytotoxic T lymphocyte escape in chronic HIV-1 infection. *J Exp Med* **199**, 905–915.

Earl, P. L., Broder, C. C., Long, D., Lee, S. A., Peterson, J., Chakrabarti, S., Doms, R. W. & Moss, B. (1994). Native oligomeric human immunodeficiency virus type 1 envelope glycoprotein elicits diverse monoclonal antibody reactivities. *J Virol* **68**, 3015–3026.

Earl, P. L., Broder, C. C., Doms, R. W. & Moss, B. (1997). Epitope map of human immunodeficiency virus type 1 gp41 derived from 47 monoclonal antibodies produced by immunization with oligomeric envelope protein. *J Virol* **71**, 2674–2684.

Earl, P. L., Sugiura, W., Montefiori, D. C., Broder, C. C., Lee, S. A., Wild, C., Lifson, J. & Moss, B. (2001). Immunogenicity and protective efficacy of oligomeric human immunodeficiency virus type 1 gp140. *J Virol* **75**, 645–653.

Ebenbichler, C. F., Thielens, N. M., Vornhagen, R., Marschang, P., Arlaud, G. J. & Dierich, M. P. (1991). Human immunodeficiency virus type 1 activates the classical pathway of complement by direct C1 binding through specific sites in the transmembrane glycoprotein gp41. *J Exp Med* **174**, 1417–1424.

Edwards, B. H., Bansal, A., Sabbaj, S., Bakari, J., Mulligan, M. J. & Goepfert, P. A. (2002). Magnitude of functional CD8+ T-cell responses to the gag protein of human immunodeficiency virus type 1 correlates inversely with viral load in plasma. *J Virol* **76**, 2298–2305.

Egan, M. A., Pavlat, W. A., Tartaglia, J., Paoletti, E., Weinhold, K. J., Clements, M. L. & Siliciano, R. F. (1995). Induction of human immunodeficiency virus type 1 (HIV-1)-specific cytolytic T lymphocyte responses in seronegative adults by a nonreplicating, host-range-restricted canarypox vector (ALVAC) carrying the HIV-1MN env gene. *J Infect Dis* **171**, 1623–1627.

Eger, K. A. & Unutmaz, D. (2004). Perturbation of natural killer cell function and receptors during HIV infection. *Trends Microbiol* **12**, 301–303.

Embretson, J., Zupancic, M., Ribas, J. L., Burke, A., Racz, P., Tenner-Racz, K. & Haase, A. T. (1993). Massive covert infection of helper T lymphocytes and macrophages by HIV during the incubation period of AIDS. *Nature* **362**, 359–362.

Emilie, D., Maillot, M. C., Nicolas, J. F., Fior, R. & Galanaud, P. (1992). Antagonistic effect of interferon-gamma on tat-induced transactivation of HIV long terminal repeat. *J Biol Chem* **267**, 20565–20570.

Emini, E. A., Schleif, W. A., Nunberg, J. H. & 9 other authors (1992). Prevention of HIV-1 infection in chimpanzees by gp120 V3 domain-specific monoclonal antibody. *Nature* **355**, 728–730.

Esser, M. T., Bess, J. W., Jr, Suryanarayana, K., Chertova, E., Marti, D., Carrington, M., Arthur, L. O. & Lifson, J. D. (2001). Partial activation and induction of apoptosis in

CD4(+) and CD8(+) T lymphocytes by conformationally authentic noninfectious human immunodeficiency virus type 1. *J Virol* **75**, 1152–1164.

Estcourt, M. J., McMichael, A. J. & Hanke, T. (2004). DNA vaccines against human immunodeficiency virus type 1. *Immunol Rev* **199**, 144–155.

Evans, D. T., O'Connor, D. H., Jing, P. & 14 other authors (1999). Virus-specific cytotoxic T-lymphocyte responses select for amino-acid variation in simian immunodeficiency virus Env and Nef. *Nat Med* **5**, 1270–1276.

Evans, T. G., Keefer, M. C., Weinhold, K. J. & 12 other authors (1999). A canarypox vaccine expressing multiple human immunodeficiency virus type 1 genes given alone or with rgp120 elicits broad and durable CD8+ cytotoxic T lymphocyte responses in seronegative volunteers. *J Infect Dis* **180**, 290–298.

Fauci, A. S. (1988). The human immunodeficiency virus: infectivity and mechanisms of pathogenesis. *Science* **239**, 617–622.

Favoreel, H. W., Van de Walle, G. R., Nauwynck, H. J. & Pensaert, M. B. (2003). Virus complement evasion strategies. *J Gen Virol* **84**, 1–15.

Feeney, M. E., Tang, Y., Roosevelt, K. A., Leslie, A. J., McIntosh, K., Karthas, N., Walker, B. D. & Goulder, P. J. (2004). Immune escape precedes breakthrough human immunodeficiency virus type 1 viremia and broadening of the cytotoxic T-lymphocyte response in an HLA-B27-positive long-term-nonprogressing child. *J Virol* **78**, 8927–8930.

Fehniger, T. A., Herbein, G., Yu, H., Para, M. I., Bernstein, Z. P., O'Brien, W. A. & Caligiuri, M. A. (1998). Natural killer cells from HIV-1+ patients produce C-C chemokines and inhibit HIV-1 infection. *J Immunol* **161**, 6433–6438.

Fehniger, T. A., Shah, M. H., Turner, M. J. & 7 other authors (1999). Differential cytokine and chemokine gene expression by human NK cells following activation with IL-18 or IL-15 in combination with IL-12: implications for the innate immune response. *J Immunol* **162**, 4511–4520.

Feldman, S., Stein, D., Amrute, S., Denny, T., Garcia, Z., Kloser, P., Sun, Y., Megjugorac, N. & Fitzgerald-Bocarsly, P. (2001). Decreased interferon-alpha production in HIV-infected patients correlates with numerical and functional deficiencies in circulating type 2 dendritic cell precursors. *Clin Immunol* **101**, 201–210.

Ferrari, G., Humphrey, W., McElrath, M. J., Excler, J. L., Duliege, A. M., Clements, M. L., Corey, L. C., Bolognesi, D. P. & Weinhold, K. J. (1997). Clade B-based HIV-1 vaccines elicit cross-clade cytotoxic T lymphocyte reactivities in uninfected volunteers. *Proc Natl Acad Sci U S A* **94**, 1396–1401.

Fidler, S. J., Dorrell, L., Ball, S., Lombardi, G., Weber, J., Hawrylowicz, C. & Rees, A. D. (1996). An early antigen-presenting cell defect in HIV-1-infected patients correlates with CD4 dependency in human T-cell clones. *Immunology* **89**, 46–53.

Finkel, T. H., Tudor-Williams, G., Banda, N. K., Cotton, M. F., Curiel, T., Monks, C., Baba, T. W., Ruprecht, R. M. & Kupfer, A. (1995). Apoptosis occurs predominantly in bystander cells and not in productively infected cells of HIV- and SIV-infected lymph nodes. *Nat Med* **1**, 129–134.

Fogli, M., Costa, P., Murdaca, G., Setti, M., Mingari, M. C., Moretta, L., Moretta, A. & De Maria, A. (2004). Significant NK cell activation associated with decreased cytolytic function in peripheral blood of HIV-1-infected patients. *Eur J Immunol* **34**, 2313–2321.

Fong, L., Mengozzi, M., Abbey, N. W., Herndier, B. G. & Engleman, E. G. (2002). Productive infection of plasmacytoid dendritic cells with human immunodeficiency virus type 1 is triggered by CD40 ligation. *J Virol* **76**, 11033–11041.

Forthal, D. N., Landucci, G. & Daar, E. S. (2001). Antibody from patients with acute human immunodeficiency virus (HIV) infection inhibits primary strains of HIV type 1 in the presence of natural-killer effector cells. *J Virol* **75**, 6953–6961.

Fouts, T. R., Tuskan, R. G., Chada, S., Hone, D. M. & Lewis, G. K. (1995). Construction and immunogenicity of *Salmonella typhimurium* vaccine vectors that express HIV-1 gp120. *Vaccine* **13**, 1697–1705.

Fowke, K. R., Kaul, R., Rosenthal, K. L. & 9 other authors (2000). HIV-1-specific cellular immune responses among HIV-1-resistant sex workers. *Immunol Cell Biol* **78**, 586–595.

Francis, M. L. & Meltzer, M. S. (1993). Induction of IFN-alpha by HIV-1 in monocyte-enriched PBMC requires gp120-CD4 interaction but not virus replication. *J Immunol* **151**, 2208–2216.

Francis, M. L., Meltzer, M. S. & Gendelman, H. E. (1992). Interferons in the persistence, pathogenesis, and treatment of HIV infection. *AIDS Res Hum Retrovir* **8**, 199–207.

Frank, I., Stoiber, H., Godar, S., Stockinger, H., Steindl, F., Katinger, H. W. & Dierich, M. P. (1996). Acquisition of host cell-surface-derived molecules by HIV-1. *AIDS* **10**, 1611–1620.

Friedrich, T. C., Dodds, E. J., Yant, L. J. & 16 other authors (2004). Reversion of CTL escape-variant immunodeficiency viruses in vivo. *Nat Med* **10**, 275–281.

Gao, X., Nelson, G. W., Karacki, P. & 9 other authors (2001). Effect of a single amino acid change in MHC class I molecules on the rate of progression to AIDS. *N Engl J Med* **344**, 1668–1675.

Garber, D. A., Silvestri, G. & Feinberg, M. B. (2004). Prospects for an AIDS vaccine: three big questions, no easy answers. *Lancet Infect Dis* **4**, 397–413.

Gauduin, M. C., Parren, P. W., Weir, R., Barbas, C. F., Burton, D. R. & Koup, R. A. (1997). Passive immunization with a human monoclonal antibody protects hu-PBL-SCID mice against challenge by primary isolates of HIV-1. *Nat Med* **3**, 1389–1393.

Gea-Banacloche, J. C., Migueles, S. A., Martino, L. & 11 other authors (2000). Maintenance of large numbers of virus-specific CD8+ T cells in HIV-infected progressors and long-term nonprogressors. *J Immunol* **165**, 1082–1092.

Geijtenbeek, T. B., Kwon, D. S., Torensma, R. & 9 other authors (2000). DC-SIGN, a dendritic cell-specific HIV-1-binding protein that enhances trans-infection of T cells. *Cell* **100**, 587–597.

Geleziunas, R., Xu, W., Takeda, K., Ichijo, H. & Greene, W. C. (2001). HIV-1 Nef inhibits ASK1-dependent death signalling providing a potential mechanism for protecting the infected host cell. *Nature* **410**, 834–838.

Geretti, A. M., Hulskotte, E. G., Dings, M. E., van Baalen, C. A., van Amerongen, G., Norley, S. G., Boers, P., Gruters, R. & Osterhaus, A. D. (1999). Decline of simian immunodeficiency virus (SIV)-specific cytotoxic T lymphocytes in the peripheral blood of long-term nonprogressing macaques infected with SIVmac32H-J5. *J Infect Dis* **180**, 1133–1141.

Gillespie, G. M., Kaul, R., Dong, T. & 8 other authors (2002). Cross-reactive cytotoxic T lymphocytes against a HIV-1 p24 epitope in slow progressors with B*57. *AIDS* **16**, 961–972.

Gloster, S. E., Newton, P., Cornforth, D., Lifson, J. D., Williams, I., Shaw, G. M. & Borrow, P. (2004). Association of strong virus-specific CD4+ T cell responses with efficient natural control of primary HIV-1 infection. *AIDS* **18**, 749–755.

Goepfert, P. A., Bansal, A., Edwards, B. H., Ritter, G. D., Jr, Tellez, I., McPherson, S. A., Sabbaj, S. & Mulligan, M. J. (2000). A significant number of human immunodeficiency virus epitope-specific cytotoxic T lymphocytes detected by tetramer binding do not produce gamma interferon. *J Virol* **74**, 10249–10255.

Goodier, M. R., Imami, N., Moyle, G., Gazzard, B. & Gotch, F. (2003). Loss of the CD56hiCD16- NK cell subset and NK cell interferon-gamma production during antiretroviral therapy for HIV-1: partial recovery by human growth hormone. *Clin Exp Immunol* **134**, 470–476.

Gougeon, M. L., Malkovsky, M., Casetti, R., Agrati, C. & Poccia, F. (2002). Innate T cell immunity to HIV-infection. Immunotherapy with phosphocarbohydrates, a novel strategy of immune intervention? *Vaccine* **20**, 1938–1941.

Goulder, P. J. & Watkins, D. I. (2004). HIV and SIV CTL escape: implications for vaccine design. *Nat Rev Immunol* **4**, 630–640.

Goulder, P. J., Bunce, M., Krausa, P. & 7 other authors (1996). Novel, cross-restricted, conserved, and immunodominant cytotoxic T lymphocyte epitopes in slow progressors in HIV type 1 infection. *AIDS Res Hum Retrovir* **12**, 1691–1698.

Goulder, P. J. R., Phillips, R. E., Colbert, R. A. & 9 other authors (1997a). Late escape from an immunodominant cytotoxic T-lymphocyte response associated with progression to AIDS. *Nat Med* **3**, 212–217.

Goulder, P. J. R., Sewell, A. K., Lalloo, D. G. & 8 other authors (1997b). Patterns of immunodominance in HIV-1-specific cytotoxic T lymphocyte responses in two human histocompatibility leukocyte antigens (HLA)-identical siblings with HLA-A*0201 are influenced by epitope mutation. *J Exp Med* **185**, 1423–1433.

Goulder, P. J., Tang, Y., Brander, C. & 16 other authors (2000). Functionally inert HIV-specific cytotoxic T lymphocytes do not play a major role in chronically infected adults and children. *J Exp Med* **192**, 1819–1832.

Goulder, P. J., Brander, C., Tang, Y. & 16 other authors (2001). Evolution and transmission of stable CTL escape mutations in HIV infection. *Nature* **412**, 334–338.

Graham, B. S., McElrath, M. J., Connor, R. I. & 16 other authors (1998). Analysis of intercurrent human immunodeficiency virus type 1 infections in phase I and II trials of candidate AIDS vaccines. AIDS Vaccine Evaluation Group, and the Correlates of HIV Immune Protection Group. *J Infect Dis* **177**, 310–319.

Gray, C. M., Lawrence, J., Schapiro, J. M. & 7 other authors (1999). Frequency of class I HLA-restricted anti-HIV CD8+ T cells in individuals receiving highly active antiretroviral therapy (HAART). *J Immunol* **162**, 1780–1788.

Greenough, T. C., Somasundaran, M., Brettler, D. B., Hesselton, R. M., Alimenti, A., Kirchhoff, F., Panicali, D. & Sullivan, J. L. (1994). Normal immune function and inability to isolate virus in culture in an individual with long-term human immunodeficiency virus type 1 infection. *AIDS Res Hum Retrovir* **10**, 395–403.

Greenough, T. C., Brettler, D. B., Somasundaran, M., Panicali, D. L. & Sullivan, J. L. (1997). Human immunodeficiency virus type 1-specific cytotoxic T lymphocytes (CTL), virus load, and CD4 T cell loss: evidence supporting a protective role for CTL in vivo. *J Infect Dis* **176**, 118–125.

Greenough, T. C., Sullivan, J. L. & Desrosiers, R. C. (1999). Declining CD4 T-cell counts in a person infected with nef-deleted HIV-1. *N Engl J Med* **340**, 236–237.

Grunfeld, C., Kotler, D. P., Shigenaga, J. K., Doerrler, W., Tierney, A., Wang, J., Pierson, R. N., Jr & Feingold, K. R. (1991). Circulating interferon-alpha levels and hypertriglyceridemia in the acquired immunodeficiency syndrome. *Am J Med* **90**, 154–162.

Gruters, R. A., van Baalen, C. A. & Osterhaus, A. D. (2002). The advantage of early recognition of HIV-infected cells by cytotoxic T-lymphocytes. *Vaccine* **20**, 2011–2015.

Gulzar, N. & Copeland, K. F. (2004). CD8+ T-cells: function and response to HIV infection. *Curr HIV Res* **2**, 23–37.

Haigwood, N. L., Watson, A., Sutton, W. F. & 9 other authors (1996). Passive immune globulin therapy in the SIV/macaque model: early intervention can alter disease profile. *Immunol Lett* **51**, 107–114.

Hamer, D. H. (2004). Can HIV be cured? Mechanisms of HIV persistence and strategies to combat it. *Curr HIV Res* **2**, 99–111.

Hanke, T., Samuel, R. V., Blanchard, T. J. & 9 other authors (1999). Effective induction of simian immunodeficiency virus-specific cytotoxic T lymphocytes in macaques by using a multiepitope gene and DNA prime-modified vaccinia virus Ankara boost vaccination regimen. *J Virol* **73**, 7524–7532.

Harcourt, G. C., Garrard, S., Davenport, M. P., Edwards, A. & Phillips, R. E. (1998). HIV-1 variation diminishes CD4 T lymphocyte recognition. *J Exp Med* **188**, 1785–1793.

Harrer, E., Harrer, T., Buchbinder, S., Mann, D. L., Feinberg, M., Yilma, T., Johnson, R. P. & Walker, B. D. (1994). HIV-1-specific cytotoxic T lymphocyte response in healthy, long-term nonprogressing seropositive persons. *AIDS Res Hum Retrovir* **10** (Suppl. 2), S77–78.

Hartshorn, K. L., Neumeyer, D., Vogt, M. W., Schooley, R. T. & Hirsch, M. S. (1987). Activity of interferons alpha, beta, and gamma against human immunodeficiency virus replication in vitro. *AIDS Res Hum Retrovir* **3**, 125–133.

Heath, S. L., Tew, J. G., Tew, J. G., Szakal, A. K. & Burton, G. F. (1995). Follicular dendritic cells and human immunodeficiency virus infectivity. *Nature* **377**, 740–744.

Heil, F., Hemmi, H., Hochrein, H., Ampenberger, F., Kirschning, C., Akira, S., Lipford, G., Wagner, H. & Bauer, S. (2004). Species-specific recognition of single-stranded RNA via toll-like receptor 7 and 8. *Science* **303**, 1526–1529.

Hendel, H., Caillat-Zucman, S., Lebuanec, H. & 9 other authors (1999). New class I and II HLA alleles strongly associated with opposite patterns of progression to AIDS. *J Immunol* **162**, 6942–6946.

Herbein, G., Mahlknecht, U., Batliwalla, F., Gregersen, P., Pappas, T., Butler, J., O'Brien, W. A. & Verdin, E. (1998). Apoptosis of CD8+ T cells is mediated by macrophages through interaction of HIV gp120 with chemokine receptor CXCR4. *Nature* **395**, 189–194.

Hess, C., Altfeld, M., Thomas, S. Y. & 9 other authors (2004). HIV-1 specific CD8+ T cells with an effector phenotype and control of virus replication. *Lancet* **363**, 863–866.

Hess, G., Rossol, S., Rossol, R. & Meyer zum Buschenfelde, K. H. (1991). Tumor necrosis factor and interferon as prognostic markers in human immunodeficiency virus (HIV) infection. *Infection* **19** (Suppl. 2), S93–97.

Hioe, C. E., Tuen, M., Chien, P. C., Jr & 7 other authors (2001). Inhibition of human immunodeficiency virus type 1 gp120 presentation to CD4 T cells by antibodies specific for the CD4 binding domain of gp120. *J Virol* **75**, 10950–10957.

Hirsch, V. M., Fuerst, T. R., Sutter, G. & 9 other authors (1996). Patterns of viral replication correlate with outcome in simian immunodeficiency virus (SIV)-infected macaques: effect of prior immunization with a trivalent SIV vaccine in modified vaccinia virus Ankara. *J Virol* **70**, 3741–3752.

Ho, D. D., Neumann, A. U., Perelson, A. S., Chen, W., Leonard, J. M. & Markowitz, M. (1995). Rapid turnover of plasma virions and CD4 lymphocytes in HIV-1 infection. *Nature* **373**, 123–126.

Hofmann-Lehmann, R., Rasmussen, R. A., Vlasak, J. & 17 other authors (2001a). Passive immunization against oral AIDS virus transmission: an approach to prevent mother-to-infant HIV-1 transmission? *J Med Primatol* **30**, 190–196.

Hofmann-Lehmann, R., Vlasak, J., Rasmussen, R. A. & 19 other authors (2001b). Postnatal passive immunization of neonatal macaques with a triple combination of human monoclonal antibodies against oral simian-human immunodeficiency virus challenge. *J Virol* **75**, 7470–7480.

Hone, D. M., Wu, S., Powell, R. J., Pascual, D. W., Van Cott, J., McGhee, J., Fouts, T. R., Tuskan, R. G. & Lewis, G. K. (1996). Optimization of live oral Salmonella-HIV-1 vaccine vectors for the induction of HIV-specific mucosal and systemic immune responses. *J Biotechnol* **44**, 203–207.

Howcroft, T. K., Strebel, K., Martin, M. A. & Singer, D. S. (1993). Repression of MHC class I gene promoter activity by two-exon Tat of HIV. *Science* **260**, 1320–1322.

Hu, P. F., Hultin, L. E., Hultin, P., Hausner, M. A., Hirji, K., Jewett, A., Bonavida, B., Detels, R. & Giorgi, J. V. (1995). Natural killer cell immunodeficiency in HIV disease is manifest by profoundly decreased numbers of CD16+CD56+ cells and expansion of a population of CD16dimCD56- cells with low lytic activity. *J Acquir Immune Defic Syndr Hum Retrovirol* **10**, 331–340.

Hu, S. L., Fultz, P. N., McClure, H. M. & 8 other authors (1987). Effect of immunization with a vaccinia-HIV env recombinant on HIV infection of chimpanzees. *Nature* **328**, 721–723.

Hufert, F. T., van Lunzen, J., Janossy, G., Bertram, S., Schmitz, J., Haller, O., Racz, P. & von Laer, D. (1997). Germinal centre CD4+ T cells are an important site of HIV replication in vivo. *AIDS* **11**, 849–857.

Ito, Y. (1994). Induction of interferon by virus glycoprotein(s) in lymphoid cells through interaction with the cellular receptors via lectin-like action: an alternative interferon induction mechanism. *Arch Virol* **138**, 187–198.

Iyasere, C., Tilton, J. C., Johnson, A. J. & 13 other authors (2003). Diminished proliferation of human immunodeficiency virus-specific CD4+ T cells is associated with diminished interleukin-2 (IL-2) production and is recovered by exogenous IL-2. *J Virol* **77**, 10900–10909.

Jakubik, J. J., Saifuddin, M., Takefman, D. M. & Spear, G. T. (1999). B lymphocytes in lymph nodes and peripheral blood are important for binding immune complexes containing HIV-1. *Immunology* **96**, 612–619.

Jeffs, S. A., McKeating, J., Lewis, S., Craft, H., Biram, D., Stephens, P. E. & Brady, R. L. (1996). Antigenicity of truncated forms of the human immunodeficiency virus type 1 envelope glycoprotein. *J Gen Virol* **77**, 1403–1410.

Jewett, A., Cavalcanti, M., Giorgi, J. & Bonavida, B. (1997). Concomitant killing in vitro of both gp120-coated CD4+ peripheral T lymphocytes and natural killer cells in the antibody-dependent cellular cytotoxicity (ADCC) system. *J Immunol* **158**, 5492–5500.

Jin, X., Bauer, D. E., Tuttleton, S. E. & 11 other authors (1999). Dramatic rise in plasma viremia after CD8(+) T cell depletion in simian immunodeficiency virus-infected macaques. *J Exp Med* **189**, 991–998.

Jin, X., Gao, X., Ramanathan, M., Jr & 7 other authors (2002). Human immunodeficiency virus type 1 (HIV-1)-specific CD8+-T-cell responses for groups of HIV-1-infected individuals with different HLA-B*35 genotypes. *J Virol* **76**, 12603–12610.

Johnson, R. P. & Desrosiers, R. C. (1998). Protective immunity induced by live attenuated simian immunodeficiency virus. *Curr Opin Immunol* **10**, 436–443.

Johnson, R. P., Lifson, J. D., Czajak, S. C. & 8 other authors (1999). Highly attenuated vaccine strains of simian immunodeficiency virus protect against vaginal challenge: inverse relationship of degree of protection with level of attenuation. *J Virol* **73**, 4952–4961.

Johnson, W. E., Sanford, H., Schwall, L., Burton, D. R., Parren, P. W., Robinson, J. E. & Desrosiers, R. C. (2003). Assorted mutations in the envelope gene of simian immunodeficiency virus lead to loss of neutralization resistance against antibodies representing a broad spectrum of specificities. *J Virol* **77**, 9993–10003.

Jones, N. A., Wei, X., Flower, D. R. & 7 other authors (2004). Determinants of HIV-1 escape from the primary CD8+ cytotoxic T lymphocyte response. *J Exp Med* **200**, 1243–1256.

Kacani, L., Stoiber, H., Speth, C., Banki, Z., Tenner-Racz, K., Racz, P. & Dierich, M. P. (2001). Complement-dependent control of viral dynamics in pathogenesis of human immunodeficiency virus and simian immunodeficiency virus infection. *Mol Immunol* **38**, 241–247.

Kalams, S. A., Johnson, R. P., Trocha, A. K., Dynan, M. J., Ngo, S., D'Aquila, R. T., Kurnick, J. T. & Walker, B. D. (1994). Longitudinal analysis of T cell receptor (TCR) gene usage by human immunodeficiency virus 1 envelope-specific cytotoxic T lymphocyte clones reveals a limited TCR repertoire. *J Exp Med* **179**, 1261–1271.

Kalams, S. A., Goulder, P. J., Shea, A. K., Jones, N. G., Trocha, A. K., Ogg, G. S. & Walker, B. D. (1999a). Levels of human immunodeficiency virus type 1-specific cytotoxic T-lymphocyte effector and memory responses decline after suppression of viremia with highly active antiretroviral therapy. *J Virol* **73**, 6721–6728.

Kalams, S. A., Buchbinder, S. P., Rosenberg, E. S., Billingsley, J. M., Colbert, D. S., Jones, N. G., Shea, A. K., Trocha, A. K. & Walker, B. D. (1999b). Association between virus-specific cytotoxic T-lymphocyte and helper responses in human immunodeficiency virus type 1 infection. *J Virol* **73**, 6715–6720.

Kamp, W., Berk, M. B., Visser, C. J. & Nottet, H. S. (2000). Mechanisms of HIV-1 to escape from the host immune surveillance. *Eur J Clin Invest* **30**, 740–746.

Kang, C. Y., Luo, L., Wainberg, M. A. & Li, Y. (1999). Development of HIV/AIDS vaccine using chimeric gag-env virus-like particles. *Biol Chem* **380**, 353–364.

Kaslow, R. A., Duquesnoy, R., VanRaden, M. & 8 other authors (1990). A1, Cw7, B8, DR3 HLA antigen combination associated with rapid decline of T-helper lymphocytes in HIV-1 infection. A report from the Multicenter AIDS Cohort Study. *Lancet* **335**, 927–930.

Kaslow, R. A., Carrington, M., Apple, R. & 12 other authors (1996). Influence of combinations of human major histocompatibility complex genes on the course of HIV-1 infection. *Nat Med* **2**, 405–411.

Kaul, R., Plummer, F. A., Kimani, J. & 8 other authors (2000). HIV-1-specific mucosal CD8+ lymphocyte responses in the cervix of HIV-1-resistant prostitutes in Nairobi. *J Immunol* **164**, 1602–1611.

Kaul, R., Rowland-Jones, S. L., Kimani, J. & 9 other authors (2001). Late seroconversion in HIV-resistant Nairobi prostitutes despite pre-existing HIV-specific CD8+ responses. *J Clin Invest* **107**, 341–349.

Kedzierska, K. & Crowe, S. M. (2002). The role of monocytes and macrophages in the pathogenesis of HIV-1 infection. *Curr Med Chem* **9**, 1893–1903.

Kelleher, A. D., Long, C., Holmes, E. C. & 15 other authors (2001). Clustered mutations in HIV-1 gag are consistently required for escape from HLA-B27-restricted cytotoxic T lymphocyte responses. *J Exp Med* **193**, 375–386.

Kerkau, T., Schmitt-Landgraf, R., Schimpl, A. & Wecker, E. (1989). Downregulation of HLA class I antigens in HIV-1-infected cells. *AIDS Res Hum Retrovir* 5, 613–620.

Kerkau, T., Bacik, I., Bennink, J. R., Yewdell, J. W., Hunig, T., Schimpl, A. & Schubert, U. (1997). The human immunodeficiency virus type 1 (HIV-1) Vpu protein interferes with an early step in the biosynthesis of major histocompatibility complex (MHC) class I molecules. *J Exp Med* 185, 1295–1305.

Klein, M. R., van Baalen, C. A., Holwerda, A. M., Kerkhof Garde, S. R., Bende, R. J., Keet, I. P. M., Eeftinck-Schattenkerk, J.-K. M. & Miedema, F. (1995). Kinetics of Gag-specific cytotoxic T lymphocyte responses during the clinical course of HIV-1 infection: a longitudinal analysis of rapid progressors and long-term asymptomatics. *J Exp Med* 181, 1365–1372.

Klenerman, P., Rowland-Jones, S., McAdam, S. & 10 other authors (1994). Cytotoxic T-cell activity antagonized by naturally occurring HIV-1 Gag variants. *Nature* 369, 403–407.

Koch, M., Pancera, M., Kwong, P. D., Kolchinsky, P., Grundner, C., Wang, L., Hendrickson, W. A., Sodroski, J. & Wyatt, R. (2003). Structure-based, targeted deglycosylation of HIV-1 gp120 and effects on neutralization sensitivity and antibody recognition. *Virology* 313, 387–400.

Kostense, S., Ogg, G. S., Manting, E. H. & 8 other authors (2001). High viral burden in the presence of major HIV-specific CD8(+) T cell expansions: evidence for impaired CTL effector function. *Eur J Immunol* 31, 677–686.

Kottilil, S., Shin, K., Planta, M., McLaughlin, M., Hallahan, C. W., Ghany, M., Chun, T. W., Sneller, M. C. & Fauci, A. S. (2004). Expression of chemokine and inhibitory receptors on natural killer cells: effect of immune activation and HIV viremia. *J Infect Dis* 189, 1193–1198.

Koup, R. A. (1994). Virus escape from CTL recognition. *J Exp Med* 180, 779–782.

Koup, R. A., Safrit, J. T., Cao, Y., Andrews, C. A., McLeod, G., Borkowsky, W., Farthing, C. & Ho, D. D. (1994). Temporal association of cellular immune responses with the initial control of viremia in primary human immunodeficiency virus type 1 syndrome. *J Virol* 68, 4650–4655.

Krown, S. E. (1998). Interferon-alpha: evolving therapy for AIDS-associated Kaposi's sarcoma. *J Interferon Cytokine Res* 18, 209–214.

Krown, S. E., Real, F. X., Cunningham-Rundles, S., Myskowski, P. L., Koziner, B., Fein, S., Mittelman, A., Oettgen, H. F. & Safai, B. (1983). Preliminary observations on the effect of recombinant leukocyte A interferon in homosexual men with Kaposi's sarcoma. *N Engl J Med* 308, 1071–1076.

Krown, S. E., Niedzwiecki, D., Bhalla, R. B., Flomenberg, N., Bundow, D. & Chapman, D. (1991). Relationship and prognostic value of endogenous interferon-alpha, beta 2-microglobulin, and neopterin serum levels in patients with Kaposi sarcoma and AIDS. *J Acquir Immune Defic Syndr* 4, 871–880.

Kunzi, M. S., Farzadegan, H., Margolick, J. B., Vlahov, D. & Pitha, P. M. (1995). Identification of human immunodeficiency virus primary isolates resistant to interferon-alpha and correlation of prevalence to disease progression. *J Infect Dis* 171, 822–828.

Labrijn, A. F., Poignard, P., Raja, A. & 16 other authors (2003). Access of antibody molecules to the conserved coreceptor binding site on glycoprotein gp120 is sterically restricted on primary human immunodeficiency virus type 1. *J Virol* 77, 10557–10565.

Lane, H. C., Kovacs, J. A., Feinberg, J. & 11 other authors (1988). Anti-retroviral effects of interferon-alpha in AIDS-associated Kaposi's sarcoma. *Lancet* **2**, 1218–1222.

Lane, H. C., Davey, V., Kovacs, J. A. & 12 other authors (1990). Interferon-alpha in patients with asymptomatic human immunodeficiency virus (HIV) infection. A randomized, placebo-controlled trial. *Ann Intern Med* **112**, 805–811.

Lang, F., Peyrat, M. A., Constant, P., Davodeau, F., David-Ameline, J., Poquet, Y., Vie, H., Fournie, J. J. & Bonneville, M. (1995). Early activation of human V gamma 9V delta 2 T cell broad cytotoxicity and TNF production by nonpeptidic mycobacterial ligands. *J Immunol* **154**, 5986–5994.

Lanier, L. L. (2001). On guard – activating NK cell receptors. *Nat Immunol* **2**, 23–27.

Lanier, L. L., Phillips, J. H., Hackett, J., Jr, Tutt, M. & Kumar, V. (1986). Natural killer cells: definition of a cell type rather than a function. *J Immunol* **137**, 2735–2739.

Learmont, J. C., Geczy, A. F., Mills, J. & 9 other authors (1999). Immunologic and virologic status after 14 to 18 years of infection with an attenuated strain of HIV-1. A report from the Sydney Blood Bank Cohort. *N Engl J Med* **340**, 1715–1722.

Le Gall, S., Erdtmann, L., Benichou, S., Berlioz-Torrent, C., Liu, L., Benarous, R., Heard, J. M. & Schwartz, O. (1998). Nef interacts with the mu subunit of clathrin adaptor complexes and reveals a cryptic sorting signal in MHC I molecules. *Immunity* **8**, 483–495.

Le Gall, S., Buseyne, F., Trocha, A., Walker, B. D., Heard, J. M. & Schwartz, O. (2000). Distinct trafficking pathways mediate Nef-induced and clathrin-dependent major histocompatibility complex class I down-regulation. *J Virol* **74**, 9256–9266.

Legrand, E., Pellegrin, I., Neau, D., Pellegrin, J. L., Ragnaud, J. M., Dupon, M., Guillemain, B. & Fleury, H. J. (1997). Course of specific T lymphocyte cytotoxicity, plasma and cellular viral loads, and neutralizing antibody titers in 17 recently seroconverted HIV type 1-infected patients. *AIDS Res Hum Retrovir* **13**, 1383–1394.

Lehner, T., Bergmeier, L., Wang, Y., Tao, L. & Mitchell, E. (1999). A rational basis for mucosal vaccination against HIV infection. *Immunol Rev* **170**, 183–196.

Lehner, T., Mitchell, E., Bergmeier, L. & 7 other authors (2000a). The role of gamma-delta T cells in generating antiviral factors and beta-chemokines in protection against mucosal simian immunodeficiency virus infection. *Eur J Immunol* **30**, 2245–2256.

Lehner, T., Wang, Y., Cranage, M. & 9 other authors (2000b). Up-regulation of beta-chemokines and down-modulation of CCR5 co-receptors inhibit simian immunodeficiency virus transmission in non-human primates. *Immunology* **99**, 569–577.

Leibson, P. J. (1997). Signal transduction during natural killer cell activation: inside the mind of a killer. *Immunity* **6**, 655–661.

Lekutis, C. & Letvin, N. L. (1998). Substitutions in a major histocompatibility complex class II-restricted human immunodeficiency virus type 1 gp120 epitope can affect CD4+ T-helper-cell function. *J Virol* **72**, 5840–5844.

Leslie, A. J., Pfafferott, K. J., Chetty, P. & 26 other authors (2004). HIV evolution: CTL escape mutation and reversion after transmission. *Nat Med* **10**, 282–289.

Letvin, N. L., Montefiori, D. C., Yasutomi, Y. & 9 other authors (1997). Potent, protective anti-HIV immune responses generated by bimodal HIV envelope DNA plus protein vaccination. *Proc Natl Acad Sci U S A* **94**, 9378–9383.

Letvin, N. L., Barouch, D. H. & Montefiori, D. C. (2002). Prospects for vaccine protection against HIV-1 infection and AIDS. *Annu Rev Immunol* **20**, 73–99.

Levy, J. A. (1993). Pathogenesis of human immunodeficiency virus infection. *Microbiol Rev* **57**, 183–289.

Levy, J. A., Mackewicz, C. E. & Barker, E. (1996). Controlling HIV pathogenesis: the role of the noncytotoxic anti-HIV response of CD8+ T cells. *Immunol Today* **17**, 217–224.

Levy, J. A., Scott, I. & Mackewicz, C. (2003). Protection from HIV/AIDS: the importance of innate immunity. *Clin Immunol* **108**, 167–174.

Lewinsohn, D. A., Lines, R., Lewinsohn, D. M., Riddell, S. R., Greenberg, P. D., Emerman, M. & Bartz, S. R. (2002). HIV-1 Vpr does not inhibit CTL-mediated apoptosis of HIV-1 infected cells. *Virology* **294**, 13–21.

Littaua, R. A., Oldstone, M. B., Takeda, A. & Ennis, F. A. (1992). A CD4+ cytotoxic T-lymphocyte clone to a conserved epitope on human immunodeficiency virus type 1 p24: cytotoxic activity and secretion of interleukin-2 and interleukin-6. *J Virol* **66**, 608–611.

Lopes, A. R., Jaye, A., Dorrell, L. & 12 other authors (2003). Greater T cell receptor heterogeneity and functional flexibility in human immunodeficiency virus type 2 (HIV-2) compared to HIV-1 infection. *J Immunol* **171**, 307–316.

Lore, K., Sonnerborg, A., Brostrom, C. & 9 other authors (2002). Accumulation of DC-SIGN$^+$ CD40$^+$ dendritic cells with reduced CD80 and CD86 expression in lymphoid tissue during acute HIV-1 infection. *AIDS Res Hum Retrovir* **16**, 683–692.

Lotti, B., Wendland, T., Furrer, H. & 10 other authors (2002). Cytotoxic HIV-1 p55gag-specific CD4$^+$ T cells produce HIV-inhibitory cytokines and chemokines. *J Clin Invest* **22**, 253–262.

Lu, W., Wu, X., Lu, Y., Guo, W. & Andrieu, J. M. (2003). Therapeutic dendritic-cell vaccine for simian AIDS. *Nat Med* **9**, 27–32.

Lubeck, M. D., Natuk, R., Myagkikh, M. & 19 other authors (1997). Long-term protection of chimpanzees against high-dose HIV-1 challenge induced by immunization. *Nat Med* **3**, 651–658.

Lum, J. J. & Badley, A. D. (2003). Resistance to apoptosis: mechanism for the development of HIV reservoirs. *Curr HIV Res* **1**, 261–274.

Lund, J., Sato, A., Akira, S., Medzhitov, R. & Iwasaki, A. (2003). Toll-like receptor 9-mediated recognition of Herpes simplex virus-2 by plasmacytoid dendritic cells. *J Exp Med* **198**, 513–520.

Lund, J. M., Alexopoulou, L., Sato, A., Karow, M., Adams, N. C., Gale, N. W., Iwasaki, A. & Flavell, R. A. (2004). Recognition of single-stranded RNA viruses by Toll-like receptor 7. *Proc Natl Acad Sci U S A* **101**, 5598–5603.

Lyles, R. H., Munoz, A., Yamashita, T. E., Bazmi, H., Detels, R., Rinaldo, C. R., Margolick, J. B., Phair, J. P. & Mellors, J. W. (2000). Natural history of human immunodeficiency virus type 1 viremia after seroconversion and proximal to AIDS in a large cohort of homosexual men. *J Infect Dis* **181**, 872–880.

Mackewicz, C. & Levy, J. A. (1992). CD8+ cell anti-HIV activity: nonlytic suppression of virus replication. *AIDS Res Hum Retrovir* **8**, 1039–1050.

Maggiorella, M. T., Baroncelli, S., Michelini, Z. & 23 other authors (2004). Long-term protection against SHIV89.6P replication in HIV-1 Tat vaccinated cynomolgus monkeys. *Vaccine* **22**, 3258–3269.

Mansky, L. M. & Temin, H. M. (1995). Lower in vivo mutation rate of human immunodeficiency virus-type 1 than that predicted from the fidelity of reverse transcriptase. *J Virol* **69**, 5087–5094.

Markowitz, M., Louie, M., Hurley, A., Sun, E., DiMascio, M., Perelson, A. S. & Ho, D. D. (2003). A novel antiviral intervention results in more accurate assessment of human immunodeficiency virus type 1 replication dynamics and T-cell decay in vivo. *J Virol* **77**, 5037–5038.

Martini, F., Urso, R., Gioia, C., De Felici, A., Narciso, P., Amendola, A., Paglia, M. G., Colizzi, V. & Poccia, F. (2000). gammadelta T-cell anergy in human immunodeficiency virus-infected persons with opportunistic infections and recovery after highly active antiretroviral therapy. *Immunology* **100**, 481–486.

Mascola, J. R., Snyder, S. W., Weislow, O. S. & 14 other authors (1996). Immunization with envelope subunit vaccine products elicits neutralizing antibodies against laboratory-adapted but not primary isolates of human immunodeficiency virus type 1. The National Institute of Allergy and Infectious Diseases AIDS Vaccine Evaluation Group. *J Infect Dis* **173**, 340–348.

Mascola, J. R., Lewis, M. G., Stiegler, G. & 11 other authors (1999). Protection of macaques against pathogenic simian/human immunodeficiency virus 89.6PD by passive transfer of neutralizing antibodies. *J Virol* **73**, 4009–4018.

Mascola, J. R., Stiegler, G., VanCott, T. C. & 8 other authors (2000). Protection of macaques against vaginal transmission of a pathogenic HIV-1/SIV chimeric virus by passive infusion of neutralizing antibodies. *Nat Med* **6**, 207–210.

Masemola, A. M., Mashishi, T. N., Khoury, G. & 18 other authors (2004). Novel and promiscuous CTL epitopes in conserved regions of Gag targeted by individuals with early subtype C HIV type 1 infection from Southern Africa. *J Immunol* **173**, 4607–4617.

Matloubian, M., Concepcion, R. J. & Ahmed, R. (1994). CD4$^+$ T cells are required to sustain CD8$^+$ cytotoxic T-cell responses during chronic viral infection. *J Virol* **68**, 8056–8063.

Mavilio, D., Benjamin, J., Daucher, M. & 8 other authors (2003). Natural killer cells in HIV-1 infection: dichotomous effects of viremia on inhibitory and activating receptors and their functional correlates. *Proc Natl Acad Sci U S A* **100**, 15011–15016.

Mazzoli, S., Trabattoni, D., Lo Caputo, S. & 13 other authors (1997). HIV-specific mucosal and cellular immunity in HIV-seronegative partners of HIV-seropositive individuals. *Nat Med* **3**, 1250–1257.

McAdam, S., Klenerman, P., Tussey, L. & 7 other authors (1995). Immunogenic HIV variant peptides that bind to HLA-B8 can fail to stimulate cytotoxic T lymphocyte responses. *J Immunol* **155**, 2729–2736.

McCune, J. M. (2001). The dynamics of CD4+ T-cell depletion in HIV disease. *Nature* **410**, 974–979.

McMichael, A. J. & Hanke, T. (2003). HIV vaccines 1983-2003. *Nat Med* **9**, 874–880.

McMichael, A. J. & Phillips, R. E. (1997). Escape of human immunodeficiency virus from immune control. *Annu Rev Immunol* **15**, 271–296.

McNeil, A. J., Yap, P. L., Gore, S. M. & 7 other authors (1996). Association of HLA types A1-B8-DR3 and B27 with rapid and slow progression of HIV disease. *QJM* **89**, 177–185.

Meddows-Taylor, S., Papathanasopoulos, M. A., Kuhn, L., Meyers, T. M. & Tiemessen, C. T. (2004). Detection of human immunodeficiency virus type 1 envelope peptide-stimulated T-helper cell responses and variations in the corresponding regions of viral isolates among vertically infected children. *Virus Genes* **28**, 311–318.

Mellors, J. W., Kingsley, L. A., Rinaldo, C. R., Jr, Todd, J. A., Hoo, B. S., Kokka, R. P. & Gupta, P. (1995). Quantitation of HIV-1 RNA in plasma predicts outcome after seroconversion. *Ann Intern Med* **122**, 573–579.

Mellors, J. W., Rinaldo, C. R., Jr, Gupta, P., White, R. M., Todd, J. A. & Kingsley, L. A. (1996). Prognosis in HIV-1 infection predicted by the quantity of virus in plasma. *Science* **272**, 1167–1170.

Meyer, H., Sutter, G. & Mayr, A. (1991). Mapping of deletions in the genome of the highly attenuated vaccinia virus MVA and their influence on virulence. *J Gen Virol* **72**, 1031–1038.

Meylan, P. R., Guatelli, J. C., Munis, J. R., Richman, D. D. & Kornbluth, R. S. (1993). Mechanisms for the inhibition of HIV replication by interferons-alpha, -beta, and -gamma in primary human macrophages. *Virology* **193**, 138–148.

Migueles, S. A., Sabbaghian, M. S., Shupert, W. L. & 8 other authors (2000). HLA B*5701 is highly associated with restriction of virus replication in a subgroup of HIV-infected long term nonprogressors. *Proc Natl Acad Sci U S A* **97**, 2709–2714.

Migueles, S. A., Laborico, A. C., Shupert, W. L. & 11 other authors (2002). HIV-specific CD8+ T cell proliferation is coupled to perforin expression and is maintained in non-progressors. *Nat Immunol* **3**, 1061–1068.

Migueles, S. A., Laborico, A. C., Imamichi, H. & 8 other authors (2003). The differential ability of HLA B*5701+ long-term nonprogressors and progressors to restrict human immunodeficiency virus replication is not caused by loss of recognition of autologous viral gag sequences. *J Virol* **77**, 6889–6898.

Montefiori, D. C. & Evans, T. G. (1999). Toward an HIV type 1 vaccine that generates potent, broadly cross-reactive neutralizing antibodies. *AIDS Res Hum Retrovir* **15**, 689–698.

Montefiori, D. C., Baba, T. W., Li, A., Bilska, M. & Ruprecht, R. M. (1996). Neutralizing and infection-enhancing antibody responses do not correlate with the differential pathogenicity of SIVmac239delta3 in adult and infant rhesus monkeys. *J Immunol* **157**, 5528–5535.

Montelaro, R. C., Parekh, B., Orrego, A. & Issel, C. J. (1984). Antigenic variation during persistent infection by equine infectious anemia virus, a retrovirus. *J Biol Chem* **259**, 10539–10544.

Moore, C. B., John, M., James, I. R., Christiansen, F. T., Witt, C. S. & Mallal, S. A. (2002). Evidence of HIV-1 adaptation to HLA-restricted immune responses at a population level. *Science* **296**, 1439–1443.

Moore, J. P., Parren, P. W. & Burton, D. R. (2001). Genetic subtypes, humoral immunity, and human immunodeficiency virus type 1 vaccine development. *J Virol* **75**, 5721–5729.

Mortara, L., Letourneur, F., Gras-Masse, H., Venet, A., Guillet, J.-G. & Bourgault-Villada, I. (1998). Selection of virus variants and emergence of virus escape mutants after immunization with an epitope vaccine. *J Virol* **72**, 1403–1410.

Mossman, S. P., Bex, F., Berglund, P. & 11 other authors (1996). Protection against lethal simian immunodeficiency virus SIVsmmPBj14 disease by a recombinant Semliki Forest virus gp160 vaccine and by a gp120 subunit vaccine. *J Virol* **70**, 1953–1960.

Motsinger, A., Haas, D. W., Stanic, A. K., Van Kaer, L., Joyce, S. & Unutmaz, D. (2002). CD1d-restricted human natural killer T cells are highly susceptible to human immunodeficiency virus 1 infection. *J Exp Med* **195**, 869–879.

Muller, S. (2004). Avoiding deceptive imprinting of the immune response to HIV-1 infection in vaccine development. *Int Rev Immunol* **23**, 423–436.

Muller, U., Steinhoff, U., Reis, L. F. L., Hemmi, S., Pavlovic, J., Zinkernagel, R. M. & Aguet, M. (1994). Functional role of type I and type II interferons in antiviral defense. *Science* **264**, 1918–1921.

Murphey-Corb, M., Martin, L. N., Davison-Fairburn, B. & 9 other authors (1989). A formalin-inactivated whole SIV vaccine confers protection in macaques. *Science* **246**, 1293–1297.

Musey, L. K., Krieger, J. N., Hughes, J. P., Schacker, T. W., Corey, L. & McElrath, M. J. (1999). Early and persistent human immunodeficiency virus type 1 (HIV-1)-specific T helper dysfunction in blood and lymph nodes following acute HIV-1 infection. *J Infect Dis* **180**, 278–284.

Mutwiri, G., Pontarollo, R., Babiuk, S. & 12 other authors (2003). Biological activity of immunostimulatory CpG DNA motifs in domestic animals. *Vet Immunol Immunopathol* **91**, 89–103.

Nathanson, N. & Mathieson, B. J. (2000). Biological considerations in the development of a human immunodeficiency virus vaccine. *J Infect Dis* **182**, 579–589.

Nelson, G. W., Kaslow, R. & Mann, D. L. (1997). Frequency of HLA allele-specific peptide motifs in HIV-1 proteins correlates with the allele's association with relative rates of disease progression after HIV-1 infection. *Proc Natl Acad Sci U S A* **94**, 9802–9807.

Nietfield, W., Bauer, M., Fevrier, M., Maier, R., Holzwarth, B., Frank, R., Maier, B., Riviere, Y. & Meyerhans, A. (1995). Sequence constraints and recognition by CTL of an HLA-B27-restricted HIV-1 gag epitope. *J Immunol* **154**, 2188–2197.

Norris, P. J., Sumaroka, M., Brander, C., Moffett, H. F., Boswell, S. L., Nguyen, T., Sykulev, Y., Walker, B. D. & Rosenberg, E. S. (2001). Multiple effector functions mediated by human immunodeficiency virus-specific CD4$^+$ T-cell clones. *J Virol* **75**, 9771–9779.

Norris, P. J., Moffett, H. F., Yang, O. O., Kaufmann, D. E., Clark, M. J., Addo, M. M. & Rosenberg, E. S. (2004). Beyond help: direct effector functions of human immunodeficiency virus type 1-specific CD4(+) T cells. *J Virol* **78**, 8844–8851.

Nowak, M. A., May, R. M., Phillips, R. E. & 8 other authors (1995). Antigenic oscillations and shifting immunodominance in HIV-1 infections. *Nature* **375**, 606–611.

O'Brien, S. J., Gao, X. & Carrington, M. (2001). HLA and AIDS: a cautionary tale. *Trends Mol Med* **7**, 379–381.

O'Connor, D., Allen, T. & Watkins, D. I. (2001). Vaccination with CTL epitopes that escape: an alternative approach to HIV vaccine development? *Immunol Lett* **79**, 77–84.

O'Connor, D. H., Allen, T. M. & Watkins, D. I. (2002a). Cytotoxic T-lymphocyte escape monitoring in simian immunodeficiency virus vaccine challenge studies. *DNA Cell Biol* **21**, 659–664.

O'Connor, D. H., Allen, T. M., Vogel, T. U. & 11 other authors (2002b). Acute phase cytotoxic T lymphocyte escape is a hallmark of simian immunodeficiency virus infection. *Nat Med* **8**, 493–499.

O'Connor, D. H., Mothe, B. R., Weinfurter, J. T. & 22 other authors (2003). Major histocompatibility complex class I alleles associated with slow simian immunodeficiency virus disease progression bind epitopes recognized by dominant acute-phase cytotoxic-T-lymphocyte responses. *J Virol* **77**, 9029–9040.

Ogg, G. S., Jin, X., Bonhoeffer, S. & 12 other authors (1998). Quantitation of HIV-1-specific cytotoxic T lymphocytes and plasma load of viral RNA. *Science* **279**, 2103–2106.

Ogg, G. S., Jin, X., Bonhoeffer, S. & 11 other authors (1999). Decay kinetics of human immunodeficiency virus-specific effector cytotoxic T lymphocytes after combination antiretroviral therapy. *J Virol* **73**, 797–800.

Oliva, A., Kinter, A. L., Vaccarezza, M. & 10 other authors (1998). Natural killer cells from human immunodeficiency virus (HIV)-infected individuals are an important source of CC-chemokines and suppress HIV-1 entry and replication in vitro. *J Clin Invest* **102**, 223–231.

Ourmanov, I., Brown, C. R., Moss, B., Carroll, M., Wyatt, L., Pletneva, L., Goldstein, S., Venzon, D. & Hirsch, V. M. (2000). Comparative efficacy of recombinant modified vaccinia virus Ankara expressing simian immunodeficiency virus (SIV) Gag-Pol and/or Env in macaques challenged with pathogenic SIV. *J Virol* **74**, 2740–2751.

Oxenius, A., Price, D. A., Easterbrook, P. J., O'Callaghan, C. A., Kelleher, A. D., Whelan, J. A., Sontag, G., Sewell, A. K. & Phillips, R. E. (2000). Early highly active antiretroviral therapy for acute HIV-1 infection preserves immune function of CD8$^+$ and CD4$^+$ T lymphocytes. *Proc Natl Acad Sci U S A* **97**, 3382–3387.

Oxenius, A., Fidler, S., Brady, M., Dawson, S. J., Ruth, K., Easterbrook, P. J., Weber, J. N., Phillips, R. E. & Price, D. A. (2001). Variable fate of virus-specific CD4$^+$ T cells during primary HIV-1 infection. *Eur J Immunol* **31**, 3782–3788.

Pantaleo, G., Demarest, J. F., Soudeyns, H. & 8 other authors (1994). Major expansion of CD8$^+$ T cells with a predominant Vβ usage during the primary immune response to HIV. *Nature* **370**, 463–467.

Pantaleo, G., Menzo, S., Vaccarezza, M. & 11 other authors (1995). Studies in subjects with long-term nonprogressive human immunodeficiency virus infection. *N Engl J Med* **332**, 209–216.

Pantaleo, G., Demarest, J. F., Schacker, T. & 14 other authors (1997). The qualitative nature of the primary immune response to HIV infection is a prognosticator of disease progression independent of the initial level of plasma viremia. *Proc Natl Acad Sci U S A* **94**, 254–258.

Patterson, S., Robinson, S. P., English, N. R. & Knight, S. C. (1999). Subpopulations of peripheral blood dendritic cells show differential susceptibility to infection with a lymphotropic strain of HIV-1. *Immunol Lett* **66**, 111–116.

Patterson, S., Rae, A., Hockey, N., Gilmour, J. & Gotch, F. (2001). Plasmacytoid dendritic cells are highly susceptible to human immunodeficiency virus type 1 infection and release infectious virus. *J Virol* **75**, 6710–6713.

Paul, N. L., Marsh, M., McKeating, J. A., Schulz, T. F., Liljestrom, P., Garoff, H. & Weiss, R. A. (1993). Expression of HIV-1 envelope glycoproteins by Semliki Forest virus vectors. *AIDS Res Hum Retrovir* **9**, 963–970.

Persaud, D., Zhou, Y., Siliciano, J. M. & Siliciano, R. F. (2003). Latency in human immunodeficiency virus type 1 infection: no easy answers. *J Virol* **77**, 1659–1665.

Petrovas, C., Mueller, Y. M. & Katsikis, P. D. (2004). HIV-specific CD8+ T cells: serial killers condemned to die? *Curr HIV Res* **2**, 153–162.

Pialoux, G., Excler, J. L., Riviere, Y. & 14 other authors (1995). A prime-boost approach to HIV preventive vaccine using a recombinant canarypox virus expressing glycoprotein 160 (MN) followed by a recombinant glycoprotein 160 (MN/LAI). The AGIS Group, and l'Agence Nationale de Recherche sur le SIDA. *AIDS Res Hum Retrovir* **11**, 373–381.

Piatak, M., Jr, Saag, M. S., Yang, L. C., Clark, S. J., Kappes, J. C., Luk, K. C., Hahn, B. H., Shaw, G. M. & Lifson, J. D. (1993). Determination of plasma viral load in HIV-1 infection by quantitative competitive polymerase chain reaction. *AIDS* **7** (Suppl. 2), S65–71.

Piguet, V. & Trono, D. (2001). Living in oblivion: HIV immune evasion. *Semin Immunol* **13**, 51–57.

Pikora, C. A. (2004). Glycosylation of the ENV spike of primate immunodeficiency viruses and antibody neutralization. *Curr HIV Res* **2**, 243–254.

Pilgrim, A. K., Pantaleo, G., Cohen, O. J., Fink, L. M., Zhou, J. Y., Zhou, J. T., Bolognesi, D. P., Fauci, A. S. & Montefiori, D. C. (1997). Neutralizing antibody responses to human immunodeficiency virus type 1 in primary infection and long-term-non-progressive infection. *J Infect Dis* **176**, 924–932.

Pinto, L. A., Sullivan, J., Berzofsky, J. A., Clerici, M., Kessler, H. A., Landay, A. L. & Shearer, G. M. (1995). ENV-specific cytotoxic T lymphocyte responses in HIV sero-negative health care workers occupationally exposed to HIV-contaminated body fluids. *J Clin Invest* **96**, 867–876.

Pircher, H., Moskophidis, D., Rohrer, U., Burki, K., Hengartner, H. & Zinkernagel, R. M. (1990). Viral escape by selection of cytotoxic T cell-resistant virus variants *in vivo*. *Nature* **346**, 629–633.

Pitcher, C. J., Quittner, C., Peterson, D. M., Connors, M., Koup, R. A., Maino, V. C. & Picker, L. J. (1999). HIV-1-specific CD4$^+$ T cells are detectable in most individuals with active HIV-1 infection, but decline with prolonged viral suppression. *Nat Med* **5**, 518–525.

Pitha, P. M. (1994). Multiple effects of interferon on the replication of human immuno-deficiency virus type 1. *Antivir Res* **24**, 205–219.

Planz, O., Ehl, S., Furrer, E., Horvath, E. & Brundler, M. A. (1997). A critical role for neutralizing antibody-producing B cells, CD4(+) T cells, and interferons in persistent and acute infections of mice with lymphocytic choriomeningitis virus: implications for adoptive immunotherapy of virus carriers. *Proc Natl Acad Sci U S A* **94**, 6874–6879.

Poccia, F., Boullier, S., Lecoeur, H., Cochet, M., Poquet, Y., Colizzi, V., Fournie, J. J. & Gougeon, M. L. (1996). Peripheral V gamma 9/V delta 2 T cell deletion and anergy to nonpeptidic mycobacterial antigens in asymptomatic HIV-1-infected persons. *J Immunol* **157**, 449–461.

Poccia, F., Battistini, L., Cipriani, B., Mancino, G., Martini, F., Gougeon, M. L. & Colizzi, V. (1999). Phosphoantigen-reactive Vgamma9Vdelta2 T lymphocytes suppress in vitro human immunodeficiency virus type 1 replication by cell-released antiviral factors including CC chemokines. *J Infect Dis* **180**, 858–861.

Poccia, F., Agrati, C., Ippolito, G., Colizzi, V. & Malkovsky, M. (2001). Natural T cell immunity to intracellular pathogens and nonpeptidic immunoregulatory drugs. *Curr Mol Med* **1**, 137–151.

Poli, G., Orenstein, J. M., Kinter, A., Folks, T. M. & Fauci, A. S. (1989). Interferon-alpha but not AZT suppresses HIV expression in chronically infected cell lines. *Science* **244**, 575–577.

Poulsen, A. G., Aaby, P., Larsen, O., Jensen, H., Naucler, A., Lisse, I. M., Christiansen, C. B., Dias, F. & Melbye, M. (1997). 9-year HIV-2-associated mortality in an urban community in Bissau, west Africa. *Lancet* **349**, 911–914.

Price, D. A., Goulder, P. J. R., Klenerman, P., Sewell, A. K., Easterbrook, P. J., Troop, M., Bangham, C. R. M. & Phillips, R. E. (1997). Positive selection of HIV-1 cytotoxic T lymphocyte escape variants during primary infection. *Proc Natl Acad Sci U S A* **94**, 1890–1895.

Price, D. A., Sewell, A. K., Dong, T., Tan, R., Goulder, P. J., Rowland-Jones, S. L. & Phillips, R. E. (1998). Antigen-specific release of beta-chemokines by anti-HIV-1 cytotoxic T lymphocytes. *Curr Biol* **8**, 355–358.

Prince, A. M., Reesink, H., Pascual, D., Horowitz, B., Hewlett, I., Murthy, K. K., Cobb, K. E. & Eichberg, J. W. (1991). Prevention of HIV infection by passive immunization with HIV immunoglobulin. *AIDS Res Hum Retrovir* **7**, 971–973.

Putkonen, P., Thorstensson, R., Ghavamzadeh, L., Albert, J., Hild, K., Biberfeld, G. & Norrby, E. (1991). Prevention of HIV-2 and SIVsm infection by passive immunization in cynomolgus monkeys. *Nature* **352**, 436–438.

Quinones-Kochs, M. I., Buonocore, L. & Rose, J. K. (2002). Role of N-linked glycans in a human immunodeficiency virus envelope glycoprotein: effects on protein function and the neutralizing antibody response. *J Virol* **76**, 4199–4211.

Rammensee, H. G. & Monaco, J. (1994). Peptimmunology. *Curr Opin Immunol* **6**, 1–2.

Rankin, R., Pontarollo, R., Ioannou, X., Krieg, A. M., Hecker, R., Babiuk, L. A. & van Drunen Littel-van den Hurk, S. (2001). CpG motif identification for veterinary and laboratory species demonstrates that sequence recognition is highly conserved. *Antisense Nucleic Acid Drug Dev* **11**, 333–340.

Ratcliffe, L. T., Lukey, P. T., MacKenzie, C. R. & Ress, S. R. (1994). Reduced NK activity correlates with active disease in HIV- patients with multidrug-resistant pulmonary tuberculosis. *Clin Exp Immunol* **97**, 373–379.

Recher, M., Lang, K. S., Hunziker, L. & 13 other authors (2004). Deliberate removal of T cell help improves virus-neutralizing antibody production. *Nat Immunol* **5**, 934–942.

Reimann, K. A., Tenner-Racz, K., Racz, P., Montefiori, D. C., Yasutomi, Y., Lin, W., Ransil, B. J. & Letvin, N. L. (1994). Immunopathogenic events in acute infection of rhesus monkeys with simian immunodeficiency virus of macaques. *J Virol* **68**, 2362–2370.

Reitter, J. N., Means, R. E. & Desrosiers, R. C. (1998). A role for carbohydrates in immune evasion in AIDS. *Nat Med* **4**, 679–684.

Richman, D. D., Wrin, T., Little, S. J. & Petropoulos, C. J. (2003). Rapid evolution of the neutralizing antibody response in HIV type 1 infection. *Proc Natl Acad Sci U S A* **100**, 4144–4149.

Rinaldo, C., Huang, X.-L., Fan, Z. & 8 other authors (1995). High levels of anti-human immunodeficiency virus type 1 (HIV-1) memory cytotoxic T-lymphocyte activity and low viral load are associated with lack of disease in HIV-1-infected long-term nonprogressors. *J Virol* **69**, 5838–5842.

Robinson, H. L., Montefiori, D. C., Johnson, R. P. & 14 other authors (1999). Neutralizing antibody-independent containment of immunodeficiency virus challenges by DNA priming and recombinant pox virus booster immunizations. *Nat Med* **5**, 526–534.

Roger, M. (1998). Influence of host genes on HIV-1 disease progression. *FASEB J* **12**, 625–632.

Roos, M. T., de Leeuw, N. A., Claessen, F. A., Huisman, H. G., Kootstra, N. A., Meeyard, L., Schellekens, P. T., Schuitemaker, H. & Miedema, F. (1994). Viro-immunological studies in acute HIV-1 infection. *AIDS* **8**, 1533–1538.

Roos, M. T., Miedema, F., Koot, M., Tersmette, M., Schaasberg, W. P., Coutinho, R. A. & Schellekens, P. T. (1995). T cell function in vitro is an independent progression marker for AIDS in human immunodeficiency virus-infected asymptomatic subjects. *J Infect Dis* **171**, 531–536.

Rose, N. F., Marx, P. A., Luckay, A. & 7 other authors (2001). An effective AIDS vaccine based on live attenuated vesicular stomatitis virus recombinants. *Cell* **106**, 539–549.

Rosenberg, E. S., Billingsley, J. M., Caliendo, A. M., Boswell, S. L., Sax, P. E., Kalams, S. A. & Walker, B. D. (1997). Vigorous HIV-1-specific CD4+ T cell responses associated with control of viremia. *Science* **278**, 1447–1450.

Rossio, J. L., Esser, M. T., Suryanarayana, K. & 10 other authors (1998). Inactivation of human immunodeficiency virus type 1 infectivity with preservation of conformational and functional integrity of virion surface proteins. *J Virol* **72**, 7992–8001.

Rowland-Jones, S., Sutton, J., Sriyoshi, K. & 7 other authors (1995). HIV-specific cytotoxic T-cells in HIV-exposed but uninfected Gambian women. *Nat Med* **1**, 59–64.

Rowland-Jones, S. L., Dong, T., Fowke, K. R. & 11 other authors (1998). Cytotoxic T cell responses to multiple conserved HIV epitopes in HIV-resistant prostitutes in Nairobi. *J Clin Invest* **102**, 1758–1765.

Ruprecht, R. M. (1999). Live attenuated AIDS viruses as vaccines: promise or peril? *Immunol Rev* **170**, 135–149.

Ruprecht, R. M., Hofmann-Lehmann, R., Rasmussen, R. A., Vlasak, J. & Xu, W. (2000). 1999: a time to re-evaluate AIDS vaccine strategies. *J Hum Virol* **3**, 88–93.

Ruprecht, R. M., Hofmann-Lehmann, R., Smith-Franklin, B. A. & 17 other authors (2001). Protection of neonatal macaques against experimental SHIV infection by human neutralizing monoclonal antibodies. *Transfus Clin Biol* **8**, 350–358.

Saifuddin, M., Hart, M. L., Gewurz, H., Zhang, Y. & Spear, G. T. (2000). Interaction of mannose-binding lectin with primary isolates of human immunodeficiency virus type 1. *J Gen Virol* **81**, 949–955.

Saksena, N. K. & Potter, S. J. (2003). Reservoirs of HIV-1 in vivo: implications for anti-retroviral therapy. *AIDS Rev* **5**, 3–18.

Salinovich, O., Payne, S. L., Montelaro, R. C., Hussain, K. A., Issel, C. J. & Schnorr, K. L. (1986). Rapid emergence of novel antigenic and genetic variants of equine infectious anemia virus during persistent infection. *J Virol* **57**, 71–80.

Sandberg, J. K., Fast, N. M., Palacios, E. H. & 8 other authors (2002). Selective loss of innate CD4(+) V alpha 24 natural killer T cells in human immunodeficiency virus infection. *J Virol* **76**, 7528–7534.

Santini, S. M., Di Pucchio, T., Lapenta, C., Parlato, S., Logozzi, M. & Belardelli, F. (2002). The natural alliance between type I interferon and dendritic cells and its role in linking innate and adaptive immunity. *J Interferon Cytokine Res* **22**, 1071–1080.

Scheppler, J. A., Nicholson, J. K., Swan, D. C., Ahmed-Ansari, A. & McDougal, J. S. (1989). Down-modulation of MHC-I in a CD4+ T cell line, CEM-E5, after HIV-1 infection. *J Immunol* **143**, 2858–2866.

Schindler, M., Wurfl, S., Benaroch, P., Greenough, T. C., Daniels, R., Easterbrook, P., Brenner, M., Munch, J. & Kirchhoff, F. (2003). Down-modulation of mature major histocompatibility complex class II and up-regulation of invariant chain cell surface expression are well-conserved functions of human and simian immunodeficiency virus nef alleles. *J Virol* **77**, 10548–10556.

Schmitz, J. E., Kuroda, M. J., Santra, S. & 13 other authors (1999). Control of viremia in simian immunodeficiency virus infection by CD8+ lymphocytes. *Science* **283**, 857–860.

Schmitz, J. E., Kuroda, M. J., Santra, S. & 14 other authors (2003). Effect of humoral immune responses on controlling viremia during primary infection of rhesus monkeys with simian immunodeficiency virus. *J Virol* **77**, 2165–2173.

Schwartz, D., Sharma, U., Busch, M. & 11 other authors (1994). Absence of recoverable infectious virus and unique immune responses in an asymptomatic HIV+ long-term survivor. *AIDS Res Hum Retrovir* **10**, 1703–1711.

Schwartz, O., Marechal, V., Le Gall, S., Lemonnier, F. & Heard, J.-M. (1996). Endocytosis of major histocompatibility complex class I molecules is induced by the HIV-1 Nef protein. *Nat Med* **2**, 338–342.

Scott, J. V., Stowring, L., Haase, A. T., Narayan, O. & Vigne, R. (1979). Antigenic variation in visna virus. *Cell* **18**, 321–327.

Scott-Algara, D. & Paul, P. (2002). NK cells and HIV infection: lessons from other viruses. *Curr Mol Med* **2**, 757–768.

Scott-Algara, D., Truong, L. X., Versmisse, P., David, A., Luong, T. T., Nguyen, N. V., Theodorou, I., Barre-Sinoussi, F. & Pancino, G. (2003). Cutting edge: increased NK cell activity in HIV-1-exposed but uninfected Vietnamese intravascular drug users. *J Immunol* **171**, 5663–5667.

Selin, L. K. & Welsh, R. M. (1994). Specificity and editing by apoptosis of virus-induced cytotoxic T lymphocytes. *Curr Opin Immunol* **6**, 553–559.

Seth, A., Ourmanov, I., Schmitz, J. E. & 9 other authors (2000). Immunization with a modified vaccinia virus expressing simian immunodeficiency virus (SIV) Gag-Pol primes for an anamnestic Gag-specific cytotoxic T-lymphocyte response and is associated with reduction of viremia after SIV challenge. *J Virol* **74**, 2502–2509.

Sethi, K. K., Naher, H. & Stroehmann, I. (1988). Phenotypic heterogeneity of cerebrospinal fluid-derived HIV-specific and HLA-restricted cytotoxic T-cell clones. *Nature* **335**, 178–181.

Shankar, P., Russo, M., Harnisch, B., Patterson, M., Skolnik, P. & Lieberman, J. (2000). Impaired function of circulating HIV-specific CD8(+) T cells in chronic human immunodeficiency virus infection. *Blood* **96**, 3094–3101.

Shearer, G. M., Pinto, L. A. & Clerici, M. (1999). Alloimmunization for immune-based therapy and vaccine design against HIV/AIDS. *Immunol Today* **20**, 66–71.

Shearer, W. T. (1998). HIV infection and AIDS. *Prim Care* **25**, 759–774.

Shen, L., Chen, Z. W., Miller, M. D., Stallard, V., Mazzara, G. P., Panicali, D. L. & Letvin, N. L. (1991). Recombinant virus vaccine-induced SIV-specific CD8+ cytotoxic T lymphocytes. *Science* **252**, 440–443.

Shibata, R., Igarashi, T., Haigwood, N., Buckler-White, A., Ogert, R., Ross, W., Willey, R., Cho, M. W. & Martin, M. A. (1999). Neutralizing antibody directed against the HIV-1 envelope glycoprotein can completely block HIV-1/SIV chimeric virus infections of macaque monkeys. *Nat Med* **5**, 204–210.

Shiver, J. W., Fu, T. M., Chen, L. & 49 other authors (2002). Replication-incompetent adenoviral vaccine vector elicits effective anti-immunodeficiency-virus immunity. *Nature* **415**, 331–335.

Siegal, F. P., Kadowaki, N., Shodell, M., Fitzgerald-Bocarsly, P. A., Shah, K., Ho, S., Antonenko, S. & Liu, Y. J. (1999). The nature of the principal type 1 interferon-producing cells in human blood. *Science* **284**, 1835–1837.

Siegal, F. P., Fitzgerald-Bocarsly, P., Holland, B. K. & Shodell, M. (2001). Interferon-alpha generation and immune reconstitution during antiretroviral therapy for human immunodeficiency virus infection. *AIDS* **15**, 1603–1612.

Siliciano, R. F. & Soloski, M. J. (1995). MHC class I-restricted processing of transmembrane proteins – mechanism and biological significance. *J Immunol* **155**, 2–5.

Siliciano, R. F., Lawton, T., Knall, C., Karr, R. W., Berman, P., Gregory, T. & Reinherz, E. L. (1988). Analysis of host-virus interactions in AIDS with anti-gp120 T cell clones: effect of HIV sequence variation and a mechanism for CD4+ cell depletion. *Cell* **54**, 561–575.

Sinicco, A., Biglino, A., Sciandra, M., Forno, B., Pollono, A. M., Raiteri, R. & Gioannini, P. (1993). Cytokine network and acute primary HIV-1 infection. *AIDS* **7**, 1167–1172.

Smith, B. A., Gartner, S., Liu, Y. & 8 other authors (2001). Persistence of infectious HIV on follicular dendritic cells. *J Immunol* **166**, 690–696.

Smyth, M. J., Kelly, J. M., Sutton, V. R., Davis, J. E., Browne, K. A., Sayers, T. J. & Trapani, J. A. (2001). Unlocking the secrets of cytotoxic granule proteins. *J Leukoc Biol* **70**, 18–29.

Soumelis, V., Scott, I., Gheyas, F., Bouhour, D., Cozon, G., Cotte, L., Huang, L., Levy, J. A. & Liu, Y. J. (2001). Depletion of circulating natural type 1 interferon-producing cells in HIV-infected AIDS patients. *Blood* **98**, 906–912.

Spear, G. T. (1993). Interaction of non-antibody factors with HIV in plasma. *AIDS* **7**, 1149–1157.

Speth, C., Stoiber, H. & Dierich, M. P. (2003). Complement in different stages of HIV infection and pathogenesis. *Int Arch Allergy Immunol* **130**, 247–257.

Stevenson, M. (2003). HIV-1 pathogenesis. *Nat Med* **9**, 853–860.

Stittelaar, K. J., Gruters, R. A., Schutten, M., van Baalen, C. A., van Amerongen, G., Cranage, M., Liljestrom, P., Sutter, G. & Osterhaus, A. D. (2002). Comparison of the efficacy of early versus late viral proteins in vaccination against SIV. *Vaccine* **20**, 2921–2927.

Stoiber, H., Thielens, N. M., Ebenbichler, C., Arlaud, G. J. & Dierich, M. P. (1994). The envelope glycoprotein of HIV-1 gp120 and human complement protein C1q bind to the same peptides derived from three different regions of gp41, the transmembrane glycoprotein of HIV-1, and share antigenic homology. *Eur J Immunol* **24**, 294–300.

Stoiber, H., Pinter, C., Siccardi, A. G., Clivio, A. & Dierich, M. P. (1996). Efficient destruction of human immunodeficiency virus in human serum by inhibiting the protective action of complement factor H and decay accelerating factor (DAF, CD55). *J Exp Med* **183**, 307–310.

Stoiber, H., Clivio, A. & Dierich, M. P. (1997). Role of complement in HIV infection. *Annu Rev Immunol* **15**, 649–674.

Stoiber, H., Kacani, L., Speth, C., Wurzner, R. & Dierich, M. P. (2001). The supportive role of complement in HIV pathogenesis. *Immunol Rev* **180**, 168–176.

Stott, E. J. (1991). Anti-cell antibody in macaques. *Nature* **353**, 393.

Stumptner-Cuvelette, P., Morchoisne, S., Dugast, M., Le Gall, S., Raposo, G., Schwartz, O. & Benaroch, P. (2001). HIV-1 Nef impairs MHC class II antigen presentation and surface expression. *Proc Natl Acad Sci U S A* **98**, 12144–12149.

Stumptner-Cuvelette, P., Jouve, M., Helft, J., Dugast, M., Glouzman, A. S., Jooss, K., Raposo, G. & Benaroch, P. (2003). Human immunodeficiency virus-1 Nef expression induces intracellular accumulation of multivesicular bodies and major histocompatibility complex class II complexes: potential role of phosphatidylinositol 3-kinase. *Mol Biol Cell* **14**, 4857–4870.

Sullivan, B. L., Knopoff, E. J., Saifuddin, M., Takefman, D. M., Saarloos, M. N., Sha, B. E. & Spear, G. T. (1996). Susceptibility of HIV-1 plasma virus to complement-mediated lysis. Evidence for a role in clearance of virus in vivo. *J Immunol* **157**, 1791–1798.

Swann, S. A., Williams, M., Story, C. M., Bobbitt, K. R., Fleis, R. & Collins, K. L. (2001). HIV-1 Nef blocks transport of MHC class I molecules to the cell surface via a PI 3-kinase-dependent pathway. *Virology* **282**, 267–277.

Swingler, S., Brichacek, B., Jacque, J.-M., Ulich, C., Zhou, J. & Stevenson, M. (2003). HIV-1 Nef intersects the macrophage CD40L signalling pathway to promote resting-cell infection. *Nature* **424**, 213–219.

Tang, J., Tang, S., Lobashevsky, E., Myracle, A. D., Fideli, U., Aldrovandi, G., Allen, S., Musonda, R. & Kaslow, R. A. (2002). Favorable and unfavorable HLA class I alleles and haplotypes in Zambians predominantly infected with clade C human immunodeficiency virus type 1. *J Virol* **76**, 8276–8284.

Tarazona, R., Casado, J. G., Delarosa, O. & 7 other authors (2002). Selective depletion of CD56(dim) NK cell subsets and maintenance of CD56(bright) NK cells in treatment-naive HIV-1-seropositive individuals. *J Clin Immunol* **22**, 176–183.

Thielens, N. M., Bally, I. M., Ebenbichler, C. F., Dierich, M. P. & Arlaud, G. J. (1993). Further characterization of the interaction between the C1q subcomponent of human C1 and the transmembrane envelope glycoprotein gp41 of HIV-1. *J Immunol* **151**, 6583–6592.

Thomsen, A. R., Johansen, J., Marker, O. & Christensen, J. P. (1996). Exhaustion of CTL memory and recrudescence of viremia in lymphocytic choriomeningitis virus-infected MHC class II-deficient mice and B cell-deficient mice. *J Immunol* **157**, 3074–3080.

Tough, D. F. (2004). Type I interferon as a link between innate and adaptive immunity through dendritic cell stimulation. *Leuk Lymphoma* **45**, 257–264.

Trimble, L. A. & Lieberman, J. (1998). Circulating CD8 T lymphocytes in human immunodeficiency virus-infected individuals have impaired function and down-modulate CD3 zeta, the signaling chain of the T-cell receptor complex. *Blood* **91**, 585–594.

Unutmaz, D. (2003). NKT cells and HIV infection. *Microbes Infect* **5**, 1041–1047.

Valentin, A., Rosati, M., Patenaude, D. J., Hatzakis, A., Kostrikis, L. G., Lazanas, M., Wyvill, K. M., Yarchoan, R. & Pavlakis, G. N. (2002). Persistent HIV-1 infection of natural killer cells in patients receiving highly active antiretroviral therapy. *Proc Natl Acad Sci U S A* **99**, 7015–7020.

van Baalen, C. A., Pontesilli, O., Huisman, R. C., Geretti, A. M., Klein, M. R., de Wolf, F., Miedema, F., Gruters, R. A. & Osterhaus, A. D. (1997). Human immunodeficiency virus type 1 Rev- and Tat-specific cytotoxic T lymphocyte frequencies inversely correlate with rapid progression to AIDS. *J Gen Virol* **78**, 1913–1918.

van Baalen, C. A., Guillon, C., van Baalen, M., Verschuren, E. J., Boers, P. H., Osterhaus, A. D. & Gruters, R. A. (2002). Impact of antigen expression kinetics on the effectiveness of HIV-specific cytotoxic T lymphocytes. *Eur J Immunol* **32**, 2644–2652.

van Baarle, D., Kostense, S., van Oers, M. H., Hamann, D. & Miedema, F. (2002). Failing immune control as a result of impaired CD8+ T-cell maturation: CD27 might provide a clue. *Trends Immunol* **23**, 586–591.

van der Vliet, H. J., von Blomberg, B. M., Hazenberg, M. D. & 10 other authors (2002). Selective decrease in circulating V alpha 24+V beta 11+ NKT cells during HIV type 1 infection. *J Immunol* **168**, 1490–1495.

Vella, C. & Daniels, R. S. (2003). CD8+ T-cell-mediated non-cytolytic suppression of human immuno-deficiency viruses. *Curr Drug Targets Infect Disord* **3**, 97–113.

von Sydow, M., Sonnerborg, A., Gaines, H. & Strannegard, O. (1991). Interferon-alpha and tumor necrosis factor-alpha in serum of patients in various stages of HIV-1 infection. *AIDS Res Hum Retrovir* **7**, 375–380.

Wagner, L., Yang, O. O., Garcia-Zepeda, E. A., Ge, Y., Kalams, S. A., Walker, B. D., Pasternack, M. S. & Luster, A. D. (1998). Beta-chemokines are released from HIV-1-specific cytolytic T-cell granules complexed to proteoglycans. *Nature* **391**, 908–911.

Wagner, R., Leschonsky, B., Harrer, E. & 7 other authors (1999). Molecular and functional analysis of a conserved CTL epitope in HIV-1 p24 recognized from a long-term nonprogressor: constraints on immune escape associated with targeting a sequence essential for viral replication. *J Immunol* **162**, 3727–3734.

Walker, D. B., Chakrabarti, S., Moss, B. & 7 other authors (1987). HIV specific cytotoxic T lymphocytes in seropositive individuals. *Nature* **328**, 345–348.

Wallace, M., Bartz, S. R., Chang, W. L., Mackenzie, D. A., Pauza, C. D. & Malkovsky, M. (1996). Gamma delta T lymphocyte responses to HIV. *Clin Exp Immunol* **103**, 177–184.

Wassef, N. M., Young, J. & Miller, R. (2003). Viral reservoirs/transient infection in HIV/AIDS: where are we now and where should we go? Summary of the June 13–14, 2002 Think Tank meeting. *AIDS Res Hum Retrovir* **19**, 333–344.

Wei, X., Ghosh, S. K., Taylor, M. E. & 9 other authors (1995). Viral dynamics in human immunodeficiency virus type 1 infection. *Nature* **373**, 117–122.

Wei, X., Decker, J. M., Wang, S. & 12 other authors (2003). Antibody neutralization and escape by HIV-1. *Nature* **422**, 307–312.

Weidt, G., Deppert, W., Utermohlen, O., Heukeshoven, J. & Lehmann-Grube, F. (1995). Emergence of virus escape mutants after immunization with epitope vaccine. *J Virol* **69**, 7147–7151.

Weissenhorn, W., Dessen, A., Calder, L. J., Harrison, S. C., Skehel, J. J. & Wiley, D. C. (1999). Structural basis for membrane fusion by enveloped viruses. *Mol Membr Biol* **16**, 3–9.

Weissman, J. D., Brown, J. A., Howcroft, T. K., Hwang, J., Chawla, A., Roche, P. A., Schiltz, L., Nakatani, Y. & Singer, D. S. (1998). HIV-1 tat binds TAFII250 and represses TAFII250-dependent transcription of major histocompatibility class I genes. *Proc Natl Acad Sci U S A* **95**, 11601–11606.

Wells, D. E., Chatterjee, S., Mulligan, M. J. & Compans, R. W. (1991). Inhibition of human immunodeficiency virus type 1-induced cell fusion by recombinant human interferons. *J Virol* **65**, 6325–6330.

Whatmore, A. M., Cook, N., Hall, G. A., Sharpe, S., Rud, E. W. & Cranage, M. P. (1995). Repair and evolution of nef in vivo modulates simian immunodeficiency virus virulence. *J Virol* **69**, 5117–5123.

Whitney, J. B. & Ruprecht, R. M. (2004). Live attenuated HIV vaccines: pitfalls and prospects. *Curr Opin Infect Dis* **17**, 17–26.

Wilson, J. D., Ogg, G. S., Allen, R. L. & 8 other authors (2000). Direct visualization of HIV-1-specific cytotoxic T lymphocytes during primary infection. *AIDS* **14**, 225–233.

Wolf, D., Witte, V., Laffert, B., Blume, K., Stromer, E., Trapp, S., d'Aloja, P., Schurmann, A. & Baur, A. S. (2001). HIV-1 Nef associated PAK and PI3-kinases stimulate Akt-independent Bad-phosphorylation to induce anti-apoptotic signals. *Nat Med* **7**, 1217–1224.

Wyand, M. S., Manson, K. H., Lackner, A. A. & Desrosiers, R. C. (1997). Resistance of neonatal monkeys to live attenuated vaccine strains of simian immunodeficiency virus. *Nat Med* **3**, 32–36.

Wyatt, R., Sullivan, N., Thali, M., Repke, H., Ho, D., Robinson, J., Posner, M. & Sodroski, J. (1993). Functional and immunologic characterization of human immunodeficiency virus type 1 envelope glycoproteins containing deletions of the major variable regions. *J Virol* **67**, 4557–4565.

Wyatt, R., Kwong, P. D., Desjardins, E., Sweet, R. W., Robinson, J., Hendrickson, W. A. & Sodroski, J. G. (1998). The antigenic structure of the HIV gp120 envelope glycoprotein. *Nature* **393**, 705–711.

Xu, X. N., Laffert, B., Screaton, G. R., Kraft, M., Wolf, D., Kolanus, W., Mongkolsapay, J., McMichael, A. J. & Baur, A. S. (1999). Induction of Fas ligand expression by HIV involves the interaction of Nef with the T cell receptor zeta chain. *J Exp Med* **189**, 1489–1496.

Yang, O. O., Kalams, S. A., Rosenzweig, M., Trocha, A., Jones, N., Koziel, M., Walker, B. D. & Johnson, R. P. (1996). Efficient lysis of human immunodeficiency virus type 1-infected cells by cytotoxic T lymphocytes. *J Virol* **70**, 5799–5806.

Yang, O. O., Sarkis, P. T. N., Trocha, A. K., Kalams, S. A., Johnson, R. P. & Walker, B. D. (2003). Impacts of avidity and specificity on the antiviral efficacy of HIV-1-specific CTL. *J Immunol* **171**, 3718–3724.

Yang, X., Lee, J., Mahony, E. M., Kwong, P. D., Wyatt, R. & Sodroski, J. (2002). Highly stable trimers formed by human immunodeficiency virus type 1 envelope glycoproteins fused with the trimeric motif of T4 bacteriophage fibritin. *J Virol* **76**, 4634–4642.

Yasutomi, Y., Koenig, S., Haun, S. S. & 7 other authors (1993). Immunization with recombinant BCG-SIV elicits SIV-specific cytotoxic T lymphocytes in rhesus monkeys. *J Immunol* **150**, 3101–3107.

Yokomaku, Y., Miura, H., Tomiyama, H. & 7 other authors (2004). Impaired processing and presentation of cytotoxic T lymphocyte (CTL) epitopes are major escape mechanisms from CTL immune pressure in human immunodeficiency virus type 1 infection. *J Virol* **78**, 1324–1332.

Yoon, K., Jeong, J. G. & Kim, S. (2001). Stable expression of human immunodeficiency virus type 1 Nef confers resistance against Fas-mediated apoptosis. *AIDS Res Hum Retrovir* **17**, 99–104.

Younes, S.-A., Bader, Y.-D., Dumont, A. R., Boulassel, R., Grossman, Z., Routy, J.-P. & Sekaly, R.-P. (2003). HIV-1 viraemia prevents the establishment of IL-2-producing HIV-specific memory CD4+ T cells endowed with proliferative capacity. *J Exp Med* **198**, 1909–1922.

Zhang, W. H., Hockley, D. J., Nermut, M. V., Morikawa, Y. & Jones, I. M. (1996). Gag-Gag interactions in the C-terminal domain of human immunodeficiency virus type 1 p24 capsid antigen are essential for Gag particle assembly. *J Gen Virol* **77**, 743–751.

Zinkernagel, R. M. (2003). On natural and artificial vaccinations. *Annu Rev Immunol* **21**, 515–546.

Persistent RNA virus infections

Catherine M. Dixon, Lucy Breakwell, Gerald Barry and
John K. Fazakerley

Centre for Infectious Diseases, College of Medicine and Veterinary Medicine,
University of Edinburgh, Edinburgh EH9 1QH, UK

Many important diseases result from persistent virus infections; examples are zoster, viral hepatitis, AIDS and ruminant pestivirus infections. Persistent virus infections are one of the most complex issues in virology. The complexity results in part from the many uses of the words 'persistent' and 'virus' such that 'persistent virus infection' ends up having many meanings. Persistence can be considered to occur at the level of the host population, at the level of the individual, within a tissue, within a culture or at the individual cell level. For example, measles virus (MV) persists within the human population and cannot be maintained in isolated populations below a critical size; in such populations, measles can be eradicated and is epidemic upon reintroduction. In larger populations with sufficient new naïve additions to the population, measles is endemic, though it only very rarely persists in any one individual. For viruses which can persist within an individual, persistence can be dynamic, with different cells infected over a period of time, or static, with the same cell or same population of cells infected over long periods of time. During human immunodeficiency virus (HIV) persistence, virus spreads dynamically between T lymphocytes, whereas following childhood chicken pox, varicella-zoster virus (VZV) persists within the same sensory nerve ganglia essentially for life to re-emerge as zoster later in life. Just as the level of persistence must be defined, so too must the nature of what is persisting; here the possibilities include infectious virus, as measured by infectivity assay, virus particles, as observed by electron microscopy, virus nucleic acid, as observed by *in situ* hybridization or PCR, or virus proteins, as observed by blotting or immunostaining (see Fig. 1). In the case of DNA viruses, a major strategy of virus persistence is to transcribe only a small and limited set of genes, a state often referred to as latency. With the exception of

SGM symposium 64: Molecular pathogenesis of virus infections.
Editors P. Digard, A. A. Nash & R. E. Randall. Cambridge University Press. ISBN 0 521 83248 9 ©SGM 2005

reversiviruses such as retroviruses, viruses with RNA genomes cannot persist in this way. An added complexity is that the language used to describe persistence all too easily falls into a form which appears to imply that viruses establish persistence for the benefit of the virus. In many cases, for example HIV or VZV, it can be argued to be a selective advantage to the virus; however, in other cases, for example measles in subacute sclerosing panencephalitis (SSPE), persistence is probably of no value to the virus; it is an accident rather than a strategy sculpted by evolutionary pressures. This chapter will consider the mechanisms by which RNA viruses become persistent within an individual. For RNA virus persistence to occur, the virus must avoid, be refractory to, or suppress one or more host responses. Firstly, there is the propensity of cells which detect a virus infection to undergo programmed cell death; secondly, there is the innate immune system, which triggers mechanisms which recognize and destroy viral RNA and initiate an antiviral state; and finally, there is the adaptive immune response, which neutralizes virus particles and destroys virus-infected cells. To understand how RNA viruses are able to persist, each of these will be considered in turn.

APOPTOSIS

It is now clear that the host response to infection by many different viruses is to initiate programmed cell death or apoptosis. Apoptosis is a form of cell death that is orchestrated according to strict and highly regulated pathways. Apoptosis, first described in 1972 (Kerr *et al.*, 1972), is now known to play a vital role in both development and survival, and a number of different pathways have been elucidated, illustrating specific activation cascades of apoptotic molecules that lead to cell death. In response to signals that increase permeability of the mitochondrial outer membrane and the subsequent release of cytochrome *c*, a complex of apoptosis-accelerating factor (Apaf)-1 molecules, caspase-9 dimers, cytochrome *c* and ATP, known as the apoptosome, forms (Hill *et al.*, 2003). This structure has the ability to trigger the activation of caspase-3, which activates a number of other pro-apoptotic molecules leading ultimately to cell death.

Altruistic cell suicide is a defence mechanism

At the individual cell level, if apoptosis can occur before the virus completes replication it will prevent new virion formation and eliminate the infection. In metazoans, this can be viewed as an altruistic response of the host cell in which death of one cell, the infected cell, allows survival of surrounding cells. In most tissues, the dead cell will be replaced and tissue damage will be minimal. In contrast, if the cell fails to undergo this suicide response and virus replicates productively, the new virus output will infect many surrounding cells. Once immune responses are primed, the fate of infected cells is anyway likely to be death, in this case not suicide but fratricide resulting from the activity of cytotoxic T cells. The fate of the infected cell is therefore the same, but an

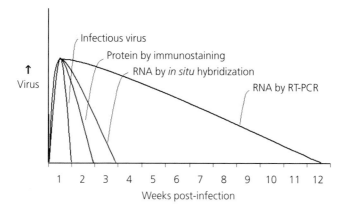

Fig. 1. Different techniques to detect the presence of Semliki Forest virus in the mouse brain give different answers for the time of clearance of the infection. Infectious virus is detected by plaque assay, viral proteins are detected by immunostaining, virus RNA positive cells are detected by *in situ* hybridization and viral RNA is detected by real-time PCR.

early death, suicide, is likely to have a selective advantage over a delayed death since it results in the demise of fewer cells with reduced tissue damage. That apoptosis is a powerful host antivirus response, as opposed to an activity which the virus has evolved to initiate to provide it with an advantage is validated by the presence of anti-apoptotic genes within the genomes of many viruses. Many viruses thus seek to suppress the apoptosis response not trigger it. Indeed, an evolutionary battle between host cell initiation of apoptosis and virus avoidance or suppression of this response can be constructed. It can be further argued that the adaptive immune response, which results in triggering of apoptosis by a cytotoxic T cell, arose to deal with virus infections in which the host cell fails to undergo apoptosis. Failure to undergo apoptosis is one factor predisposing to a persistent infection and it can arise in three ways. Firstly, the virus may avoid triggering host cell detection mechanisms; secondly, the virus may actively suppress the apoptosis response; and thirdly, the cell may not have the propensity to undergo apoptosis.

Avoidance of apoptosis

To address the first of these possibilities, the mechanisms by which cells detect RNA viruses are far from clear. It has been suggested that virus entry and fusion result in the release of ceramide and that this triggers apoptosis (Jan *et al.*, 2000). For RNA viruses, replication of virus RNA in the cytoplasm results in at least partially double-stranded RNA (dsRNA), a molecular form that does not exist in uninfected cells. There are at least two cellular detection systems for dsRNA, protein kinase R (PKR) and Toll-like receptor (TLR) 3 (Wu & Kaufman, 1997; Alexopoulou *et al.*, 2001). Activation of these two systems is required for initiation of the interferon (IFN) response (discussed in detail later) and it would make much sense if they were also the mechanisms by which cells initiated apoptosis. Priming by IFN may well increase the propensity of cells to undergo apoptosis. IFN up-regulates PKR and 2′,5′-oligoadenylate synthetase (OAS) and the latter activates RNase-L. PKR and RNase-L are known inducers of apoptosis (Wu & Kaufman, 1997; Tan & Katze, 1999). TLR3 activation can trigger apoptosis through the downstream protein Toll/IL-1R domain-containing adapter inducing IFN-β (TRIF) (Ruckdeschel *et al.*, 2004).

Many viruses shut off host cell protein synthesis and this may be a trigger for apoptosis. Another possible trigger for apoptosis is activation of endoplasmic reticulum (ER) stress by the unfolded protein response. This is triggered by proteins accumulating in the ER. The ER chaperone BiP serves as a negative regulator of ER stress proteins such as PKR-like ER kinase (PERK) and the ER-resident transmembrane kinase/endoribonuclease IRE1. BiP sequesters PERK and IRE1 to the ER membrane. However, upon binding unfolded proteins, BiP releases these ER stress proteins. Upon reaching a threshold, the released kinases trigger the stress response which includes phosphorylation of eIF2α and activation of caspase-12 (Rao *et al.*, 2001). Infection of Madin–Darby bovine kidney cells with cytopathic strains of the RNA pestivirus bovine viral diarrhoea virus (BVDV) results in activation of PERK, hyperphosphorylation of eIF2α, activation of caspase-12, down-regulation of Bcl-2 and apoptosis (Jordan *et al.*, 2002).

Another well-described pathway initiating apoptosis is activation of cell surface death receptors. Ligation of the tumour necrosis factor (TNF) receptor, the TNF-related apoptosis-inducing ligand (TRAIL) receptor or Fas lead to the activation of caspase-8 and cleavage of Bid, causing it to translocate to the mitochondrion and activate apoptosis (Locksley *et al.*, 2001; Luo *et al.*, 1998). Cells infected with influenza virus initiate synthesis of Fas-L, which acts in an autocrine manner to initiate cell death (Wurzer *et al.*, 2004).

RNA replication of positive-stranded RNA viruses generally occurs within membrane vesicles. This may have evolved to exclude host cell RNA detection and degradation proteins from areas of virus RNA replication and in particular from viral dsRNA. Replication of poliovirus (PV) takes place in vesicular structures with double membranes (Bienz *et al.*, 1992). Coronaviruses synthesize RNA on endosomal membranes and multivesicular bodies (van der Meer *et al.*, 1999), whereas flaviviruses use the intracellular packets of membrane vesicles derived from the membranes of the *trans*-Golgi network (Mackenzie *et al.*, 1999). For alphaviruses, RNA replication occurs in small vesicular invaginations on the surface of large vacuoles, known respectively as spherules and cytopathic vacuoles (Grimley *et al.*, 1968, 1972; Kujala *et al.*, 2001). Electron-dense material protrudes from the spherules into the cytoplasm and these structures contain ribosomes, indicating that newly synthesized viral RNA is translated within or just outside these spherules.

Unlike the positive-stranded viruses, replication and transcription of negative-stranded RNA viruses generally takes place free in the cytoplasm. However, the viral RNA is encapsidated by nucleocapsid protein. In the paramyxoviruses, a nucleocapsid-binding domain of the viral phosphoprotein, P, allows binding of the viral polymerase to the nucleocapsid (reviewed by Curran & Kolakofsky, 1999). Translocation of the polymerase along the RNA is believed to involve the continuous reattachment of the polymerase to the nucleocapsid proteins, and unencapsidated RNA cannot act as a template for the polymerase (Kingston *et al.*, 2004). This strategy may have evolved to prevent the formation of long dsRNA molecules likely to trigger and be susceptible to host cell defences. In this nucleocapsid strategy of RNA replication and transcription, the elongating stand of RNA is only attached to the template RNA (forming dsRNA) at the point where the polymerase binds.

Reoviruses with dsRNA genomes have the most elaborate strategy to avoid triggering host cell defences. After entering the cell, the segmented dsRNA genomes remain within the virus core particle (Nibert, 1998; Reinisch *et al.*, 2000). Within this containment unit, positive-stranded mRNA is made (Banerjee & Shatkin, 1970). These mRNAs leave the core particle and are used to make viral proteins. These proteins assemble into new core particles, which synthesize either additional mRNAs or genomic dsRNA which are incorporated into new virus particles (Schwartz *et al.*, 2002).

While dsRNA-binding proteins may be activated during RNA virus infections, presumably through exposure to small amounts of dsRNA, by sequestering replicating RNA in vesicles or virus core particles or by covering the RNA with nucleoproteins, RNA viruses appear to have evolved strategies to try to avoid triggering host cell defences.

Inhibition of apoptosis

A second explanation for the failure of virus-infected cells to undergo apoptosis is active virus inhibition of this process. Many DNA viruses carry anti-apoptotic genes. These include vFLIPs (virus FLICE-like inhibitory proteins) which inhibit caspase-8 (FLICE) activation, for example the K13 gene of human herpesvirus 8; Bcl-2 homologues which suppress pro-apoptotic Bcl-2 family members such as Bax and Bid, for example BHRF1 of Epstein–Barr virus or 5HL of African swine fever virus; suppressors of p53 activation, for example E6 of human papillomavirus; and inhibitors of caspases, for example p35 of baculovirus (Cho *et al.*, 2002; Clem, 2001). In contrast to DNA viruses, fewer RNA viruses are known with anti-apoptotic activities. The V proteins of paramyxoviruses suppress apoptosis (Wansley & Parks, 2002; Wansley *et al.*, 2003). Cells infected with Simian virus 5 do not undergo apoptosis but when the conserved C-terminal domain of the V protein is deleted infection is cytopathic. This can be prevented by *trans* expression of the V protein (Sun *et al.*, 2004). BVDV induces apoptosis in cells when the NS2-3 protein is cleaved but mutations in the cleavage site which result in uncleaved NS2-3 result in an infection which does not trigger apoptosis (Corapi *et al.*, 1988; Moennig *et al.*, 1993). The mechanism of this remains unclear. *In vitro* models of PV infection show a loss of TRAIL receptors on the cell surface resulting in a diminished sensitivity of infected cells to pro-apoptotic factors (Neznanov *et al.*, 2002). HIV prevents Fas-mediated apoptosis by the Nef-dependent inhibition of apoptosis signal-regulating kinase 1 (ASK1). Furthermore, Nef inactivates the pro-apoptotic protein Bad, thereby blocking the mitochondria-induced apoptotic pathway (Wolf *et al.*, 2001). The Tax protein of human T-lympho-tropic virus type 1 also appears to be anti-apoptotic (Saggioro *et al.*, 2003).

Relative predisposition to apoptosis

The third possibility for the failure of virus-infected cells to undergo apoptosis is that the cell does not have the propensity to do so. For replaceable cell populations, as argued above, altruistic cell suicide is a highly effective antiviral strategy. However, in a population of non-renewable cells this strategy could be less advantageous, and, if the cell population were non-renewable and vitally important, highly disadvantageous. The latter situation pertains in many populations of neurons in the central nervous system (CNS). In both mice and humans, the adult CNS contains mostly post-mitotic fully differentiated neurons. Many of these cells are vital, for example neurons in the respiratory centre, and must survive a lifetime without further cell division. Altruistic death of these cells upon infection could compromise the survival of the individual and it can be speculated that these cells are a specialized case, an exception to the principle of altruistic suicide (Allsopp & Fazakerley, 2000).

The situation is well illustrated by Semliki Forest virus (SFV) infection of the mouse brain. Immature differentiating neuronal populations in the developing mouse brain remain highly susceptible to apoptosis, and during the sculpting of neuronal circuits, large numbers die by this process in the first 2 weeks after birth (reviewed by Fazakerley, 2001). Intranasal inoculation of SFV in neonatal mice results in virus tracking along the circuits of the developing olfactory system, and, in these highly apoptosis-susceptible neuronal cell populations, a wave of virus infection is followed by a wave of apoptosis (Oliver & Fazakerley, 1998; Fazakerley & Allsopp, 2001). In contrast, in the adult mouse brain after neuronal circuits have been sculpted, the majority of neurons are fully differentiated, post-mitotic and relatively resistant to apoptosis. SFV infection of these neurons can result in virus persistence (Fazakerley & Webb, 1987; Amor *et al.*,

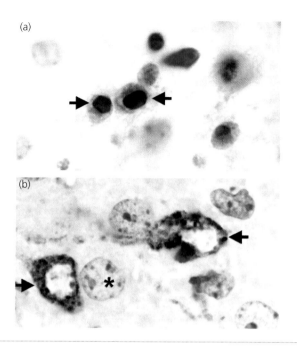

Fig. 2. Transition from apoptosis-susceptible to apoptosis-resistant neurons occurs as the postnatal mouse brain develops. (a) Cells in the olfactory bulb of a mouse inoculated intranasally at 8 days of age with SFV A7[74]. Nearly all the cells in this brain region are infected. By 24 h post-infection, the cells have condensed darkly staining (haematoxylin) pycnotic nuclei and have activated caspase-3 (brown immunostaining, arrows) indicative of apoptosis. (b) Two virus-positive neurons (brown immunostaining, arrows) in the brain of an adult mouse with severe combined immunodeficiency, 3 weeks after SFV A7[74] infection. Compared to the nuclei of adjacent uninfected cells (*), infected cells have pale swollen nuclei indicative of necrosis. No activated caspase-3 immunostaining is seen in these brains. This virus can persist for many months in the adult mouse brain.

1996). In the adult mouse, the only population of CNS neurons remaining highly susceptible to apoptosis are the immature neural progenitor cells of the rostral migratory stream (A. Boyd & J. K. Fazakerley, unpublished). Fig. 2 illustrates the difference in the outcome of SFV infection between immature apoptosis-susceptible neonatal neurons and mature apoptosis-resistant adult mouse neurons.

INNATE IMMUNE RESPONSE

IFNs are an antiviral defence induced by the presence of dsRNA, a required inter-mediate of RNA virus replication. Upon infection, dsRNA can activate PKR and TLR3, which induce phosphorylation and activation of several transcription factors, such as nuclear factor (NF)-κB, ATF-2/c-Jun and IFN regulatory factor (IRF)-3 through phosphorylation cascades. These bind to the IFN-β promoter inducing gene expression (Maniatis *et al.*, 1998). Pathways that contribute to transcription factor activation are still being unravelled and new components which may be involved in IFN induction are continually being elucidated (Servant *et al.*, 2002). Activated IRF-3 is also thought to initiate expression of IFN-α4 (Au *et al.*, 1993). This is the initial wave of the IFN response. The response is maintained and augmented by IRF-7 (Marie *et al.*, 1998). Released IFNs act in both an autocrine and paracrine manner to induce an antiviral state by binding to the type I IFN receptor. Signal transduction via the signal transducers and activators of transcription (STAT) signalling pathway initiates transcription of many antiviral genes, including those encoding PKR, RNase-L, 2'-5'-OAS, RNA-specific adenosine deaminase (ADAR) and Mx proteins.

Avoidance of IFN triggering

To establish and maintain a persistent infection, RNA viruses must either not trigger the IFN response, suppress it or not be affected by it. Virus replication strategies to avoid dsRNA-binding proteins that may either trigger cell responses or degrade virus RNA were discussed in the previous section. In addition, some viruses have proteins to sequester any intermediary RNA replication products that are released from sites of virus replication. These proteins are effective against PKR, OAS and ADAR, which all possess dsRNA-binding motifs. The outer σ3 capsid protein of reovirus binds dsRNA and variations in the sensitivity of reovirus strains to IFN can be related to the affinity of σ3 for dsRNA (Bergeron *et al.*, 1998). The N-terminal domain of the influenza virus NS1 protein binds dsRNA and probably functions to sequester this away from cellular detection proteins (Lu *et al.*, 1995). BVDV-infected cells release a virus-encoded glycoprotein, Erns, which has RNase activity and degrades any dsRNA released from infected cells (Iqbal *et al.*, 2004). This probably acts to prevent TLR3-mediated establishment of an antiviral state in surrounding cells.

A number of viruses have evolved methods to inactivate PKR. The non-structural NS5A protein of hepatitis C virus (HCV) directly binds and inhibits PKR (Gale *et al.*, 1997). In addition, the HCV E2 protein possesses a sequence containing phosphorylation sites similar to that in PKR and eIF-2α which may inhibit PKR activity (Taylor *et al.*, 1999). PKR degradation has been observed during PV infection (Black *et al.*, 1989). The OAS–RNase-L system is particularly effective against picornaviruses; however, encephalomyocarditis virus induces the cellular inhibitor of RNase-L, RLI, which disrupts 2'-5'-oligoadenylate binding and activation of RNase-L (Martinand *et al.*, 1998).

Suppression of IFN responses

Active suppression of the IFN response has been observed for a number of RNA viruses. The leader (L) protein of cardioviruses and aphthoviruses inhibits induction of the initial IFN response most probably through interaction with IRF-3 (Pesch *et al.*,

2001). The V protein of paramyxoviruses has a diverse range of functions, including apoptosis prevention but also cell cycle alterations (Lin & Lamb, 2000) and suppression of IFN synthesis and dsRNA signalling (He *et al.*, 2002). The V protein encoded by MV associates with STAT-1, STAT-2 and IRF-9. Binding of STATs by measles or henipa-virus V proteins prevents their nuclear accumulation (Rodriguez *et al.*, 2002; Palosaari *et al.*, 2003). In the *Rubulavirus* species simian virus 5, mumps virus and human parainfluenza virus type II, the V protein targets STATs for ubiquitin-dependent proteasome degradation (Didcock *et al.*, 1999; Ulane & Horvath, 2002). In addition to dsRNA binding, the NS1 protein of influenza virus inhibits IRF-3 nuclear translocation and therefore inhibits transcription of IFN-β (Lin *et al.*, 1998). The Ebola virus VP35, which has identity to influenza virus NS1, blocks phosphorylation of IRF-3 (Hartman *et al.*, 2004). The non-structural NSs protein of Bunyamwera virus also antagonizes the transcription factors IRF-3 and NF-κB inhibiting IFN-β expression (Weber *et al.*, 2002). Interestingly, NSs shows strong sequence similarity to the pro-apoptotic *Drosophila* protein Reaper (Colon-Ramos *et al.*, 2003).

ACQUIRED IMMUNE RESPONSE

Many viruses that persist have developed mechanisms to avoid detection and/or clearance by cellular and humoral immune responses. This can occur either by replicating in an area where such responses do not occur, or do not readily occur, or by actively suppressing these responses.

Immunoprivileged site

It is often considered that virus persistence is more likely to occur in an immuno-privileged site. Foot-and-mouth disease virus (FMDV) persistence has been localized to specialized cells in the pharynx, non-cornified cells of the dorsal soft palate (Zhang & Kitching, 2001). The efficient immune response to FMDV that results in clearance of virus by antibodies within the rest of the body does not extend to the pharyngeal area. FMDV infection also results in rapid down-regulation of MHC class I (Sanz-Parra *et al.*, 1998), a phenomenon also observed with the related PV (Doedens & Kirkegaard, 1995). This avoids recognition by the CD8$^+$ T-cell response. Animals with FMDV detectable in their saliva 28 days post-infection are designated 'carriers' (Salt, 1993) and such animals have been identified both post-infection and post-vaccination and may be important sources of infection for new outbreaks (Dawe *et al.*, 1994).

The CNS is frequently stated to be an immunoprivileged site on the basis that the resting uninfected CNS has no lymphocytes, no antibody, and no complement proteins, only low levels of MHC expression and high levels of anti-inflammatory TGF-β. Clearly, this would seem to be an environment in which a virus might be able to establish persistence. However, the concept of an immunoprivileged site must be considered with caution since the CNS is capable of mounting florid immune responses with large numbers of activated T and B lymphocytes, strong MHC-I expression and high titres of antibodies. CNS virus infections nearly always arrive via other tissues and replication at extraneural sites initiates immune responses. On entering the CNS, infected cells make IFNs and chemokines and these initiate CNS responses and recruit activated lymphocytes from the vasculature. The CNS is better viewed as an immuno-specialized site. Nevertheless, something about immune responses in this organ does seem to allow the persistence of many viruses despite an active immune response. Many viruses including Borna virus, MV, rubella virus, mouse hepatitis virus and Theiler's murine encephalomyelitis virus (TMEV) are able to persist in the CNS.

Inadequate immune responses

TMEV is a natural endemic picornavirus infection of mice. How the virus persists in wild mouse populations remains unclear. Experimental infection of the CNS of mice has been much studied, originally as a model of poliomyelitis since the virus can infect and destroy motor neurons in the spinal cord. More recently this virus has been studied to understand the mechanism by which it induces demyelinating disease. Demyelin-ation is associated with virus persistence in the spinal cord and occurs only in some strains of mice. The genetics of this has been studied and three major loci have been defined. The MHC haplotype has a major effect and in particular the *H2D* gene (Aubagnac *et al.*, 2001). Two other loci have been defined, *Tmevp2* and *Tmevp3*, both on mouse chromosome 10 close to the *IFN-γ* locus (Bureau *et al.*, 1993; Bihl *et al.*, 1999). All mouse strains infected with TMEV make T- and B-cell responses and cytotoxic T-cell activity and antibodies can be readily measured. However, there appear to be differences between susceptible and resistant mouse strains in the magnitude of these responses (Dethlefs *et al.*, 1997). As with the *H2D* locus it seems likely that the *Tmevp2* and *Tmevp3* loci encode genes, probably cytokines, which influence immune responses. Quantification of spleen and brain cytokine transcripts in SJL/J (persistence susceptible) and C57BL/6 (persistence resistant) mice has demonstrated changes in cytokine gene expression (R. Walker & J. K. Fazakerley, unpublished). It seems that inadequate or inappropriate immune responses, perhaps the type or magnitude of responses, allow TMEV to persist in the CNS in certain genetic backgrounds.

In SSPE, an almost invariably fatal late childhood complication post-MV infection, MV establishes a persistent infection of neurons and glial cells. Persistence is associated with changes in virus gene expression; in particular, changes in expression of the matrix and envelope proteins which result in no or minimal release of virus particles (Schneider-Schaulies *et al.*, 1993). However, these mutations may not be causal to virus persistence but consequent, since mutations in these regions of the genome can accumulate if there is no selective pressure to maintain the original sequence. In cells in which the virus RNA replicates but is not spread by the production of virus particles, there is selective pressure on replication-associated genes but not on virus structural genes. Despite the activities (described in detail earlier) of the MV N and V proteins in down-regulating innate immune responses, infection does trigger CNS innate immune responses as demonstrated by the expression in cells around foci of infection of the IFN-induced MxA protein (Ogata *et al.*, 2004). Measles is also known to be immuno-suppressive but there are generally high titres of antibodies to virus in both blood and cerebrospinal fluid. *In vitro* studies indicate that antibody may drive persistence by down-regulating virus gene expression (Fujinami & Oldstone, 1979). The neuro-pathology of SSPE indicates that this can be an inflammatory disease, so neither the changes in gene expression nor replication within the so-called 'immunoprivileged site' of the CNS avoid triggering adaptive immune responses. It is likely that some cells are productively infected or with time undergo cell death and that release of virus antigens drives the CNS immune response but this response is inadequate to eliminate the virus. Perhaps replication within neurons without expression of virus envelope glycoproteins on the cell surface avoids removal of infected cells by antibodies, or perhaps V protein inhibits IFN-γ, while absence of MHC-I expression on neurons avoids cytotoxic T-cell surveillance. Most likely it is the combination of these factors along with the relative resistance of mature neurons to initiate apoptosis which result in the failure to eliminate MV from the CNS.

Antigenic variation

Antigenic variation of virus proteins allows viruses to 'escape' specific immune responses. RNA viruses rapidly generate mutations due to the infidelity of non-proofreading RNA polymerases. Mutations in the viral glycoproteins of equine infectious anaemia virus and the selection pressure of evolving antibody responses generate a succession of antibody-resistant viruses resulting in a persistent infection (Kono *et al.*, 1973). Similarly, the E2 glycoprotein of HCV contains two hypervariable regions which undergo frequent changes in individuals with persistent HCV infection (Farci *et al.*, 2000). The rapid generation of mutations in HIV leads to drug resistance, altered cell tropism, cytopathogenicity and variants that 'escape' the CD8[+] T-cell response. Viruses with mutations in key CD8[+] T-cell epitopes are no longer recognized by these lymphocytes and therefore persist (Dai *et al.*, 1992; Klenerman *et al.*, 1994). HIV not only possesses immune evasion strategies but also induces fatal immuno-suppression; HIV infects CD4[+] T cells, which are destroyed by CD8[+] T cells (Dalgleish *et al.*, 1984; Safrit *et al.*, 1994).

Induction of tolerance

A highly effective way of avoiding adaptive immune responses is to induce tolerance. Total immune tolerance is induced by BVDV infection *in utero*; the virus is recognized

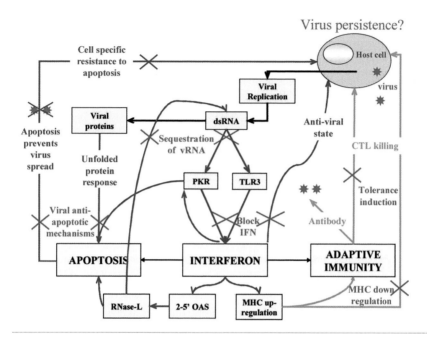

Fig. 3. Host cell responses to virus and virus mechanisms to evade them. Blue arrows indicate innate immune mechanisms, green arrows indicate adaptive immune mechanisms and red crosses are points at which viruses can suppress these responses predisposing towards the establishment of persistence.

as 'self', and is not targeted by T cells (Coria & McClurkin, 1978). Two biotypes of BVDV, cytopathic (cp) and non-cytopathic (ncp), are defined by the outcome of infections *in vitro* (Akkina, 1983). The BVDV genome encodes a single open reading frame that is cleaved post-translation. NcpBVDV has incomplete cleavage of the NS2-3 protein. Persistent infections develop following ncpBVDV infection *in utero* at a time before development of functional lymphocytes (Orban *et al.*, 1983). This results in central deletion of any lymphocytes with antiviral specificities. Studies investigating the levels of IFN in the amniotic fluid of pregnant dams could not detect any IFN when the dam had been infected with ncpBVDV; it is possible that ncpBVDV also suppresses IFN responses (Charleston *et al.*, 2001). Persistently infected animals have a subclinical infection and constantly excrete virus, becoming a source of infection in the herd. Persistently infected animals can spontaneously develop rapidly fatal mucosal disease. This results either from infection by cpBVDV, persistently infected animals have no immunity, or from mutation of the ncpBVDV to give rise to cpBVDV by restoration of the cleavage of NS2-3 (Corapi *et al.*, 1988; Moennig *et al.*, 1993).

Neonatal tolerance is also induced by lymphocytic choriomeningitis virus (LCMV). This and other arenaviruses establish asymptomatic, chronic infections of rodents worldwide. Transmission of these viruses to humans varies from asymptomatic infection to severe haemorrhagic fevers, for example Lassa fever or Bolivian haemorrhagic fever. Infection of adult mice results in a biphasic immune response, with initial NK cell expansion giving way to a massive CD8$^+$ T-cell response. At the peak of the immune response, 50 % of all CD8$^+$ T cells are specific for LCMV (Murali-Krishna *et al.*, 1998). In contrast, infection of the neonatal rodent results in negative selection of virus-specific T cells and tolerance (Pircher *et al.*, 1989; Ciurea *et al.*, 1999). Other immune mechanisms are not affected by this specific tolerance as shown by rejection of skin grafts and high levels of antibody against LCMV in carrier mice (Buchmeier & Oldstone, 1978). Cells infected with LCMV do not undergo apoptosis and in the absence of specific immunity the outcome is persistence.

In conclusion, although the mechanisms by which RNA viruses can persist are not fully understood, several factors are important; Fig. 3 summarizes some of these. RNA virus persistence results from a combination of the avoidance or prevention of apoptosis, innate immune responses and adaptive immune responses.

REFERENCES

Akkina, R. K. (1983). *Analysis of immunogenic determinants of bovine viral diarrhea virus.* PhD thesis, University of Minnesota, St Paul, MN 1982, USA.

Alexopoulou, L., Holt, A. C., Medzhitov, R. & Flavell, R. A. (2001). Recognition of double-stranded RNA and activation of NF-kappaB by Toll-like receptor 3. *Nature* **413**, 732–738.

Allsopp, T. E. & Fazakerley, J. K. (2000). Altruistic cell suicide and the specialized case of the virus-infected nervous system. *Trends Neurosci* **23**, 284–290.

Amor, S., Scallan, M. F., Morris, M. M., Dyson, H. & Fazakerley, J. K. (1996). Role of immune responses in protection and pathogenesis during Semliki Forest virus encephalitis. *J Gen Virol* **77**, 281–291.

Au, W. C., Su, Y., Raj, N. B. & Pitha, P. M. (1993). Virus-mediated induction of interferon A gene requires cooperation between multiple binding factors in the interferon alpha promoter region. *J Biol Chem* **268**, 24032–24040.

Aubagnac, S., Brahic, M. & Bureau, J. F. (2001). Viral load increases in SJL/J mice persistently infected by Theiler's virus after inactivation of the beta(2)m gene. *J Virol* **75**, 7723–7726.

Banerjee, A. K. & Shatkin, A. J. (1970). Transcription in vitro by reovirus-associated ribonucleic acid-dependent polymerase. *J Virol* **6**, 1–11.

Bergeron, J., Mabrouk, T., Garzon, S. & Lemay, G. (1998). Characterization of the thermosensitive ts453 reovirus mutant: increased dsRNA binding of sigma 3 protein correlates with interferon resistance. *Virology* **246**, 199–210.

Bienz, K., Egger, D., Pfister, T. & Troxler, M. (1992). Structural and functional characterization of the poliovirus replication complex. *J Virol* **66**, 2740–2747.

Bihl, F., Brahic, M. & Bureau, J. F. (1999). Two loci, Tmevp2 and Tmevp3, located on the telomeric region of chromosome 10, control the persistence of Theiler's virus in the central nervous system of mice. *Genetics* **152**, 385–392.

Black, T. L., Safer, B., Hovanessian, A. & Katze, M. G. (1989). The cellular 68,000-Mr protein kinase is highly autophosphorylated and activated yet significantly degraded during poliovirus infection: implications for translational regulation. *J Virol* **63**, 2244–2251.

Buchmeier, M. J. & Oldstone, M. B. (1978). Virus-induced immune complex disease: identification of specific viral antigens and antibodies deposited in complexes during chronic lymphocytic choriomeningitis virus infection. *J Immunol* **120**, 1297–1304.

Bureau, J. F., Montagutelli, X., Bihl, F., Lefebvre, S., Guenet, J. L. & Brahic, M. (1993). Mapping loci influencing the persistence of Theiler's virus in the murine central nervous system. *Nat Genet* **5**, 87–91.

Charleston, B., Fray, M. D., Baigent, S., Carr, B. V. & Morrison, W. I. (2001). Establishment of persistent infection with non-cytopathic bovine viral diarrhoea virus in cattle is associated with a failure to induce type I interferon. *J Gen Virol* **82**, 1893–1897.

Cho, N. H., Kim, Y. T. & Kim, J. W. (2002). Alteration of cell cycle in cervical tumor associated with human papillomavirus: cyclin-dependent kinase inhibitors. *Yonsei Med J* **43**, 722–728.

Ciurea, A., Klenerman, P., Hunziker, L., Horvath, E., Odermatt, B., Ochsenbein, A. F., Hengartner, H. & Zinkernagel, R. M. (1999). Persistence of lymphocytic choriomeningitis virus at very low levels in immune mice. *Proc Natl Acad Sci U S A* **96**, 11964–11969.

Clem, R. J. (2001). Baculoviruses and apoptosis: the good, the bad, and the ugly. *Cell Death Differ* **8**, 137–143.

Colon-Ramos, D. A., Irusta, P. M., Gan, E. C. & 9 other authors (2003). Inhibition of translation and induction of apoptosis by bunyaviral nonstructural proteins bearing sequence similarity to Reaper. *Mol Biol Cell* **14**, 4162–4172.

Corapi, W. V., Donis, R. O. & Dubovi, E. J. (1988). Monoclonal antibody analyses of cytopathic and noncytopathic viruses from fatal bovine viral diarrhea virus infections. *J Virol* **62**, 2823–2827.

Coria, M. F. & McClurkin, A. W. (1978). Specific immune tolerance in an apparently healthy bull persistently infected with bovine viral diarrhea virus. *J Am Vet Med Assoc* **172**, 449–451.

Curran, J. & Kolakofsky, D. (1999). Replication of paramyxoviruses. *Adv Virus Res* **54**, 403–422.

Dai, L. C., West, K., Littaua, R., Takahashi, K. & Ennis, F. A. (1992). Mutation of human immunodeficiency virus type 1 at amino acid 585 on gp41 results in loss of killing by CD8+ A24-restricted cytotoxic T lymphocytes. *J Virol* **66**, 3151–3154.

Dalgleish, A. G., Beverley, P. C., Clapham, P. R., Crawford, D. H., Greaves, M. F. & Weiss, R. A. (1984). The CD4 (T4) antigen is an essential component of the receptor for the AIDS retrovirus. *Nature* **312**, 763–767.

Dawe, P. S., Flanagan, F. O., Madekurozwa, R. L., Sorensen, K. J., Anderson, E. C., Foggin, C. M., Ferris, N. P. & Knowles, N. J. (1994). Natural transmission of foot-and-mouth disease virus from African buffalo (*Syncerus caffer*) to cattle in a wildlife area of Zimbabwe. *Vet Rec* **134**, 230–232.

Dethlefs, S., Brahic, M. & Larsson-Sciard, E. L. (1997). An early, abundant cytotoxic T-lymphocyte response against Theiler's virus is critical for preventing viral persistence. *J Virol* **71**, 8875–8878.

Didcock, L., Young, D. F., Goodbourn, S. & Randall, R. E. (1999). The V protein of simian virus 5 inhibits interferon signalling by targeting STAT1 for proteasome-mediated degradation. *J Virol* **73**, 9928–9933.

Doedens, J. R. & Kirkegaard, K. (1995). Inhibition of cellular protein secretion by poliovirus proteins 2B and 3A. *EMBO J* **14**, 894–907.

Farci, P., Shimoda, A., Coiana, A. & 9 other authors (2000). The outcome of acute hepatitis C predicted by the evolution of the viral quasispecies. *Science* **288**, 339–344.

Fazakerley, J. K. (2001). Neurovirology and developmental neurobiology. *Adv Virus Res* **56**, 73–124.

Fazakerley, J. K. & Allsopp, T. E. (2001). Programmed cell death in virus infections of the nervous system. *Curr Top Microbiol Immunol* **253**, 95–119.

Fazakerley, J. K. & Webb, H. E. (1987). Semliki Forest virus-induced, immune-mediated demyelination: adoptive transfer studies and viral persistence in nude mice. *J Gen Virol* **68**, 377–385.

Fujinami, R. S. & Oldstone, M. B. (1979). Antiviral antibody reacting on the plasma membrane alters measles virus expression inside the cell. *Nature* **279**, 529–530.

Gale, M. J., Jr, Korth, M. J., Tang, N. M., Tan, S. L., Hopkins, D. A., Dever, T. E., Polyak, S. J., Gretch, D. R. & Katze, M. G. (1997). Evidence that hepatitis C virus resistance to interferon is mediated through repression of the PKR protein kinase by the non-structural 5A protein. *Virology* **230**, 217–227.

Grimley, P. M., Berezesky, I. K. & Friedman, R. M. (1968). Cytoplasmic structures associated with an arbovirus infection: loci of viral ribonucleic acid synthesis. *J Virol* **2**, 1326–1338.

Grimley, P. M., Levin, J. G., Berezesky, I. K. & Friedman, R. M. (1972). Specific membranous structures associated with the replication of group A arboviruses. *J Virol* **10**, 1326–1338.

Hartman, A. L., Towner, J. S. & Nichol, S. T. (2004). A C-terminal basic amino acid motif of Zaire ebolavirus VP35 is essential for type I interferon antagonism and displays high identity with the RNA-binding domain of another interferon antagonist, the NS1 protein of influenza A virus. *Virology* **328**, 177–184.

He, B., Paterson, R. G., Stock, N., Durbin, J. E., Durbin, R. K., Goodbourn, S., Randall, R. E. & Lamb, R. A. (2002). Recovery of paramyxovirus simian virus 5 with a V protein lacking the conserved cysteine-rich domain: the multifunctional V protein blocks both interferon-beta induction and interferon signaling. *Virology* **303**, 15–32.

Hill, M. M., Adrain, C. & Martin, S. J. (2003). Portrait of a killer: the mitochondrial apoptosome emerges from the shadows. *Mol Interv* **3**, 19–26.

Iqbal, M., Poole, E., Goodbourn, S. & McCauley, J. W. (2004). Role for bovine viral diarrhea virus Erns glycoprotein in the control of activation of beta interferon by double-stranded RNA. *J Virol* **78**, 136–145.

Jan, J. T., Chatterjee, S. & Griffin, D. E. (2000). Sindbis virus entry into cells triggers apoptosis by activating sphingomyelinase, leading to the release of ceramide. *J Virol* **74**, 6425–6432.

Jordan, R., Wang, L., Graczyk, T. M., Block, T. M. & Romano, P. R. (2002). Replication of a cytopathic strain of bovine viral diarrhea virus activates PERK and induces endoplasmic reticulum stress-mediated apoptosis of MDBK cells. *J Virol* **76**, 9588–9599.

Kerr, J. F., Wyllie, A. H. & Currie, A. R. (1972). Apoptosis: a basic biological phenomenon with wide-ranging implications in tissue kinetics. *Br J Cancer* **26**, 239–257.

Kingston, R. L., Hamel, D. J., Gay, L. S., Dahlquist, F. W. & Matthews, B. W. (2004). Structural basis for the attachment of a paramyxoviral polymerase to its template. *Proc Natl Acad Sci U S A* **101**, 8301–8306.

Klenerman, P., Rowland-Jones, S., McAdam, S. & 10 other authors (1994). Cytotoxic T-cell activity antagonized by naturally occurring HIV-1 Gag variants. *Nature* **369**, 403–407.

Kono, Y., Kobayashi, K. & Fukunaga, Y. (1973). Antigenic drift of equine infectious anemia virus in chronically infected horses. *Arch Gesamte Virusforsch* **41**, 1–10.

Kujala, P., Ikaheimonen, A., Ehsani, N., Vihinen, H., Auvinen, P. & Kaariainen, L. (2001). Biogenesis of the Semliki Forest virus RNA replication complex. *J Virol* **75**, 3873–3884.

Lin, G. Y. & Lamb, R. A. (2000). The paramyxovirus simian virus 5 V protein slows progression of the cell cycle. *J Virol* **74**, 9152–9166.

Lin, R., Heylbroeck, C., Pitha, P. M. & Hiscott, J. (1998). Virus-dependent phosphorylation of the IRF-3 transcription factor regulates nuclear translocation, transactivation potential, and proteasome-mediated degradation. *Mol Cell Biol* **18**, 2986–2996.

Locksley, R. M., Killeen, N. & Lenardo, M. J. (2001). The TNF and TNF receptor super-families: integrating mammalian biology. *Cell* **104**, 487–501.

Lu, Y., Wambach, M., Katze, M. G. & Krug, R. M. (1995). Binding of the influenza virus NS1 protein to double-stranded RNA inhibits the activation of the protein kinase that phosphorylates the eIF-2 translation initiation factor. *Virology* **214**, 222–228.

Luo, X., Budihardjo, I., Zou, H., Slaughter, C. & Wang, X. (1998). Bid, a Bcl2 interacting protein, mediates cytochrome c release from mitochondria in response to activation of cell surface death receptors. *Cell* **94**, 481–490.

Mackenzie, J. M., Jones, M. K. & Westaway, E. G. (1999). Markers for trans-Golgi membranes and the intermediate compartment localize to induced membranes with distinct replication functions in flavivirus-infected cells. *J Virol* **73**, 9555–9567.

Maniatis, T., Falvo, J. V., Kim, T. H., Kim, T. K., Lin, C. H., Parekh, B. S. & Wathelet, M. G. (1998). Structure and function of the interferon-beta enhanceosome. *Cold Spring Harb Symp Quant Biol* **63**, 609–620.

Marie, I., Durbin, J. E. & Levy, D. E. (1998). Differential viral induction of distinct interferon-alpha genes by positive feedback through interferon regulatory factor-7. *EMBO J* **17**, 6660–6669.

Martinand, C., Salehzada, T., Silhol, M., Lebleu, B. & Bisbal, C. (1998). RNase L inhibitor (RLI) antisense constructions block partially the down regulation of the 2-5A/RNase L pathway in encephalomyocarditis-virus-(EMCV)-infected cells. *Eur J Biochem* **254**, 248–255.

Moennig, V., Greiser-Wilke, I., Frey, H. R., Haas, L., Liebler, E., Pohlenz, J. & Liess, B. (1993). Prolonged persistence of cytopathogenic bovine viral diarrhea virus (BVDV) in a persistently viremic cattle. *Zentralbl Veterinarmed B* **40**, 371–377.

Murali-Krishna, K., Altman, J. D., Suresh, M., Sourdive, D. J., Zajac, A. J., Miller, J. D., Slansky, J. & Ahmed, R. (1998). Counting antigen-specific CD8 T cells: a re-evaluation of bystander activation during viral infection. *Immunity* **8**, 177–187.

Neznanov, N., Chumakov, K. P., Ullrich, A., Agol, V. I. & Gudkov, A. V. (2002). Unstable receptors disappear from cell surface during poliovirus infection. *Med Sci Monit* **8**, BR391–BR396.

Nibert, M. L. (1998). Structure of mammalian orthoreovirus particles. *Curr Top Microbiol Immunol* **233**, 1–30.

Ogata, S., Ogata, A., Schneider-Schaulies, S. & Schneider-Schaulies, J. (2004). Expression of the interferon-alpha/beta-inducible MxA protein in brain lesions of subacute sclerosing panencephalitis. *J Neurol Sci* **223**, 113–119.

Oliver, K. R. & Fazakerley, J. K. (1998). Transneuronal spread of Semliki Forest virus in the developing mouse olfactory system is determined by neuronal maturity. *Neuroscience* **82**, 867–877.

Orban, S., Liess, B., Hafez, S. M., Frey, H. R., Blindow, H. & Sasse-Patzer, B. (1983). Studies on transplacental transmissibility of a Bovine Virus Diarrhoea (BVD) vaccine virus. I. Inoculation of pregnant cows 15 to 90 days before parturition (190th to 265th day of gestation). *Zentralbl Veterinarmed B* **30**, 619–634.

Palosaari, H., Parisien, J. P., Rodriguez, J. J., Ulane, C. M. & Horvath, C. M. (2003). STAT protein interference and suppression of cytokine signal transduction by measles virus V protein. *J Virol* **77**, 7635–7644.

Pesch, V., van Eyll, O. & Michiels, T. (2001). The leader protein of Theiler's virus inhibits immediate-early alpha/beta interferon production. *J Virol* **75**, 7811–7817.

Pircher, H., Burki, K., Lang, R., Hengartner, H. & Zinkernagel, R. M. (1989). Tolerance induction in double specific T-cell receptor transgenic mice varies with antigen. *Nature* **342**, 559–561.

Rao, R. V., Hermel, E., Castro-Obregon, S., del Rio, G., Ellerby, L. M., Ellerby, H. M. & Bredesen, D. E. (2001). Coupling endoplasmic reticulum stress to the cell death program. Mechanism of caspase activation. *J Biol Chem* **276**, 33869–33874.

Reinisch, K. M., Nibert, M. L. & Harrison, S. C. (2000). Structure of the reovirus core at 3·6 Å resolution. *Nature* **404**, 960–967.

Rodriguez, J. J., Parisien, J. P. & Horvath, C. M. (2002). Nipah virus V protein evades alpha and gamma interferons by preventing STAT1 and STAT2 activation and nuclear accumulation. *J Virol* **76**, 11476–11483.

Ruckdeschel, K., Pfaffinger, G., Haase, R., Sing, A., Weighardt, H., Hacker, G., Holzmann, B. & Heesemann, J. (2004). Signaling of apoptosis through TLRs critically involves toll/IL-1 receptor domain-containing adapter inducing IFN-beta, but not MyD88, in bacteria-infected murine macrophages. *J Immunol* **173**, 3320–3328.

Safrit, J. T., Andrews, C. A., Zhu, T., Ho, D. D. & Koup, R. A. (1994). Characterization of human immunodeficiency virus type 1-specific cytotoxic T lymphocyte clones isolated during acute seroconversion: recognition of autologous virus sequences within a conserved immunodominant epitope. *J Exp Med* **179**, 463–472.

Saggioro, D., Acquasaliente, L., Daprai, L. & Chieco-Bianchi, L. (2003). Inhibition of apoptosis by human T-lymphotropic virus type-1 tax protein. *Ann N Y Acad Sci* **1010**, 591–597.

Salt, J. S. (1993). The carrier state in foot and mouth disease – an immunological review. *Br Vet J* **149**, 207–223.

Sanz-Parra, A., Sobrino, F. & Ley, V. (1998). Infection with foot-and-mouth disease virus results in a rapid reduction of MHC class I surface expression. *J Gen Virol* **79**, 433–436.

Schneider-Schaulies, S., Schneider-Schaulies, J., Bayer, M., Loffler, S. & ter Meulen, V. (1993). Spontaneous and differentiation-dependent regulation of measles virus gene expression in human glial cells. *J Virol* **67**, 3375–3383.

Schwartz, M., Chen, J., Janda, M., Sullivan, M., den Boon, J. & Ahlquist, P. (2002). A positive-strand RNA virus replication complex parallels form and function of retrovirus capsids. *Mol Cell* **9**, 505–514.

Servant, M. J., Grandvaux, N. & Hiscott, J. (2002). Multiple signaling pathways leading to the activation of interferon regulatory factor 3. *Biochem Pharmacol* **64**, 985–992.

Sun, M., Rothermel, T. A., Shuman, L., Aligo, J. A., Xu, S., Lin, Y., Lamb, R. A. & He, B. (2004). Conserved cysteine-rich domain of paramyxovirus simian virus 5 V protein plays an important role in blocking apoptosis. *J Virol* **78**, 5068–5078.

Tan, S. L. & Katze, M. G. (1999). The emerging role of the interferon-induced PKR protein kinase as an apoptotic effector: a new face of death? *J Interferon Cytokine Res* **19**, 543–554.

Taylor, D. R., Shi, S. T., Romano, P. R., Barber, G. N. & Lai, M. M. (1999). Inhibition of the interferon-inducible protein kinase PKR by HCV E2 protein. *Science* **285**, 107–110.

Ulane, C. M. & Horvath, C. M. (2002). Paramyxoviruses SV5 and HPIV2 assemble STAT protein ubiquitin ligase complexes from cellular components. *Virology* **304**, 160–166.

van der Meer, Y., Snijder, E. J., Dobbe, J. C., Schleich, S., Denison, M. R., Spaan, W. J. & Locker, J. K. (1999). Localization of mouse hepatitis virus nonstructural proteins and RNA synthesis indicates a role for late endosomes in viral replication. *J Virol* **73**, 7641–7657.

Wansley, E. K. & Parks, G. D. (2002). Naturally occurring substitutions in the P/V gene convert the noncytopathic paramyxovirus simian virus 5 into a virus that induces alpha/beta interferon synthesis and cell death. *J Virol* **76**, 10109–10121.

Wansley, E. K., Grayson, J. M. & Parks, G. D. (2003). Apoptosis induction and interferon signaling but not IFN-beta promoter induction by an SV5 P/V mutant are rescued by coinfection with wild-type SV5. *Virology* **316**, 41–54.

Weber, F., Bridgen, A., Fazakerley, J. K., Streitenfeld, H., Kessler, N., Randall, R. E. & Elliott, R. M. (2002). Bunyamwera bunyavirus nonstructural protein NSs counteracts the induction of alpha/beta interferon. *J Virol* **76**, 7949–7955.

Wolf, D., Witte, V., Laffert, B., Blume, K., Stromer, E., Trapp, S., d'Aloja, P., Schurmann, A. & Baur, A. S. (2001). HIV-1 Nef associated PAK and PI3-kinases stimulate Akt-independent Bad-phosphorylation to induce anti-apoptotic signals. *Nat Med* **7**, 1217–1224.

Wu, S. & Kaufman, R. J. (1997). A model for the double-stranded RNA (dsRNA)-dependent dimerization and activation of the dsRNA-activated protein kinase PKR. *J Biol Chem* **272**, 1291–1296.

Wurzer, W. J., Ehrhardt, C., Pleschka, S., Berberich-Siebelt, F., Wolff, T., Walczak, H., Planz, O. & Ludwig, S. (2004). NF-kappaB-dependent induction of tumor necrosis factor-related apoptosis-inducing ligand (TRAIL) and Fas/FasL is crucial for efficient influenza virus propagation. *J Biol Chem* **279**, 30931–30937.

Zhang, Z. D. & Kitching, R. P. (2001). The localization of persistent foot and mouth disease virus in the epithelial cells of the soft palate and pharynx. *J Comp Pathol* **124**, 89–94.

Pathogenesis of Ebola and Marburg viruses

Amy L. Hartman, Jonathan S. Towner and Stuart Nichol

Special Pathogens Branch, Division of Viral and Rickettsial Diseases, Centers for Disease Control and Prevention, Atlanta, GA 30306, USA

EBOLA AND MARBURG VIRUS EMERGENCE AND RE-EMERGENCE

Ebola virus has captured the public imagination as something like the Darth Vader of the microbial world due in part to the associated rapidly progressing high fatality haemorrhagic disease and in part fuelled by sensational movies, books and magazine articles. Ebola virus was first discovered during investigation of severe haemorrhagic fever (HF) outbreaks in southern Sudan and the Democratic Republic of Congo (DRC, formerly Zaire) in 1976 (World Health Organization, 1978a, b). It was quickly determined that each outbreak was caused by newly discovered negative-stranded RNA viruses, which became known as Ebola-Sudan and Ebola-Zaire viruses, respectively (Buchmeier et al., 1983; Cox et al., 1983; McCormick et al., 1983). These viruses are now referred to as Zaire ebolavirus (ZEBOV) and Sudan ebolavirus (SEBOV). The outbreak in northern DRC resulted in 318 cases and a case fatality of 88 %, while the Sudan outbreak was somewhat less severe, with 284 cases and a case fatality of 53 % (World Health Organization, 1978a, b). Almost a decade prior to these outbreaks, Marburg virus (MARV) had been identified as the cause of a HF in humans and non-human primates in vaccine facilities in Europe in 1967 (Kissling et al., 1968; World Health Organization, 1967). These outbreaks were associated with tissues from African green monkeys imported from the Lake Victoria region of Uganda. Over the following decades, Ebola virus and MARV have been associated with several sporadic HF outbreaks with high case fatality in humans in various parts of sub-Saharan Africa (Table 1 and Fig. 1). In addition, in 1989 Reston ebolavirus was discovered to be the cause of HF among non-human primates in research facilities in the USA associated

SGM symposium 64: Molecular pathogenesis of virus infections.
Editors P. Digard, A. A. Nash & R. E. Randall. Cambridge University Press. ISBN 0 521 83248 9

Table 1. High mortality HF outbreaks associated with ZEBOV, SEBOV and MARV

Date	Location	Cases	Fatality (%)
ZEBOV			
1976	DRC (formerly Zaire)	318	88
1977	DRC	1	100
1994–1995	Gabon	49	65
1995	DRC	315	81
1996 spring	Gabon	37	57
1996 autumn	Gabon	60	75
2001–2002	Gabon	123	79
2003	Republic of Congo	143	90
SEBOV			
1976	Southern Sudan	284	53
1979	Southern Sudan	34	65
2000–2001	Northern Uganda	425	53
2004	Southern Sudan	17	42
MARV			
1967	Uganda > Germany/Yugoslavia	31	23
1975	Zimbabwe > South Africa	3	33
1980	Kenya	2	50
1987	Kenya	1	100
1998	Northeastern DRC	141	82

with importation of infected rhesus monkeys from export facilities in the Philippines (Jahrling *et al.*, 1990; Peters *et al.*, 1992). Although seroconversion was seen in humans, no human disease has been observed associated with Reston ebolavirus. Ivory Coast ebolavirus was implicated in chimpanzee deaths and one non-fatal human case in 1994 in the Tai Forest area of Côte d'Ivoire (Le Guenno *et al.*, 1995). These four different species of ebolavirus, together with MARV, make up the family *Filoviridae* of negative-stranded non-segmented RNA viruses. The pathogenesis of Reston and Ivory Coast ebolaviruses in humans will not be considered further in this review due to the limited data available.

CLINICAL DISEASE

Vascular dysfunction is a central feature of the disease associated with all filoviruses, and the main difference between the species is the speed and severity of disease progression in humans and other primates. Of the ebolaviruses, the best studied and most pathogenic is ZEBOV. The most well-documented ZEBOV disease outbreak was in Kikwit, DRC, in 1995 (Centers for Disease Control and Prevention, 1995; World Health Organization, 1995). Based on the observation of 315 cases, the incubation period was approximately 4–10 days. The case fatality was 81 %, and disease progression was rapid with death occurring within 7–11 days post onset of illness (Khan *et al.*, 1999). The largest and most well-documented SEBOV outbreak occurred in northern Uganda in 2000. This outbreak resulted in 425 cases with a case fatality of 53 %. Onset of illness and time to death were similar to those for ZEBOV (Towner

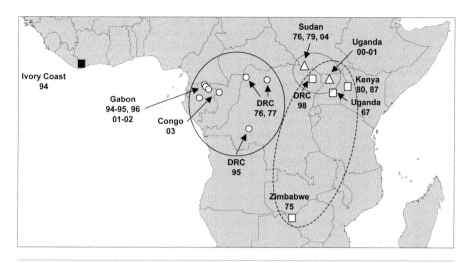

Fig. 1. Distribution of ZEBOV, SEBOV and MARV. The year and location of disease outbreaks (or virus source, if different) are shown for ZEBOV (open circles), SEBOV (open triangles) and MARV (open boxes). Location of the single Ivory Coast ebolavirus case is also indicated (closed box). Suspected ranges of ZEBOV (solid line), SEBOV (dotted line) and MARV (dashed line) are also shown.

et al., 2004). The only large MARV outbreak to have occurred in a natural setting was in Durba, DRC, in 1998, where 141 cases occurred with a case fatality of 82 %, similar to that of ZEBOV (Bausch *et al.*, 2003). Within this outbreak, time to onset of illness and death were also similar to those for ZEBOV.

The general course of clinical disease is similar with ZEBOV, SEBOV and MARV infections, with minor differences occurring in severity as noted above. Initial symptoms can be rather non-specific, including fever with generalized body pains, nausea/vomiting, headache, diarrhoea, anorexia, malaise/fatigue, abdominal pain, dysphagia or hiccups, which can make initial diagnosis based solely on clinical symptoms rather difficult. This can lead to initial misdiagnosis and virus transmission to family members and health care workers. A non-productive cough may appear particularly in EBOV-infected cases. The more distinctive petechial haemorrhagic rash is visible on the trunk and shoulders in only approximately 50 % of MARV- or ZEBOV-infected individuals, and even less with SEBOV-infected cases. Virus antigen and RNA are easily detectable in the sera of nearly all cases following onset of symptoms, and high viraemias frequently develop as the disease progresses (Towner *et al.*, 2004).

Even at these early stages of disease, differences are evident in those patients that will die versus those that will survive. For instance, approximately 2–3 log higher virus RNA levels are seen in patients who will progress to a fatal outcome (Towner *et al.*, 2004). The disease progresses rapidly in most patients. Conjunctival haemorrhage, signs of easy bruising and prolonged bleeding at needle puncture sites may occur. Patients that will go on to a fatal outcome begin to display other prominent signs of bleeding and coagulation defects, such as blood in the urine and faeces and/or massive gastro-intestinal or vaginal bleeding. Signs of disseminated intravascular coagulation (DIC) and multi-organ failure become increasingly evident. Severe nausea, vomiting, tachypnoea, anuria, delirium and coma are common in the terminal stages of disease. Most deaths occur within 7–10 days post onset of illness.

Marked immunosuppression is a hallmark of filovirus infections. Most strikingly, little or no virus-specific humoral antibody responses are detected in fatally infected patients (Baize *et al.*, 1999; Ksiazek *et al.*, 1999a, b; Towner *et al.*, 2004). Antibody responses are observed in most cases that survive, and general clearance of virus occurs rapidly in these patients. However, even in survivors, infectious virus and detectable virus RNA can persist, particularly in semen, for several weeks in some patients (Rodriguez *et al.*, 1999). Ebola virus infection induces cytokine responses, but again distinct differences can be seen between fatally and non-fatally infected patients. Early detection of anti-inflammatory cytokines such as IL-10 in patient plasma and high levels of neopterin and IL-1 receptor A (IL-1RA) are common features of fatal infections. Conversely, early detection of IL-1β and elevated IL-6 in patient plasma is commonly found in survivors (Baize *et al.*, 2002; Leroy *et al.*, 2000). At the time of death, virus-specific immuno-histochemistry staining reveals the presence of large amounts of virus in most of the major organs, endothelial cell layers, and at mucosal surfaces of ZEBOV- and SEBOV-infected patients (Zaki & Goldsmith, 1999). Prominent involvement of spleen and liver can also be seen.

FACTORS CONTRIBUTING TO HIGHLY LETHAL DISEASE

Given the relatively small number of human cases and the remote resource-poor locations where outbreaks occur, most of what is known regarding the pathogenesis of EBOV and MARV pathogenesis comes from animal model studies. Several disease models are available and each generally approximates the human disease, although some significant differences have been noted. The most studied and perhaps most accurate disease model is ZEBOV infection of cynomolgus (*Macaca fascicularis*) and rhesus (*Macaca mulatta*) macaques. While these infected primates do develop HF, it differs from the human disease in that disease develops more rapidly and is uniformly fatal. Passage of ZEBOV in guinea pigs or adult mice can result in 'heating up' of these viruses to become highly lethal in these rodents (Bray *et al.*, 1998; Volchkov *et al.*, 2000). However, HF is not a prominent feature of rodent lethal infections, and coagulopathy similar to that observed in primates is not observed (Bray *et al.*, 2001; Geisbert *et al.*, 2002b). In addition, the lymphocyte apoptosis which is so evident in human and non-human infections is absent from the rodent disease models (Bray *et al.*, 1998;

Connally *et al.*, 1999). Comparison of the results from such animal disease models with the known features of the disease in humans can produce a general picture of the pathogenesis of ZEBOV infection in humans and provide insights into the extreme pathogenicity of the virus. Overall, the high replication efficiency and broad tropism of the virus together with the induced immune and vascular dysfunction are the features of ZEBOV infections that are likely to be most important for the high case fatality observed.

Early events: importance of DC-SIGN and dysfunction of virus-infected dendritic cells and macrophages

Epidemiological evidence indicates that close contact with an individual or their body fluids during the symptomatic stage of their infection is generally required for transmission during an Ebola HF outbreak (Dowell *et al.*, 1999; World Health Organization, 1978b). There is no evidence to suggest that aerosol transmission plays a significant role during outbreaks. Presumably virus enters the body via the mucous membranes or through cuts or abrasions in the skin. Local virus replication then takes place at the site of entry. The actual cellular receptor involved in binding and internalization of virus has not yet been identified for any of the filoviruses. It has been suggested that perhaps the asialoglycoprotein receptor folate receptor-α may be utilized by MARV (Chan *et al.*, 2001), but subsequent studies indicated that this was not the case (Simmons *et al.*, 2003b). Tissue culture cell infections and pathology observations of infected human or animal models have shown that filoviruses have the ability to enter and replicate in a wide range of cell types, and therefore the receptor is most likely a commonly expressed protein or lectin. Regardless of the specific cellular receptor, lectins such as DC-SIGN (dendritic-cell specific intercellular adhesion molecule-grabbing nonintegrin), DC-SIGNR (DC-SIGN-related) and L-SIGN (liver/lymph node-SIGN) have been shown to play an important role in increasing filovirus binding to target cells (Baribaud *et al.*, 2002; Simmons *et al.*, 2003a), thereby strongly influencing the tropism of the virus during the initial stages of infection.

Following local virus replication, infection of immature dendritic cells (DCs) in peripheral locations is greatly enhanced due to virus interaction with DC-SIGN expressed in abundance on these cells (Baribaud *et al.*, 2002; Simmons *et al.*, 2003a). In non-human primate models, DCs and macrophages have been shown to be the early targets of ZEBOV (Geisbert *et al.*, 2003b). Infection of highly migratory DCs and macrophages provides a transport mechanism to facilitate virus movement to the local lymph nodes and other lymphoid tissue where further rounds of virus replication can take place. Perhaps more importantly, virus infection of the DCs may contribute to much of the immunosuppression that is characteristic of filovirus infections because DCs play a central role in linking the innate and adaptive immune response.

Following infection or interaction with an invading virus, immature DCs normally mature and initiate immune responses via secretion of chemokines and proinflammatory cytokines as well as increasing expression of a variety of costimulatory and chemokine receptors. DCs are critical for the initial innate immune response because of their rapid production of type I interferons (IFN-α/β), which alert neighbouring cells to the presence of an invading virus and elicit antiviral gene expression. Subsequently, DC maturation results in migration of the cell from the periphery to local lymph nodes, where the DC presents virus peptide antigens on the major histocompatibility complexes (MHCs) to naïve T cells, thereby initiating their differentiation into effector T cells. Thus DCs serve as a critical link between the innate immune responses at the site of infection and induction of adaptive immune responses within the lymph nodes. Not surprisingly, several viruses such as human immunodeficiency virus type 1, cytomegalovirus and measles virus, as well as bacteria such as *Mycobacterium tuberculosis*, specifically target DC function to effectively subvert host immune defences (van Kooyk *et al.*, 2003). Pathogens that target DCs via DC-SIGN may alter the balance of the Th1–Th2 response (resulting in a less efficient antiviral response) by the pathogen interaction with cell surface expressed DC-SIGN altering cell signalling events.

Cell infections with filoviruses are frequently very cytolytic and result in release of high virus titres. However, with respect to EBOV and MARV, virus infection of DCs results in high levels of virus replication and release, but not cytolysis of the cells. It has been shown *in vitro* that human monocyte-derived DCs can be infected by EBOV or MARV and that infection results in inhibition of secretion of proinflammatory cytokines, prevention of the up-regulation of costimulatory molecules, disruption of their maturation and poor T-cell stimulation (Bosio *et al.*, 2003; Mahanty *et al.*, 2003). Thus EBOV and MARV infection of DCs appears to render them dysfunctional, which most likely contributes to the poor induction of adaptive immune responses seen later in infection with these viruses.

The precise mechanism by which filoviruses inhibit early innate immune responses and normal DC function is unclear, but studies suggest that the virus VP35 protein plays an important role. Several studies have shown that ZEBOV and MARV infection inhibits production of IFN-α in dsRNA-stimulated cells, and only limited amounts of IFN-α are produced in ZEBOV-infected human macrophages (Bosio *et al.*, 2003; Gupta *et al.*, 2001). The IFN response is likely antagonized by virus production of VP35 (Basler *et al.*, 2000, 2003). The suppression mechanism has not been fully elucidated, but VP35 has been shown to block phosphorylation and nuclear translocation of IRF-3, resulting in reduced transcription of IFN-β as well as a number of other antiviral response genes (Basler *et al.*, 2003; Hartman *et al.*, 2004). Interestingly, VP35 and the NS1 protein of influenza virus share a common basic domain which is critical for IFN antagonism

(Hartman *et al*., 2004). For VP35, the C-terminal basic amino acids R305, K309 and R312 were shown to be important for IFN antagonism, and all three residues are highly conserved among all known filovirus VP35 sequences (Hartman *et al*., 2004).

Intermediate events: spread of infection, bystander lymphocyte apoptosis, multifocal necrosis, and tissue factor involvement

From the initial site of infection and local lymph nodes, virus then spreads as either free or monocyte-associated virus via the lymphatic system and the bloodstream to infect DCs and macrophages within other lymphoid tissues, parenchymal cells, spleen, liver and other tissues. In the liver, virus is first found in Kupffer cells, followed by hepatocytes. In many cells types, EBOV or MARV are highly cytolytic, resulting in necrosis in a number of tissues and organs. While the virus is rapidly spreading, dysfunction of infected DCs and macrophages (including VP35-mediated effects) continues to undermine the development of adaptive immune responses that could potentially curtail virus replication.

In addition to direct cytolysis of infected cells, data from ZEBOV-infected humans (Baize *et al*., 1999) and non-human primates (Geisbert *et al*., 2000, 2003b) suggest that while most lymphocytes are not infected by these viruses, massive bystander apoptosis can occur in fatally infected individuals. Increased expression of FAS and TRAIL on macrophages may contribute to caspase-independent apoptosis of uninfected lymphocytes (Geisbert *et al*., 2000; Hensley *et al*., 2002), which can further exacerbate the virus-induced immunosuppression and lymphopenia. Consistent with this, loss of $CD8^+$ T cells and plasma cells can be seen in both fatally ZEBOV- and SEBOV-infected patients (Sanchez *et al*., 2004).

Coagulation abnormalities become increasingly evident as filovirus infection progresses. Coagulation normally plays two important roles in the body. First, coagulation stops bleeding through haemostatic mechanisms, and, secondly, it can limit spread of invading pathogens by encapsulation. Once infection has progressed to the point where significant numbers of macrophages have become infected, then tissue factor (TF), a member of the cytokine receptor family, likely begins to play an important pathogenic role. It has long been known that TF activation of DIC contributes to lethal outcome in bacterial septicaemia, and there are numerous similarities between bacterial sepsis and the DIC and systemic shock observed in severe viral HFs such as Ebola (Ruf, 2004). Virus or bacterial infections can cause rapid induction of cells of the innate immune system to synthesize TF and initiate the coagulation cascade (Broze, 1995). TF binds enzyme coagulation factor VIIa and substrate factor X, which is then converted to

Xa. This complex (TF–VIIa–Xa) interacts with TF pathway inhibitor (TFPI) to form the inactive complex (TF–VIIa–Xa–TFPI). Coagulation on vascular cells is initiated by release of Xa, leading to thrombin formation, fibrin deposition and platelet activation. This coagulation cascade is balanced by an anticoagulant cascade involving protein C (Esmon, 2001). Protein C (in association with endothelial cell protein C receptor – EPCR) is activated by thrombin bound to thrombomodulin. There are links between the cellular receptor families involved in coagulation pathways and adaptive immune responses, leading to cross-talk between these pathways during infection and inflammation.

Ebolavirus replication in macrophages and monocytes results in increased synthesis of TF (Geisbert et al., 2003c). In addition, early IL-6 synthesis is seen to be elevated in fatally infected patients and IL-6 up-regulates TF expression on monocytes. Also, an increase in TF and a corresponding decrease in protein C can be observed in macrophages and plasma from ZEBOV-infected non-human primates, indicative of dysregulation of the fibrinolytic system (Ruf, 2004). Increased TF leads to marked fibrin encasement of EBOV-infected cells (lymphoid tissue, macrophages, Kupffer cells and DCs), which is seen early in infected non-human primates (Geisbert et al., 2003d). In addition, D-dimers, which are a marker of fibrinolysis, can be detected 1 day post ZEBOV infection of non-human primates (Geisbert et al., 2003c). The importance of TF relative to high mortality HF is further illustrated by the finding that treatment of ZEBOV-infected rhesus macaques with rNAPc2, a recombinant inhibitor of the factor VIIa–TF complex, prolonged survival time and resulted in 33 % survival in an otherwise 100 % lethal infection (Geisbert et al., 2003a). This finding is particularly encouraging given the current absence of any clinically proven effective intervention for filovirus HF patients.

Late events: virus glycoprotein, endothelial cell leakage, nitric oxide and antibody

The role of filovirus glycoproteins in pathogenesis remains unclear despite considerable study. The primary product of the glycoprotein gene of ebolaviruses is a small secreted non-structural glycoprotein termed sGP. An mRNA editing event occurs in approximately 25 % of glycoprotein gene transcripts to allow synthesis of the full-length precursor of the virus spike glycoprotein (preGP), which is post-translationally cleaved by furin to GP1 and GP2, which become disulphide linked (Volchkov et al., 1995; Sanchez et al., 1996). GP1–GP2 homotrimers make up the mature virus surface protein responsible for receptor binding and fusion (Malashkevich et al., 1999; Weissenhorn et al., 1998a, b). The MARV glycoprotein gene directly encodes the preGP without any RNA editing, and the virus does not synthesize sGP (Feldmann et al., 2001; Sanchez *et al.*, 1998). Considerable amounts of sGP (EBOV only), GP1 and soluble GP1–GP2 are released into the medium of filovirus-infected cells, and, in addition, sGP can be detected in ZEBOV-infected patient serum (Dolnik *et al.*, 2004; Sanchez *et al.*, 1996). Such observations have led to investigation of the various pathogenic roles of the mature G1–G2, and speculation that virus soluble glycoproteins may play a role in activation of mononuclear phagocytic cells and endothelial cells and as potential immune decoys (Feldmann *et al.*, 2003; Ito *et al.*, 2001; Sullivan *et al.*, 2003b).

Filovirus infection of many cell types results in considerable cytotoxicity, and several groups have shown cytotoxicity or cell detachment associated with expression of ZEBOV glycoprotein (Chan *et al.*, 2000; Takada *et al.*, 2000; Yang *et al.*, 2000). The highly O-glycosylated mucin-like region within GP1 may contribute to the cytotoxicity of the expressed GP1–GP2 (Sanchez *et al.*, 1996; Yang *et al.*, 2000). GP1–GP2 synthesis in ebolavirus-infected cells is limited by sGP production and the need for RNA editing to allow full-length GP1–GP2 production (Sanchez *et al.*, 1996; Volchkov *et al.*, 1995). Consistent with the amount of GP1–GP2 expressed being important in cytopathic effect, more rapid cell cytotoxicity has been observed with ZEBOV which contains an additional nucleotide at the editing site, resulting in greatly increased production of GP1–GP2 and only small amounts of sGP (Calain *et al.*, 1999; Volchkov *et al.*, 2001). However, such mutant ZEBOVs are highly pathogenic in animal models (Volchkov *et al.*, 2000). Also, comparable cytotoxicity and case fatality are seen with MARV, which does not limit GP1–GP2 synthesis by the need for RNA editing, or produce any sGP. Similarly, Reston ebolavirus produces large quantities of sGP, but does not appear to cause disease in infected humans.

Variation in cleavability of virus surface glycoproteins by furin or furin-like convertases has been shown to correlate with altered pathogenicity of many enveloped RNA viruses, such as influenza or Newcastle disease virus (Klenk & Garten, 1994). Such findings led to the expectation that a correlation may exist between filovirus pathogenicity and GP1–GP2 cleavage. Surprisingly, it was found that GP1–GP2 cleavage was not essential for virus release or infectivity in tissue culture (Neumann *et al.*, 2002; Wool-Lewis & Bates, 1999).

Interestingly, sequence identity has been found between a C-terminal region of the GP2 protein of EBOV and MARV and an immunosuppressive domain found in retrovirus glycoproteins (Ignatiev, 1999; Volchkov *et al.*, 1992). Synthetic peptides encompassing these regions have been shown to inhibit blastogenic lymphocyte proliferation and human NK cell activity (Ignatiev, 1999), suggesting an immunosuppressive activity similar to that seen in retroviruses (Cianciolo *et al.*, 1985; Harris *et al.*, 1987; Kadota *et al.*, 1991).

Endothelial cell leakage becomes increasingly evident as infection progresses and can be quite marked even early in infection prior to actual virus infection of the cells (Geisbert *et al.*, 2003d). This suggests that, until relatively late in the course of infection, much of the impairment of the endothelial barrier is indirect via virus-induced cytokines and chemokines rather that a direct effect associated with virus replication in these cells. As endothelial cell dysfunction occurs prior to actual infection by virus, it has also been speculated that circulating virus glycoproteins (sGP, or soluble GP1 or GP1–GP2) produced early in infection by infected mononuclear phagocytic cells could bind to endothelial cells and contribute to the endothelial cell leakage which is a major factor in the shock and haemorrhage observed in fatally infected patients (Feldmann *et al.*, 2003). However, little evidence has been found to date to support this conjecture.

Increased levels of nitric oxide (NO) have been detected in fatally ZEBOV-infected non-human primates (Geisbert *et al.*, 2003b). Also, higher blood levels of NO have been shown to correlate with increasing disease severity in SEBOV-infected cases (Sanchez *et al.*, 2004). The NO levels detected in fatal cases are similar to those seen associated with vascular disruption, cardiac stress and heart failure (Beckman & Koppenol, 1996). These data suggest that NO may play a role in the acute phase hypotension and late phase shock observed in infected patients.

The role of the humoral immune response in control of filovirus infection in humans is unclear. Virus-specific IgM and IgG are usually (but not always) detectable in convalescent phase sera of survivors and even in some fatal cases (Ksiazek *et al.*, 1999a, b; Towner *et al.*, 2004). Virus-specific antibody is absent from most fatally ZEBOV- and SEBOV-infected humans and all fatally infected non-human primates, suggesting a role of antibody in virus clearance. However, immediate or even repeated administration of high-titre anti-ZEBOV hyperimmune IgG following infection did not prevent death in ZEBOV-infected cynomolgus monkeys (Jahrling *et al.*, 1999), suggesting that antibody is not playing a major role in recovery.

Antibody-dependent enhancement (ADE) of ZEBOV infection has been demonstrated *in vitro* and has been suggested to play a role in the high mortality disease associated with filoviruses (Takada *et al.*, 2003). However, other investigators failed to identify any ADE of infection on testing of several ZEBOV-specific antisera (Geisbert *et al.*, 2002a). In addition, no early deaths have been observed in animals immunized with a wide variety of ZEBOV preparations or glycoprotein-expressing vectors (Chepurnov *et al.*, 1995; Geisbert *et al.*, 2002a; Jahrling *et al.*, 1999; Sullivan *et al.*, 2000, 2003a). Also, antibody is not detected in most fatally ZEBOV-infected humans and all fatally infected non-human primates. Taken together, these observations would suggest little role of ADE in filovirus pathogenesis.

Terminal events: extensive cell death and multi-organ failure

Based on immunohistochemisty studies of autopsy specimens, it is clear that EBOV or MARV is present in many organs and tissues by the later stages of infection, and there are few organs or tissues of the body where virus antigen is not detectable (Zaki & Goldsmith, 1999). It is likely that direct virus cytopathic effect, together with indirect effects of a dysregulated cytokine cascade and general DIC, all contribute to the last stage massive multifocal organ injury that ensues in fatal cases.

CONCLUDING REMARKS

HFs associated with ZEBOV, SEBOV and MARV are rapid highly lethal diseases. It would appear that no single virulence factor is responsible for the highly pathogenic nature of these viruses. While these viruses have the ability to efficiently replicate and cause cytopathology in a wide range of cells and tissues, they also efficiently target DCs and induce immune and vascular dysfunction. It is clear that critical early events predetermine disease outcome, but what are the more important factors determining why some ZEBOV-, SEBOV- or MARV-infected patients die and others survive? Why was there 23 % case fatality among the 31 MARV-infected European individuals in the 1967 outbreak in Germany and Yugoslavia, but 82 % case fatality among 141 MARV-infected African patients in 1998 in DRC? Why are case-fatality levels observed in recent EBOV and MARV HF outbreaks no better than those observed in original outbreaks almost 30–40 years ago (Bausch et al., 2003; Towner et al., 2004), despite improvements in patient supportive care? The precise answers to these questions are unknown, but two major factors likely play an important role: route or extent of virus exposure and host genetics.

Observations from outbreak settings suggest that those individuals with obvious high EBOV or MARV exposure are more likely to go on to fatal outcome. Such patients would include medical staff with known intensive unprotected exposure to end stage patients with florid disease symptoms or their body fluids, or patients exposed via reuse of virus-contaminated needles. Differences in MHC class I alleles have been reported to correlate with SEBOV disease outcome (Sanchez et al., 2004). Differences in host genetics are likely reflected in differing innate immune responses in infected individuals. Rapid and robust proinflammatory immune responses appear to correlate with favourable outcome and orchestrate effective adaptive immune responses. Recent promising results in non-human primates in the areas of treatment using inhibitors of TF-initiated coagulation (Geisbert et al., 2003a) and successful accelerated vaccination strategies (Sullivan et al., 2003a) represent significant advances toward the goal of preventing filovirus HF-related deaths.

REFERENCES

Baize, S., Leroy, E. M., Georges-Courbot, M.-C., Capron, M., Lansoud-Soukate, J., Debre, P., Fisher-Hoch, S. P., McCormick, J. B. & Georges, A. J. (1999). Defective humoral responses and extensive intravascular apoptosis are associated with fatal outcome in Ebola virus-infected patients. *Nat Med* **5**, 423–426.

Baize, S., Leroy, E. M., Georges, A. J., Georges-Courbot, M.-C., Capron, M., Bedjabaga, I., Lansoud-Soukate, J. & Mavoungou, E. (2002). Inflammatory responses in Ebola virus-infected patients. *Clin Exp Immunol* **128**, 163–168.

Baribaud, F., Pohlmann, S., Leslie, G., Mortari, F. & Doms, R. W. (2002). Quantitative expression and virus transmission analysis of DC-SIGN on monocyte-derived dendritic cells. *J Virol* **76**, 9135–9142.

Basler, C. F., Wang, X., Muhlberger, E., Volchkov, V., Paragas, J., Klenk, H.-D., Garcia-Sastre, A. & Palese, P. (2000). The Ebola virus VP35 protein functions as a type I IFN antagonist. *Proc Natl Acad Sci U S A* **97**, 12289–12294.

Basler, C. F., Mikulasova, A., Martinez-Sobrido, L., Paragas, J., Muhlberger, E., Bray, M., Klenk, H.-D., Palese, P. & Garcia-Sastre, A. (2003). The Ebola virus VP35 protein inhibits activation of interferon regulatory factor 3. *J Virol* **77**, 7945–7956.

Bausch, D. G., Borchert, M., Grein, T. & 16 other authors (2003). Risk factors for Marburg hemorrhagic fever, Democratic Republic of the Congo. *Emerg Infect Dis* **9**, 1531–1537.

Beckman, J. S. & Koppenol, W. H. (1996). Nitric oxide, superoxide, and peroxynitrite: the good, the bad, and ugly. *Am J Physiol* **271**, C1424–1437.

Bosio, C. M., Aman, M. J., Grogan, C., Hogan, R., Ruthel, G., Negley, D., Mohamadzadeh, M., Bavari, S. & Schmaljohn, A. (2003). Ebola and Marburg viruses replicate in monocyte-derived dendritic cells without inducing the production of cytokines and full maturation. *J Infect Dis* **188**, 1630–1638.

Bray, M., Davis, K., Geisbert, T., Schmaljohn, C. & Huggins, J. (1998). A mouse model for evaluation of prophylaxis and therapy of Ebola hemorrhagic fever. *J Infect Dis* **178**, 651–661.

Bray, M., Hatfill, S., Hensley, L. & Huggins, J. W. (2001). Haematological, biochemical and coagulation changes in mice, guinea-pigs and monkeys infected with a mouse-adapted variant of Ebola Zaire virus. *J Comp Pathol* **125**, 243–253.

Broze, G. J., Jr (1995). Tissue factor pathway inhibitor and the revised theory of coagulation. *Annu Rev Med* **46**, 103–112.

Buchmeier, M. J., DeFries, R. U., McCormick, J. B. & Kiley, M. P. (1983). Comparative analysis of the structural polypeptides of Ebola viruses from Sudan and Zaire. *J Infect Dis* **147**, 276–281.

Calain, P., Monroe, M. C. & Nichol, S. T. (1999). Ebola virus defective interfering particles and persistent infection. *Virology* **262**, 114–128.

Centers for Disease Control and Prevention (1995). Outbreak of Ebola viral hemorrhagic fever – Zaire 1995. *Morb Mortal Wkly Rep* **44**, 381–382.

Chan, S. Y., Ma, M. C. & Goldsmith, M. A. (2000). Differential induction of cellular detachment by envelope glycoproteins of Marburg and Ebola (Zaire) viruses. *J Gen Virol* **81**, 2155–2159.

Chan, S. Y., Empig, C. J., Welte, F. J., Speck, R. F., Schmaljohn, A., Kreisberg, J. F. & Goldsmith, M. A. (2001). Folate receptor-alpha is a cofactor for cellular entry by Marburg and Ebola viruses. *Cell* **106**, 117–126.

Chepurnov, A. A., Chernukhin, I. V., Ternovoi, V. A., Kudoyarova, N. M., Makhova, N. M., Azayev, M. S. & Smolina, M. O. (1995). Attempts at creating a vaccine against Ebola fever. *Vopr Virusol* **40**, 257–260.

Cianciolo, G. J., Copeland, T. D., Oroszlan, S. & Snyderman, R. (1985). Inhibition of lymphocyte proliferation by a synthetic peptide homologous to retroviral envelope proteins. *Science* **230**, 453–455.

Connally, B. M., Steele, K. E., Davis, K. J., Geisbert, T. W., Kell, W. M., Jaax, N. K. & Jahrling, P. B. (1999). Pathogenesis of experimental Ebola virus infection in guinea pigs. *J Infect Dis* **179** (Suppl. 1), S203–S217.

Cox, N. J., McCormick, J. B., Johnson, K. M. & Kiley, M. P. (1983). Evidence for two subtypes of Ebola virus based on oligonucleotide mapping of RNA. *J Infect Dis* **147**, 272–275.

Dolnik, O., Volchkova, V., Garten, W., Carbonnelle, C., Becker, S., Kahn, J., Ströher, U., Klenk, H.-D. & Volchkov, V. V. (2004). Ectodomain shedding of the glycoprotein GP of Ebola virus. *EMBO J* **23**, 2175–2184.

Dowell, S. F., Mukunu, R., Ksiazek, T. G., Khan, A. S., Rollin, P. E. & Peters, C. J. (1999). Transmission of Ebola hemorrhagic fever: a study of risk factors in family members, Kikwit, Democratic Republic of the Congo, 1995. Commission de Lutte contre les Epidémies à Kikwit. *J Infect Dis* **179** (Suppl. 1), S87–S91.

Esmon, C. T. (2001). Role of coagulation inhibitors in inflammation. *Thromb Haemostasis* **86**, 51–56.

Feldmann, H., Volchkov, V. E., Volchkova, V. A., Stroeher, U. & Klenk, H.-D. (2001). Biosynthesis and role of filoviral glycoproteins. *J Gen Virol* **82**, 2839–2848.

Feldmann, H., Jones, S., Klenk, H.-D. & Schnittler, H.-J. (2003). Ebola virus: from discovery to vaccine. *Nat Rev Immunol* **3**, 677–685.

Geisbert, T. W., Hensley, L. E., Gibb, T. R., Steele, K. E., Jaax, N. K. & Jahrling, P. B. (2000). Apoptosis induced in vitro and in vivo during infection by Ebola and Marburg viruses. *Lab Invest* **80**, 171–186.

Geisbert, T. W., Hensley, L. E., Geisbert, J. B. & Jahrling, P. B. (2002a). Evidence against an important role for infectivity-enhancing antibodies in Ebola virus infections. *Virology* **293**, 15–19.

Geisbert, T. W., Pushko, P., Anderson, K., Smith, J., Davis, K. J. & Jahrling, P. B. (2002b). Evaluation in nonhuman primates of vaccines against Ebola virus. *Emerg Infect Dis* **8**, 503–507.

Geisbert, T. W., Hensley, L. E., Jahrling, P. B. & 7 other authors (2003a). Treatment of Ebola virus infection with a recombinant inhibitor of factor VIIa/tissue factor: a study in rhesus monkeys. *Lancet* **362**, 1953–1958.

Geisbert, T. W., Hensley, L. E., Larsen, T. & 7 other authors (2003b). Pathogenesis of Ebola hemorrhagic fever in cynomolgus macaques: evidence that dendritic cells are early and sustained targets of infection. *Am J Pathol* **163**, 2347–2370.

Geisbert, T. W., Young, H. A., Jahrling, P. B., Davis, K. J., Kagan, E. & Hensley, L. E. (2003c). Mechanisms underlying coagulation abnormalities in Ebola hemorrhagic fever: overexpression of tissue factor in primate monocytes/macrophages is a key event. *J Infect Dis* **188**, 1618–1629.

Geisbert, T. W., Young, H. A., Jahrling, P. B., Davis, K. J., Larsen, T., Kagan, E. & Hensley, L. E. (2003d). Pathogenesis of Ebola hemorrhagic fever in primate models: evidence that hemorrhage is not a direct effect of virus-induced cytolysis of endothelial cells. *Am J Pathol* **163**, 2371–2382.

Gupta, M., Mahanty, S., Ahmed, R. & Rollin, P. E. (2001). Monocyte-derived human macrophages and peripheral blood mononuclear cells infected with Ebola virus secrete MIP-1alpha and TNF-alpha and inhibit poly-IC-induced IFN-alpha in vitro. *Virology* **284**, 20–25.

Harris, D. T., Cianciolo, G. J., Snyderman, R., Argov, S. & Koren, H. S. (1987). Inhibition of human natural killer cell activity by a synthetic peptide homologous to a conserved region in the retroviral protein, p15E. *J Immunol* **138**, 889–894.

Hartman, A. L., Towner, J. S. & Nichol, S. T. (2004). A C-terminal basic amino acid motif of Zaire ebolavirus VP35 is essential for type I interferon antagonism and displays high identity with the RNA-binding domain of another interferon antagonist, the NS1 protein of influenza A virus. *Virology* **328**, 177–184.

Hensley, L. E., Young, H. A., Jahrling, P. B. & Geisbert, T. W. (2002). Proinflammatory response during Ebola virus infection of primate models: possible involvement of the tumor necrosis factor receptor superfamily. *Immunol Lett* **80**, 169–179.

Ignatiev, G. M. (1999). Immune response to filovirus infections. *Curr Top Microbiol Immunol* **235**, 205–217.

Ito, H., Watanabe, S., Takada, A. & Kawaoka, Y. (2001). Ebola virus glycoprotein: proteolytic processing, acylation, cell tropism, and detection of neutralizing antibodies. *J Virol* **75**, 1576–1580.

Jahrling, P. B., Geisbert, T. W., Dalgard, D. W., Johnson, E. D., Ksiazek, T. G., Hall, W. C. & Peters, C. J. (1990). Preliminary report: isolation of Ebola virus from monkeys imported to USA. *Lancet* **335**, 502–505.

Jahrling, P. B., Geisbert, T. W., Geisbert, J. B., Swearengen, J. R., Bray, M., Jaax, N. K., Huggins, J. W., LeDuc, J. W. & Peters, C. J. (1999). Evaluation of immune globulin and recombinant interferon α-2b for treatment of experimental Ebola virus infections. *J Infect Dis* **179** (Suppl. 1), S224–S234.

Kadota, J., Cianciolo, G. J. & Snyderman, R. (1991). A synthetic peptide homologous to retroviral transmembrane envelope proteins depresses protein kinase C mediated lymphocyte proliferation and directly inactivated protein kinase C: a potential mechanism for immunosuppression. *Microbiol Immunol* **35**, 443–459.

Khan, A. S., Kweteminga, T. F., Heymann, D. H. & 16 other authors (1999). The reemergence of Ebola hemorrhagic fever (EHF), Democratic Republic of the Congo, 1995. *J Infect Dis* **179** (Suppl. 1), S76–S86.

Kissling, R. E., Robinson, R. Q., Murphy, F. A. & Whitfield, S. G. (1968). Agent of disease contracted from green monkeys. *Science* **160**, 888–890.

Klenk, H.-D. & Garten, W. (1994). Host cell proteases controlling virus pathogenicity. *Trends Microbiol* **2**, 39–43.

Ksiazek, T. G., Rollin, P. E., Williams, A. J. & 9 other authors (1999a). Clinical virology of Ebola hemorrhagic fever (EHF): virus, virus antigen, IgG and IgM antibody findings among EHF patients in Kikwit, 1995. *J Infect Dis* **179** (Suppl. 1), S177–S187.

Ksiazek, T. G., West, C. P., Rollin, P. E., Jahrling, P. B. & Peters, C. J. (1999b). ELISA for the detection of antibodies to Ebola viruses. *J Infect Dis* **179** (Suppl. 1), S191–S198.

Le Guenno, B., Formenty, P., Wyers, M., Gounon, P., Walker, F. & Boesch, C. (1995). Isolation and partial characterisation of a new strain of Ebola virus. *Lancet* **345**, 1271–1274.

Leroy, E. M., Baize, S., Volchkov, V. E. & 7 other authors (2000). Human asymptomatic Ebola infection and strong inflammatory response. *Lancet* **355**, 2210–2215.

Mahanty, S., Hutchinson, K., Agarwal, S., McRae, M., Rollin, P. E. & Pulendran, B. (2003). Cutting edge: impairment of dendritic cells and adaptive immunity by Ebola and Lassa viruses. *J Immunol* **170**, 2797–2801.

Malashkevich, V. N., Schneider, B. J., McNally, M. L., Milhollen, M. A., Pang, J. X. & Kim, P. S. (1999). Core structure of the envelope glycoprotein GP2 from Ebola virus at 1·9-Å resolution. *Proc Natl Acad Sci U S A* **96**, 2662–2667.

McCormick, J. B., Bauer, S. P., Elliott, L. H., Webb, P. A. & Johnson, K. M. (1983). Biologic differences between strains of Ebola virus from Zaire and Sudan. *J Infect Dis* **147**, 264–267.

Neumann, G., Feldmann, H., Watanabe, S., Lukashevich, I. & Kawaoka, Y. (2002). Reverse genetics demonstrates that proteolytic processing of the Ebola virus glycoprotein is not essential for replication in cell culture. *J Virol* **76**, 406–410.

Peters, C. J., Johnson, E. D., Jahrling, P. B., Ksiazek, T. G., Rollin, P. E., White, J., Hall, W., Trotter, R. & Jaax, N. (1992). Filoviruses. In *Emerging Viruses*, pp. 159–175. Edited by Stephen S. Morse. Oxford: Oxford University Press.

Rodriguez, L., De Roo, A., Guimard, Y. & 7 other authors (1999). Persistence and genetic stability of Ebola virus during the outbreak in Kikwit, Zaire 1995. *J Infect Dis* **179** (Suppl. 1), S170–S176.

Ruf, W. (2004). Emerging roles of tissue factor in viral hemorrhagic fever. *Trends Immunol* **25**, 461–464.

Sanchez, A., Trappier, S. G., Mahy, B. W. J., Peters, C. J. & Nichol, S. T. (1996). The virion glycoproteins of Ebola viruses are encoded in two reading frames and are expressed through transcriptional editing. *Proc Natl Acad Sci U S A* **93**, 3602–3607.

Sanchez, A., Trappier, S. G., Ströher, U., Nichol, S. T., Bowen, M. D. & Feldmann, H. (1998). Variation in the glycoprotein and VP35 genes of Marburg virus strains. *Virology* **240**, 138–146.

Sanchez, A., Lukwiya, M., Bausch, D., Mahanty, S., Sanchez, A. J., Wagoner, K. D. & Rollin, P. E. (2004). Analysis of human peripheral blood samples from fatal and nonfatal cases of Ebola (Sudan) hemorrhagic fever: cellular responses, virus load, and nitric oxide levels. *J Virol* **78**, 10370–10377.

Simmons, G., Reeves, J. D., Grogan, C. C. & 10 other authors (2003a). DC-SIGN and DC-SIGNR bind Ebola glycoproteins and enhance infection of macrophages and endothelial cells. *Virology* **305**, 115–123.

Simmons, G., Rennekamp, A. J., Chai, N., Vandenberghe, L. H., Riley, J. L. & Bates, P. (2003b). Folate receptor alpha and caveolae are not required for Ebola virus glycoprotein-mediated viral infection. *J Virol* **77**, 13433–13438.

Sullivan, N. J., Sanchez, A., Rollin, P. E., Yang, Z.-Y. & Nabel, G. J. (2000). Development of a preventive vaccine for Ebola virus infection in primates. *Nature* **408**, 605–609.

Sullivan, N. J., Geisbert, T. W., Geisbert, J. B., Xu, L., Yang, Z.-Y., Roederer, M., Koup, R. A., Jahrling, P. B. & Nabel, G. J. (2003a). Accelerated vaccination for Ebola virus haemorrhagic fever in non-human primates. *Nature* **424**, 681–684.

Sullivan, N., Yang, Z.-Y. & Nabel, G. J. (2003b). Ebola virus pathogenesis: implications for vaccines and therapies. *J Virol* **77**, 9733–9737.

Takada, A., Watanabe, S., Ito, H., Okazaki, K., Kida, H. & Kawaoka, Y. (2000). Downregulation of beta1 integrins by Ebola virus glycoprotein: implication for virus entry. *Virology* **278**, 20–26.

Takada, A., Feldmann, H., Ksiazek, T. G. & Kawaoka, Y. (2003). Antibody-dependent enhancement of Ebola virus infection. *J Virol* **77**, 7539–7544.

Towner, J. S., Rollin, P. E., Bausch, D. G. & 10 other authors (2004). Rapid diagnosis of Ebola hemorrhagic fever by reverse transcription-PCR in an outbreak setting and assessment of patient viral load as a predictor of outcome. *J Virol* **78**, 4330–4341.

van Kooyk, Y., Appelmelk, B. & Geijtenbeek, T. B. (2003). A fatal attraction: *Mycobacterium tuberculosis* and HIV-1 target DC-SIGN to escape immune surveillance. *Trends Mol Med* **9**, 153–159.

Volchkov, V. E., Blinov, V. M. & Netesov, S. V. (1992). The envelope glycoprotein of Ebola virus contains an immunosuppressive-like domain similar to oncogenic retroviruses. *FEBS Lett* **305**, 181–184.

Volchkov, V. E., Becker, S., Volchkova, V. A., Ternovoj, V. A., Kotov, A. N., Netesov, S. V. & Klenk, H.-D. (1995). GP mRNA of Ebola virus is edited by the Ebola virus polymerase and by T7 and vaccinia virus polymerase. *Virology* **214**, 421–430.

Volchkov, V. E., Chepurnov, A. A., Volchkova, V. A., Ternovoj, V. A. & Klenk, H. D. (2000). Molecular characterization of guinea pig-adapted variants of Ebola virus. *Virology* **277**, 147–155.

Volchkov, V. E., Volchkova, V. A., Muhlberger, E., Kolesnikova, L. V., Weik, M., Dolnik, O. & Klenk, H.-D. (2001). Recovery of infectious Ebola virus from complementary DNA: RNA editing of the GP gene and viral cytotoxicity. *Science* **291**, 1965–1969.

Weissenhorn, W., Calder, L. J., Wharton, S. A., Skehel, J. J. & Wiley, D. C. (1998a). The central structural feature of the membrane fusion protein subunit from the Ebola virus glycoprotein is a long triple-stranded coiled coil. *Proc Natl Acad Sci U S A* **95**, 6032–6036.

Weissenhorn, W., Carfi, A., Lee, K.-H., Skehel, J. J. & Wiley, D. C. (1998b). Crystal structure of the Ebola virus membrane fusion subunit, GP2, from the envelope glycoprotein ectodomain. *Mol Cell* **2**, 605–616.

Wool-Lewis, R. J. & Bates, P. (1999). Endoproteolytic processing of the Ebola virus envelope glycoprotein: cleavage is not required for function. *J Virol* **73**, 1419–1426.

World Health Organization (1967). Outbreaks in laboratory personnel working with Cercopithecus monkeys from East Africa – Europe. *Wkly Epidemiol Rec* **42**, 479–480.

World Health Organization (1978a). Ebola haemorrhagic fever in Sudan, 1976. Report of a WHO/International Study Team. *Bull W H O* **56**, 247–270.

World Health Organization (1978b). Ebola haemorrhagic fever in Zaire, 1976. *Bull W H O* **56**, 271–293.

World Health Organization (1995). Ebola haemorrhagic fever – Zaire. *Wkly Epidemiol Rec* **70**, 149–150.

Yang, Z.-Y., Duckers, H. J., Sullivan, N. J., Sanchez, A., Nabel, E. G. & Nabel, G. J. (2000). Identification of the Ebola virus glycoprotein as the main viral determinant of vascular cell cytotoxicity and injury. *Nat Med* **6**, 886–889.

Zaki, S. R. & Goldsmith, C. S. (1999). Pathologic features of filovirus infections in humans. *Curr Top Microbiol Immunol* **235**, 97–116.

Molecular approaches to the pathogenesis of feline coronaviruses

Charlotte Dye and Stuart Siddell

Department of Molecular and Cellular Medicine, University of Bristol, Bristol BS8 1TD, UK

INTRODUCTION

Feline coronavirus (FCoV) infection is extremely common in cats. For example, in the United Kingdom, approximately 40 % of the domestic cat population has been infected. In multi-cat households, this figure increases to around 90 % (Addie, 2000; Addie & Jarrett, 1992; Sparkes *et al.*, 1992). Natural infections with FCoV are usually transient, although a significant percentage of infections may become persistent (Addie & Jarrett, 2001). FCoV infections may be asymptomatic or may result in mild, self-limiting gastrointestinal disease. In these cases, the causative agent is known as feline enteric coronavirus (FECV). In a small percentage of animals, a fatal multi-systemic, immune-mediated disease occurs and this is known as feline infectious peritonitis (FIP) (Pedersen, 1995). The virus associated with FIP is referred to as feline infectious peritonitis virus (FIPV) and it is proposed that cats acquire FIPV by mutation of an endogenous FECV, or occasionally through excreted virus from other FIPV-infected animals (Pedersen & Hawkins, 1995; Poland *et al.*, 1996; Vennema *et al.*, 1998). Any genetic differences between FECV and FIPV that can account for the different disease patterns remain to be identified.

There are two types of FCoV that can be distinguished by serology and by sequence analysis. Type I viruses are most prevalent in the field and account for approximately 80 % of all infections (Hohdatsu *et al.*, 1992). Type II viruses are less prevalent and are characterized by recombination events that result in the replacement of the FCoV spike glycoprotein gene with the equivalent gene of the canine enteric coronavirus (CCoV)

SGM symposium 64: Molecular pathogenesis of virus infections.
Editors P. Digard, A. A. Nash & R. E. Randall. Cambridge University Press. ISBN 0 521 83248 9 ©SGM 2005

(Herrewegh *et al.*, 1998). There is no evidence that either type is more commonly associated with FIP in natural infections (Benetka *et al.*, 2004).

NATURAL HISTORY OF FIP

Clinical features

Two forms of FIP exist with very different clinical presentations. The so-called 'wet' or 'effusive' form of FIP is believed to develop when the host fails to mount any form of effective immune response. In contrast, the 'dry' or 'granulatomous' form of FIP develops when a partial immune response is mounted. The 'wet' form of FIP is typified by rapidly progressive, effusive disease. Ascites is the most common finding but pleural effusions and fluid accumulation in other body cavities can also be seen. In the early stages of infection, cats present with non-specific signs such as pyrexia, lethargy and anorexia. Recurrent diarrhoea or constipation may also be a feature. Jaundice is a major finding and is often present irrespective of the degree of hepatic involvement. Progressive abdominal distension over a period of days to weeks ensues together with marked weight loss. Dyspnoea is a feature in around 25 % of cases where there is pleural involvement. The disease course is short and invariably fatal. The 'dry' form of FIP results from granuloma formation on the surface of tissues. Non-specific signs of chronic weight loss, pyrexia, lethargy and anorexia may be present for weeks before the appearance of any organ-specific signs. The central nervous system is commonly involved, eventually resulting in neurological signs. However, any organ can be affected and renal, hepatic, pancreatic and ocular changes are also commonly seen. As with the 'wet' form, the 'dry' form of the disease is invariably fatal, although with the 'dry' form the course is often more protracted (Haagmans *et al.*, 1996; Hoskins, 1993; Pedersen, 1987).

Transmission

FCoVs are highly infectious. They are spread predominantly by the faeco-oral route but also in oro-nasal secretions and in urine (Hoskins, 1993). Cats recovering from infection will shed virus in their faeces and are a potential risk to other susceptible cats. Most cats will shed virus for a period of a few weeks to months either continuously or transiently. Occasionally persistent carriers are found which will shed virus indefinitely (Addie & Jarrett, 2001). In multi-cat households, a large proportion of cats will be shedding FCoV at any given time. Each individual cat will undergo cycles of infection, shedding, recovery and re-infection (Foley *et al.*, 1997a). Close contact between infected and susceptible cats facilitates the most effective transmission. However, FCoV can persist in the environment (Hoskins, 1993) and there is a risk of transmission via fomites. There is no known vertical transmission of FCoV.

Susceptibility

Pure-bred cats are predisposed to the development of FIP. Age is also an important risk factor with the majority of cats developing disease before 2 years (Foley *et al.*, 1997a). A second, smaller disease peak is seen in older cats of more than 12 years. The majority of disease is seen in cats suffering from stress or concurrent disease. The classical presentation is kittens from multi-cat households or shelter situations where shedding rates are maximal, intercurrent disease is common and stress levels are high. FIP disease is usually sporadic, a finding that suggests that horizontal transmission of virulent virus is infrequent. However, a number of clustered outbreaks have been described.

Pathogenesis

Following ingestion FCoV binds to the apical surface of gut epithelial cells in the feline gastrointestinal tract. The small intestine is preferentially attacked in acute infections but the large intestine often becomes involved during more chronic carrier states. The tips of intestinal villi are the primary site of virus entry. Attachment of the serotype II virus to the host cell is mediated by the metalloprotease aminopeptidase N (APN) (Hohdatsu *et al.*, 1998a; Tresnan & Holmes, 1998). APN is a 150 kDa glycoprotein that is also expressed on the plasma membranes of granulocytes, lymphocytes and monocytes, as well as on other non-haematopoietic tissue, including fibroblasts, synaptic membranes in the central nervous system, epithelial cells from the renal proximal tubules and the intestinal brush border, and endothelial and epithelial cells of the respiratory tract. The host receptor for serotype I virus is unknown (Hohdatsu *et al.*, 1998a).

Once within the enterocyte, FCoV replicates rapidly. Virus particles released from the dying cells attach to other non-infected enterocytes in the vicinity and the cycle continues. Clinical signs of enteritis develop if sufficient enterocytes are affected before effective immunity develops. Host immunity consists of both humoral and cell-mediated responses. The humoral response is polyclonal but the anti-spike (S) protein response is dominant and is responsible for the induction of neutralizing antibodies. The FCoV S, membrane (M) and nucleocapsid (N) proteins are believed to elicit cell-mediated immune responses but their significance for protective immunity is not yet clear (Welsh *et al.*, 1986).

One hypothesis is that FIP occurs when a cat is exposed to variants of FECV that have mutated within the host and are now able to multiply in macrophages (Pedersen & Hawkins, 1995; Poland *et al.*, 1996; Vennema *et al.*, 1998). However, both FECV and FIPV are able to pass from the gut into the bloodstream (Herrewegh *et al.*, 1995) and FCoV particles are found within monocytes even in benign FECV infections. Nevertheless, classical FIP lesions are typified by pyogranulomatous plaques and vasculitis, which emphasizes that monocyte-associated viraemia is a direct prerequisite of granuloma formation. FCoV-infected monocytes attach to activated MHC class II venous endothelial cells and exit from the vessels, forming perivenous accumulations. Intense inflammatory infiltration is associated with destruction of the basal laminae and this leads to leakage of fluid, resulting in effusion accumulation in body cavities.

The host adaptive immune system plays an important role in the pathogenesis of FIP. Briefly, it is thought that virus-specific antibodies can influence disease development (Addie et al., 1995; Herrewegh et al., 1997; Pedersen, 1976) by facilitating virus uptake into macrophages (Hohdatsu et al., 1993, 1994). During experimental infection, seropositive cats often develop an accelerated, fulminating course of disease, which has been explained by the idea of 'antibody-dependent enhancement' (ADE). However, ADE has seldom been observed in natural FIP infections (Hohdatsu et al., 1998b) and there is a possibility that it may simply be an interesting laboratory artefact.

Lack of an effective cell-mediated response (CMI) is believed to play a central role in disease development (Pedersen, 1987). However, few studies have explored the role of the FIPV-specific CMI due to lack of reagents and suitable assays. FIPV-specific CD4+ve and CD8+ve T cells can be identified using intracellular cytokine staining and flow cytometry assays. Studies comparing the cellular immune responses in cats that survived or succumbed to FIPV challenge are still preliminary.

Diagnosis

Diagnosis of FIP is difficult and a definitive diagnosis can only be made by histo-pathological examination of tissue biopsy or post-mortem material. Routine laboratory tests may increase the suspicion of FIP but do not provide conclusive evidence. Haematology classically reveals lymphopenia, mature neutrophilia and mild anaemia, but their absence does not exclude a diagnosis of FIP. Similarly, serum biochemical changes will frequently show hyperbilirubinaemia and hyperglobulin-aemia with a polyclonal gammopathy, but these changes are not always present (Paltrinieri et al., 1998; Sparkes et al., 1991). $\alpha2$-Globulins tend to be prevalent in early disease, switching to a γ-globulin response as clinical signs develop (Gunn-Moore et al., 1998; Stoddart et al., 1988). Assays for FCoV specific antibody are widely available and frequently used. Immunofluorescent antibody techniques are most commonly used in the commercial laboratory setting and have good sensitivity and specificity. However, coronavirus antibody titre results are often confusing in the clinical situation (Barr, 1996). An antibody titre simply confirms exposure to FCoV and conveys no information about whether or not this is the benign FECV or the pathogenic FIPV. Cats with FIP can present with both very high and very low antibody titres thus the test has little value in either the diagnosis or prognosis of FIP.

Control

Effective treatments for FIP are not available. Environmental controls are currently the most powerful and widely used tool for the control of FCoV infection. Good hygiene and disinfection techniques are essential. Ideally group sizes should be small with no more than five cats housed together. This reduces shedding rates and levels of intercurrent disease. Stress levels should be kept to a minimum. Quantitative real-time PCR methods are now available, enabling the level as well as the frequency of virus shedding to be assessed (Gut *et al.*, 1999). This is a useful tool since high-level shedders pose a greater threat of disease transmission (Foley *et al.*, 1997a). Separation of cats within a household according to their level of faecal shedding has been advocated (Addie & Jarrett, 2001; Foley *et al.*, 1997b; Lutz *et al.*, 2002). This may help to accelerate the rate of viral load reduction but is impractical in many cattery situations. Isolation and early weaning protocols are similarly difficult to instigate, but may be beneficial.

There have been many attempts to produce a vaccine for FCoV but its development is fraught with problems. The first vaccination attempts were with sub-lethal doses of live virus. Although effective it was impossible to accurately calculate suitable doses for each individual and many of the trial cats developed FIP. Avirulent live vaccine did not offer any protection to cats previously exposed to FCoV, making it of little use in the field (Gerber *et al.*, 1990). Live heterologous CCoV and human coronavirus (HCoV) immunization also offered no protection, and CCoV vaccination, in fact, resulted in disease enhancement (Barlough *et al.*, 1985). At the present time, there is only one licensed FIP vaccine on the market (Primucell; Pfizer). This is based on a modified, live, temperature-sensitive strain of FIPV. It is licensed for use in kittens from 16 weeks of age upwards (which means that any benefits it may convey are lost for the younger kitten population, who are at greatest risk) and, indeed, the efficacy of this vaccine is still a matter of debate (Fehr *et al.*, 1997; Gerber, 1995; McArdle *et al.*, 1995; Scott *et al.*, 1995). Recently a two-stage process enabling genetic modification of FCoV has been described (Haijema *et al.*, 2003) and, in an initial study, different combinations of accessory (i.e. non-essential) genes have been deleted from the FIPV genome. The genetically modified viruses replicated efficiently and grew to high titres in cell culture and no clinical signs of FIP were seen in inoculated cats. When challenged with the pathogenic FIPV 79-1146 strain, the cats previously vaccinated with the modified viruses remained healthy compared to non-vaccinated control cats, which showed clear symptoms of FIP (Haijema *et al.*, 2004). These genetically modified viruses make promising candidates for vaccine development.

Vectored and non-vectored subunit vaccines for FIP have also been tested, but still with only moderate success. For example, a recombinant vaccinia virus expressing the FIPV

S protein did not prevent infection, and indeed appeared to enhance the onset of FIP during challenge (Vennema *et al.*, 1990). Similarly, recombinant vaccinia or raccoon poxviruses expressing FIPV N or M proteins were found to provide only limited protection against low-dose challenge (Vennema *et al.*, 1991; Wasmoen *et al.*, 1995). At the present time, the most promising vectored subunit vaccine appears to be a protein antigen derived from recombinant baculovirus-expressed N protein (Hohdatsu *et al.*, 2003). This cell lysate is effective in preventing the progression of FIP without inducing ADE of FIPV infection in cats. Finally, attempts to produce DNA-based subunit vaccines have been initiated but are as yet unsuccessful (Glansbeek *et al.*, 2002).

CORONAVIRUSES

Genome organization and expression

Coronavirus particles contain a genomic RNA of approximately 27 000–30 000 nucleotides and four structural proteins: namely the spike glycoprotein S, the membrane protein M, the small envelope protein E and the nucleocapsid protein N. Three of these four proteins are embedded in the viral envelope. These are the S protein, which mediates binding of the virus particle to the target cell and the subsequent fusion of viral and cellular membranes, the M protein, which has a crucial role in the incorporation of the virus nucleocapsid into virus particles, and the E protein, which facilitates virus assembly, possibly by inducing curvature into pre-Golgi membranes, the site at which coronaviruses assemble by budding. The fourth structural protein, N, is associated with the viral RNA genome to form a ribonucleoprotein complex (Siddell *et al.*, 2005).

Coronavirus replication takes place in the cytoplasm of infected cells. The virus particle enters the cell by receptor-mediated endocytosis and, following disassembly, the virus replicase proteins are translated from the genomic RNA. The replicase proteins are initially synthesized as large polyproteins, which are then processed by virus-encoded proteinases to produce a functional replicase–transcriptase complex (Ziebuhr *et al.*, 2000). The coronavirus replicase gene has been shown, or is predicted, to encode multiple enzymic functions, including polymerase, proteases, helicase, ribonucleases, methyltransferase and phosphatase (Ivanov *et al.*, 2004; Snijder *et al.*, 2003). Subsequently, the replicase–transcriptase complex is responsible for the synthesis of a set of subgenomic mRNAs. These mRNAs are produced by a unique mechanism that involves discontinuous transcription during negative strand RNA synthesis. The subgenomic mRNAs are structurally polycistronic but, with some exceptions, functionally monocistronic. The subgenomic mRNAs are translated to produce the structural proteins of the virus and a number of accessory proteins. The accessory

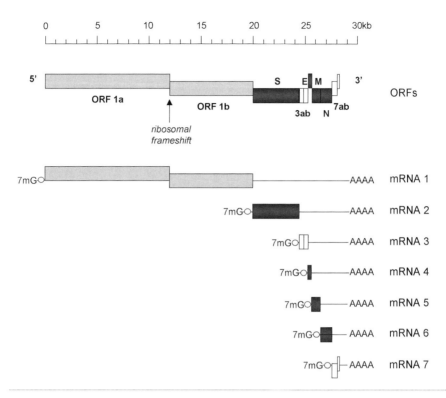

Fig. 1. Organization and expression of the FCoV genome. The genomic ORFs encoding structural proteins are solid; those encoding non-structural proteins are lightly shaded. ORFs encoding accessory proteins are unshaded. The ORFs are drawn to scale in the correct relative reading frames. The structural relationships of the genomic and subgenomic mRNAs are shown. 7mG represents a 5′ cap structure and AAAA represents a 3′ polyadenylate tract. The leader sequence present at the 5′-end of each mRNA is represented by an ellipse. The ribosomal frameshifting element found in genome-length mRNA1 and the translated region for each mRNA is also shown.

proteins are thought to be non-essential for replication in tissue culture but clearly provide a selective advantage *in vivo* (Siddell *et al.*, 2005). Fig. 1 illustrates this replication strategy with specific reference to FCoV.

Reverse genetics

The development of systems for manipulating the coronavirus genome using traditional reverse genetic approaches has been hampered by both the genome size and the instability of specific coronavirus cDNA sequences in bacterial systems. However, recently, reverse genetic systems have been established using non-traditional approaches that are based on the use of bacterial artificial chromosomes (Almazan *et al.*, 2000), the *in vitro* ligation of coronavirus cDNA fragments (Yount *et al.*, 2000), and the use of vaccinia virus as a vector for the propagation of coronavirus genomic cDNAs (Thiel *et al.*, 2001). Recombinant viruses with gene inactivations, deletions or attenuating modifications can now be generated and used to study the role of specific gene products in virus replication or pathogenesis. Also, genetically attenuated viruses can be produced which are potential vaccine candidates. The reverse genetic system based upon the use of vaccinia virus cloning vectors is a generic approach that has been described for the generation of recombinant human coronavirus 229E (Casais *et al.*, 2001), recombinant avian infectious bronchitis coronavirus (Casais *et al.*, 2001) and recombinant mouse hepatitis virus (Coley *et al.*, 2005). The basic strategy can be divided into three phases.

The assembly of a full-length coronavirus genomic cDNA. This normally involves the generation of numerous subgenomic cDNA fragments that are either amplified as bacterial plasmid DNA or prepared in large amounts by preparative reverse-transcriptase PCR (RT-PCR). The cDNAs are then ligated sequentially, *in vitro*, to produce a small number of cDNAs that encompass the entire genome. The specific ligation strategy is dictated by the sequence of the coronavirus in question but a common feature is the use of convenient, naturally occurring or engineered restriction sites, especially if they cleave, for example, interrupted palindromic sequences. It is also necessary to modify the cDNAs that represent the 5′- and 3′-ends of the coronavirus genome. Normally, a transcription promoter sequence for the bacteriophage T7 RNA polymerase is positioned upstream of the coronavirus genome, and a (unique) restriction site followed by the hepatitis δ ribozyme is placed downstream of the poly-A tail of the coronavirus genome. The terminal cDNA constructs must also have appropriate *Eag*I or *Bsp*120I restriction sites to facilitate cloning into a unique *Not*I restriction site present in the genomic DNA of vaccinia virus, strain v*Not*I/tk (Merchlinsky & Moss, 1992).

The cloning and propagation of the coronavirus genomic cDNA in vaccinia virus vectors. The next stage is to ligate, *in vitro*, the coronavirus cDNA fragments and the long and short arms of *Not*I-cleaved v*Not*I/tk genomic DNA. This ligation is done in the presence of *Not*I to prevent religation of the vaccinia virus DNA. Subsequently, the ligation reaction is transfected into mammalian cells that have been previously infected with fowlpox virus. Recombinant vaccinia virus, the genome of which includes a full-length copy of the coronavirus genome, is then rescued.

Rescue of recombinant coronaviruses. Essentially, recombinant coronaviruses are rescued by generating genomic-length RNA transcripts from the coronavirus component of the recombinant vaccinia virus DNA template. These transcripts are then transfected into permissive cells. The transcription reaction is normally done *in vitro* but it is also possible to rescue recombinant coronaviruses via the transcription of template DNA in the permissive cell itself. The ability to rescue recombinant coronaviruses is significantly enhanced by (but not dependent upon) the directed expression of the coronavirus nucleocapsid protein in the transfected cells.

In addition to use of recombinant vaccinia viruses as cloning vectors, vaccinia virus-mediated homologous recombination can be used to introduce specific mutations into the coronavirus genomic cDNA during its propagation in vaccinia virus (Coley *et al.*, 2005). Once an infectious coronavirus cDNA has been obtained, this element of the reverse genetic approach is actually the rate-limiting step, and it is therefore imperative that a rapid and easy procedure is available.

CONCLUSIONS

Recent advances in understanding the molecular and cellular biology of coronaviruses, together with the development of robust reverse genetic systems, herald a new era in the application of molecular approaches to the prevention and control of coronavirus-related disease. This will include diseases of humans, for example severe acute respiratory syndrome (SARS), diseases of livestock, for example infectious bronchitis in chickens, and diseases of companion animals, for example FIP. Clearly, there is a long way to go, but at least the route is now well sign-posted.

REFERENCES

Addie, D. D. (2000). Clustering of feline coronaviruses in multicat households. *Vet J* **159**, 8–9.

Addie, D. D. & Jarrett, J. O. (1992). Feline coronavirus antibodies in cats. *Vet Rec* **131**, 202–203.

Addie, D. D. & Jarrett, O. (2001). Use of a reverse-transcriptase polymerase chain reaction for monitoring the shedding of feline coronavirus by healthy cats. *Vet Rec* **148**, 649–653.

Addie, D. D., Toth, S., Murray, G. D. & Jarrett, O. (1995). Risk of feline infectious peritonitis in cats naturally infected with feline coronavirus. *Am J Vet Res* **56**, 429–434.

Almazan, F., Gonzalez, J. M., Penzes, Z., Izeta, A., Calvo, E., Plana-Duran, J. & Enjuanes, L. (2000). Engineering the largest RNA virus genome as an infectious bacterial artificial chromosome. *Proc Natl Acad Sci U S A* **97**, 5516–5521.

Barlough, J. E., Johnson-Lussenburg, C. M., Stoddart, C. A., Jacobson, R. H. & Scott, F. W. (1985). Experimental inoculation of cats with human coronavirus 229E and subsequent challenge with feline infectious peritonitis virus. *Can J Comp Med* **49**, 303–307.

Barr, M. C. (1996). FIV, FeLV, and FIPV: interpretation and misinterpretation of serological test results. *Semin Vet Med Surg (Small Anim)* **11**, 144–153.

Benetka, V., Kubber-Heiss, A., Kolodziejek, J., Nowotny, N., Hofmann-Parisot, M. & Mostl, K. (2004). Prevalence of feline coronavirus types I and II in cats with histopathologically verified feline infectious peritonitis. *Vet Microbiol* **99**, 31–42.

Casais, R., Thiel, V., Siddell, S. G., Cavanagh, D. & Britton, P. (2001). Reverse genetics system for the avian coronavirus infectious bronchitis virus. *J Virol* **75**, 12359–12369.

Coley, S. E., Lavi, E., Sawicki, S. G., Fu, L., Schelle, B., Karl, N., Siddell, S. G. & Thiel, V. (2005). Recombinant mouse hepatitis virus strain A59 from cloned, full-length cDNA replicates to high titers in vitro and is fully pathogenic in vivo. *J Virol* (in press).

Fehr, D., Holznagel, E., Bolla, S., Hauser, B., Herrewegh, A. A., Horzinek, M. C. & Lutz, H. (1997). Placebo-controlled evaluation of a modified life virus vaccine against feline infectious peritonitis: safety and efficacy under field conditions. *Vaccine* **15**, 1101–1109.

Foley, J. E., Poland, A., Carlson, J. & Pedersen, N. C. (1997a). Patterns of feline coronavirus infection and fecal shedding from cats in multiple-cat environments. *J Am Vet Med Assoc* **210**, 1307–1312.

Foley, J. E., Poland, A., Carlson, J. & Pedersen, N. C. (1997b). Risk factors for feline infectious peritonitis among cats in multiple-cat environments with endemic feline enteric coronavirus. *J Am Vet Med Assoc* **210**, 1313–1318.

Gerber, J. D. (1995). Overview of the development of a modified, live temperature-sensitive FIP vaccine. *Feline Pract* **23**, 62–66.

Gerber, J. D., Ingersoll, J. D., Gast, A. M., Christianson, K. K., Selzer, N. L., Landon, R. M., Pfeiffer, N. E., Sharpee, R. L. & Beckenhauer, W. H. (1990). Protection against feline infectious peritonitis by intranasal inoculation of a temperature-sensitive FIPV vaccine. *Vaccine* **8**, 536–542.

Glansbeek, H. L., Haagmans, B. L., te Lintelo, E. G., Egberink, H. F., Duquesne, V., Aubert, A., Horzinek, M. C. & Rottier, P. J. (2002). Adverse effects of feline IL-12 during DNA vaccination against feline infectious peritonitis virus. *J Gen Virol* **83**, 1–10.

Gunn-Moore, D. A., Caney, S. M., Gruffydd-Jones, T. J., Helps, C. R. & Harbour, D. A. (1998). Antibody and cytokine responses in kittens during the development of feline infectious peritonitis (FIP). *Vet Immunol Immunopathol* **65**, 221–242.

Gut, M., Leutenegger, C. M., Huder, J. B., Pedersen, N. C. & Lutz, H. (1999). One-tube fluorogenic reverse transcription-polymerase chain reaction for the quantitation of feline coronaviruses. *J Virol Methods* **77**, 37–46.

Haagmans, B. L., Egberink, H. F. & Horzinek, M. C. (1996). Apoptosis and T-cell depletion during feline infectious peritonitis. *J Virol* **70**, 8977–8983.

Haijema, B. J., Volders, H. & Rottier, P. J. (2003). Switching species tropism: an effective way to manipulate the feline coronavirus genome. *J Virol* **77**, 4528–4538.

Haijema, B. J., Volders, H. & Rottier, P. J. (2004). Live, attenuated coronavirus vaccines through the directed deletion of group-specific genes provide protection against feline infectious peritonitis. *J Virol* **78**, 3863–3871.

Herrewegh, A. A., de Groot, R. J., Cepica, A., Egberink, H. F., Horzinek, M. C. & Rottier, P. J. (1995). Detection of feline coronavirus RNA in feces, tissues, and body fluids of naturally infected cats by reverse transcriptase PCR. *J Clin Microbiol* **33**, 684–689.

Herrewegh, A. A., Mahler, M., Hedrich, H. J., Haagmans, B. L., Egberink, H. F., Horzinek, M. C., Rottier, P. J. & de Groot, R. J. (1997). Persistence and evolution of feline coronavirus in a closed cat-breeding colony. *Virology* **234**, 349–363.

Herrewegh, A. A., Smeenk, I., Horzinek, M. C., Rottier, P. J. & de Groot, R. J. (1998). Feline coronavirus type II strains 79-1683 and 79-1146 originate from a double recombination between feline coronavirus type I and canine coronavirus. *J Virol* **72**, 4508–4514.

Hohdatsu, T., Okada, S., Ishizuka, Y., Yamada, H. & Koyama, H. (1992). The prevalence of types I and II feline coronavirus infections in cats. *J Vet Med Sci* **54**, 557–562.

Hohdatsu, T., Yamada, H., Ishizuka, Y. & Koyama, H. (1993). Enhancement and neutralization of feline infectious peritonitis virus infection in feline macrophages by neutralizing monoclonal antibodies recognizing different epitopes. *Microbiol Immunol* **37**, 499–504.

Hohdatsu, T., Tokunaga, J. & Koyama, H. (1994). The role of IgG subclass of mouse monoclonal antibodies in antibody-dependent enhancement of feline infectious peritonitis virus infection of feline macrophages. *Arch Virol* **139**, 273–285.

Hohdatsu, T., Izumiya, Y., Yokoyama, Y., Kida, K. & Koyama, H. (1998a). Differences in virus receptor for type I and type II feline infectious peritonitis virus. *Arch Virol* **143**, 839–850.

Hohdatsu, T., Yamada, M., Tominaga, R., Makino, K., Kida, K. & Koyama, H. (1998b). Antibody-dependent enhancement of feline infectious peritonitis virus infection in feline alveolar macrophages and human monocyte cell line U937 by serum of cats experimentally or naturally infected with feline coronavirus. *J Vet Med Sci* **60**, 49–55.

Hohdatsu, T., Yamato, H., Ohkawa, T., Kaneko, M., Motokawa, K., Kusuhara, H., Kaneshima, T., Arai, S. & Koyama, H. (2003). Vaccine efficacy of a cell lysate with recombinant baculovirus-expressed feline infectious peritonitis (FIP) virus nucleocapsid protein against progression of FIP. *Vet Microbiol* **97**, 31–44.

Hoskins, J. D. (1993). Coronavirus infection in cats. *Vet Clin North Am Small Anim Pract* **23**, 1–16.

Ivanov, K. A., Hertzig, T., Rozanov, M., Bayer, S., Thiel, V., Gorbalenya, A. E. & Ziebuhr, J. (2004). Major genetic marker of nidoviruses encodes a replicative endoribonuclease. *Proc Natl Acad Sci U S A* **101**, 12694–12699.

Lutz, H., Machler, M. R., Gut, M., Leutenegger, C. & Meli, M. (2002). FCoV shedding pattern of privately owned cats under field conditions. In *Second International FCoV/FIP Symposium*. Edited by D. D. Addie, M. Horzinek, O. Jarrett, H. Lutz & N. Pedersen. Glasgow: Concorde Services.

McArdle, F., Tennant, B., Bennett, M., Kelly, D. F., Gaskell, C. J. & Gaskell, R. M. (1995). Independent evaluation of a modified, live FIPV vaccine under experimental conditions. *Feline Pract* **23**, 67–72.

Merchlinsky, M. & Moss, B. (1992). Introduction of foreign DNA into the vaccinia virus genome by in vitro ligation: recombination-independent selectable cloning vectors. *Virology* **190**, 522–526.

Paltrinieri, S., Cammarata, M. P., Cammarata, G. & Comazzi, S. (1998). Some aspects of humoral and cellular immunity in naturally occuring feline infectious peritonitis. *Vet Immunol Immunopathol* **65**, 205–220.

Pedersen, N. C. (1976). Serologic studies of naturally occurring feline infectious peritonitis. *Am J Vet Res* **37**, 1449–1453.

Pedersen, N. C. (1987). Virologic and immunologic aspects of feline infectious peritonitis virus infection. *Adv Exp Med Biol* **218**, 529–550.

Pedersen, N. C. (1995). An overview of feline enteric coronavirus and infectious peritonitis virus infections. *Feline Pract* **23**, 7–20.

Pedersen, N. C. & Hawkins, K. F. (1995). Mechanisms for persistence of acute and chronic feline calicivirus infections in the face of vaccination. *Vet Microbiol* **47**, 141–156.

Poland, A. M., Vennema, H., Foley, J. E. & Pedersen, N. C. (1996). Two related strains of feline infectious peritonitis virus isolated from immunocompromised cats infected with a feline enteric coronavirus. *J Clin Microbiol* **34**, 3180–3184.

Scott, F. W., Corapi, W. V. & Olsen, C. W. (1995). Independent evaluation of a live, modified FIPV vaccine under experimental conditions. *Feline Pract* **23**, 74–76.

Siddell, S. G., Ziebuhr, J. & Snijder, E. J. (2005). Coronaviruses, toroviruses and arteriviruses. In *Topley and Wilson's Microbiology and Microbial Infections*, 10th edn. Edited by B. W. J. Mahy & V. ter Meulen. London: Edward Arnold (in press).

Snijder, E. J., Bredenbeek, P. J., Dobbe, J. C. & 7 other authors (2003). Unique and conserved features of genome and proteome of SARS-coronavirus, an early split-off from the coronavirus group 2 lineage. *J Mol Biol* **331**, 991–1004.

Sparkes, A. H., Gruffydd-Jones, T. J. & Harbour, D. A. (1991). Feline infectious peritonitis: a review of clinicopathological changes in 65 cases, and a critical assessment of their diagnostic value. *Vet Rec* **129**, 209–212.

Sparkes, A. H., Gruffydd-Jones, T. J. & Harbour, D. A. (1992). Feline coronavirus antibodies in UK cats. *Vet Rec* **131**, 223–224.

Stoddart, M. E., Whicher, J. T. & Harbour, D. A. (1988). Cats inoculated with feline infectious peritonitis virus exhibit a biphasic acute phase plasma protein response. *Vet Rec* **123**, 622–624.

Thiel, V., Herold, J., Schelle, B. & Siddell, S. G. (2001). Infectious RNA transcribed *in vitro* from a cDNA copy of the human coronavirus genome cloned in vaccinia virus. *J Gen Virol* **82**, 1273–1281.

Tresnan, D. B. & Holmes, K. V. (1998). Feline aminopeptidase N is a receptor for all group I coronaviruses. *Adv Exp Med Biol* **440**, 69–75.

Vennema, H., de Groot, R. J., Harbour, D. A., Dalderup, M., Gruffydd-Jones, T., Horzinek, M. C. & Spaan, W. J. (1990). Early death after feline infectious peritonitis virus challenge due to recombinant vaccinia virus immunization. *J Virol* **64**, 1407–1409.

Vennema, H., de Groot, R. J., Harbour, D. A., Horzinek, M. C. & Spaan, W. J. (1991). Primary structure of the membrane and nucleocapsid protein genes of feline infectious peritonitis virus and immunogenicity of recombinant vaccinia viruses in kittens. *Virology* **181**, 327–335.

Vennema, H., Poland, A., Foley, J. & Pedersen, N. C. (1998). Feline infectious peritonitis viruses arise by mutation from endemic feline enteric coronaviruses. *Virology* **243**, 150–157.

Wasmoen, T. L., Kadakia, N. P., Unfer, R. C., Fickbohm, B. L., Cook, C. P., Chu, H. J. & Acree, W. M. (1995). Protection of cats from infectious peritonitis by vaccination with a recombinant raccoon poxvirus expressing the nucleocapsid gene of feline infectious peritonitis virus. *Adv Exp Med Biol* **380**, 221–228.

Welsh, R. M., Haspel, M. V., Parker, D. C. & Holmes, K. V. (1986). Natural cytotoxicity against mouse hepatitis virus-infected cells. II. A cytotoxic effector cell with a B lymphocyte phenotype. *J Immunol* **136**, 1454–1460.

Yount, B., Curtis, K. M. & Baric, R. S. (2000). Strategy for systematic assembly of large RNA and DNA genomes: transmissible gastroenteritis virus model. *J Virol* **74**, 10600–10611.

Ziebuhr, J., Snijder, E. J. & Gorbalenya, A. E. (2000). Virus-encoded proteinases and proteolytic processing in the *Nidovirales*. *J Gen Virol* **81**, 853–879.

The transmissible spongiform encephalopathies

Jean C. Manson and Rona M. Barron

Institute for Animal Health, Neuropathogenesis Unit, Ogston Building, West Mains Road,
Edinburgh EH9 3JF, UK

INTRODUCTION

The transmissible spongiform encephalopathies (TSEs) have been a major focus for concern and for research activity since the emergence of bovine spongiform encephalopathy (BSE) in the late 1980s. Nearly two decades later, how has our understanding of these diseases progressed?

The TSEs are neurodegenerative diseases which affect a variety of mammalian species. The most studied of these diseases are scrapie in sheep, BSE in cattle, Creutzfeldt–Jakob disease (CJD) in humans and more recently chronic wasting disease (CWD) in deer. They are all characterized by long asymptomatic phases post-infection which can last for months or years. While incubation times of several years have been observed with CJD which has been acquired as a result of inoculation with contaminated growth hormone, incubation times of up to 40 years have been observed in some cases of Kuru, a TSE acquired through ritualistic cannibalistic practices of the Fore tribe in New Guinea. The long asymptomatic phase is followed by a short clinical phase displaying a variety of neurological abnormalities, which include ataxia and dementia, finally resulting in death, which is thought to be due to neuronal loss. However, while neuronal loss is clearly evident in microscopic examination of many animals terminally affected with TSEs, in others it is not always observed. Another hallmark of these diseases is the presence of vacuoles in the brain of an affected individual, from which the term spongiosis is derived. The exact origin of these vacuoles is not clear although it has been hypothesized that they are the product of apoptotic neurons.

SGM symposium 64: Molecular pathogenesis of virus infections.
Editors P. Digard, A. A. Nash & R. E. Randall. Cambridge University Press. ISBN 0 521 83248 9 ©SGM 2005

One of the most distinctive features of these diseases, however, is the deposition in the brain of the PrP protein (Prusiner, 1996). In the uninfected individual, PrP is a protease-sensitive sialoglycoprotein anchored to the cell membrane through a GPI (glycosylphosphatidylinositol) anchor (PrP^C) (Stahl *et al.*, 1990). During the course of disease there is a conformational change in PrP^C, resulting in the accumulation of an abnormal, protease-resistant isoform (PrP^{Sc}). It has been proposed that PrP^{Sc} is the infectious agent (Prusiner, 1982) and is also responsible for the neurotoxicity in these diseases, but the precise role of PrP^{Sc} in disease still remains to be established.

The accumulation of PrP^{Sc} during disease can vary from diffuse deposits to amyloid plaques in the brain of an infected individual. With some TSE agents, e.g. sheep scrapie and variant CJD (vCJD), deposition of PrP^{Sc} can also be seen in a variety of peripheral organs (Keulen *et al.*, 1996; Hilton *et al.*, 2002), whereas with other agents and hosts (e.g. BSE in cattle) PrP^{Sc} deposition appears more restricted, while BSE in sheep shows a similar widespread distribution of PrP^{Sc} to that seen in vCJD (Foster *et al.*, 2001). While PrP^{Sc} deposition, vacuolar pathology and neuronal loss are often found in the same region of the infected brain, this is not always the case, and in some instances, PrP^{Sc} deposition can be detected early on in the disease process before the appearance of neuronal damage (Jeffrey *et al.*, 2001). In other cases, PrP^{Sc} deposition is not always detected in a terminally infected brain despite the presence of vacuolar pathology and neuronal loss (Lasmezas *et al.*, 1997; Manson *et al.*, 1999). The exact relationship between PrP deposition, vacuolation and neuronal loss therefore remains to be established.

Disease-associated forms of PrP can also be detected by protease treatment of an infectious tissue homogenate followed by immunoblot analysis using PrP-specific antibodies. These protease-resistant forms of PrP are detected in most tissues which carry infectivity and are generally referred to as PrP-res to denote their protease resistance. The three bands of PrP-res detected correspond to the three different glycoforms of PrP, produced by differential *N*-linked glycosylation at amino acids 180 and 196 (Figs 1 and 2).

DECLINING AND EMERGING TSEs

BSE

Since 1986, when BSE was first recognized, around 180 000 cattle in the UK have developed the disease and many more animals are predicted to have been infected with BSE but slaughtered prior to the onset of the clinical phase (Anderson *et al.*, 1996; Donnelly *et al.*, 2002). The origin of BSE is not known, although the epidemic was likely to have been maintained and spread through the practice of recycling infected

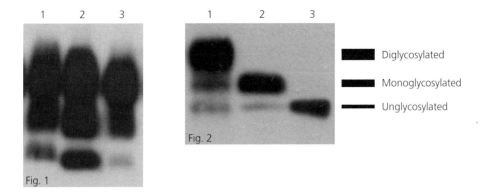

Fig. 1. Identification of BSE profile by immunostaining. The PrP glycoform of BSE is consistent when analysed by SDS-PAGE and immunostaining and is different from that obtained with isolates of natural scrapie. Lanes: 1, natural scrapie isolate; 2, BSE isolate; 3, experimental BSE in sheep. Blot probed with monoclonal antibody 6H4.

Fig. 2. Glycosylation of PrP. *N*-Linked glycosylation of PrP at amino acids 180 and 196 results in the production of three glycoforms of PrP: diglycosylated, where both sites are occupied; monoglycosylated, where one of the sites is occupied; and unglycosylated, where neither site is occupied. The glycosylation of PrP can be altered in transgenic mice by targeting polymorphisms into the *N*-linked glycosylation sites of the endogenous murine PrP gene. This results in the production of PrP lacking one or both sets of glycans. Lanes: 1, wild-type PrP; 2, PrP lacking glycosylation at amino acid 196; 3, PrP lacking glycosylation at both amino acids 180 and 196.

bovine tissues back to cows through cattle feed (Wilesmith *et al.*, 1991). The epidemic peaked in 1992, but since that time, due to the control measures which were put in place banning the feeding of ruminant protein to ruminants, the epidemic has consistently declined. However, some cases continued to occur in animals born after the ban was put in place, leading to a more extensive ban on feeding mammalian protein to any farmed animal in 1996. A small number of cases in animals born after the 1996 ban have been attributed to leakage of contaminated feed, although other routes of transmission cannot be ruled out (Wilesmith, 2002).

BSE has now been identified in most European countries and in a number of non-European countries. The highest numbers of cases outside the UK have been reported in the Republic of Ireland, France and Portugal. A number of countries, including Canada, USA, Finland, Greece and Israel, have reported only a single case. Since control methods were enforced later in other European countries, the peaks of these epidemics appeared later than in the UK, although most European countries are now showing a decline in the annual number of cases (Smith & Bradley, 2003).

Of considerable concern is the possibility that BSE may have been passed to sheep. Several studies have concluded from limited data available that if BSE is present in UK sheep it will be at a low level (Baylis, 2002; Ferguson *et al.*, 2002; Kao *et al.*, 2002). If it has been transmitted to sheep, however, it is important to establish whether it is likely to behave like sheep scrapie, which is not apparently transmissible to humans, or like BSE, for which there is considerable evidence of transmission to humans (Bruce *et al.*, 1997; Hill *et al.*, 1997). There is thus extensive surveillance being carried out for the presence of BSE in sheep. One method being used in this surveillance is immunoblot analysis of infected brain using PrP-specific antibodies, since PrP-res associated with BSE has a remarkably consistent size and ratio of glycoforms which differs from that associated with known sheep scrapie strains (Fig. 1).

vCJD

In 1996, the first case of vCJD was announced (Will *et al.*, 1996) and the linkage of vCJD with BSE was subsequently confirmed through techniques based on strain typing experiments in mice and the gel banding patterns of disease-associated PrP (Bruce *et al.*, 1997; Hill *et al.*, 1997). vCJD differs from sporadic CJD (sCJD) most notably on its early age of onset (Will *et al.*, 1996). An age-related susceptibility has been demonstrated for some TSE agents in mice although the basis for this susceptibility is not yet understood (Outram *et al.*, 1973). Thus factors other than diet may contribute to the unusual early onset of vCJD. The clinical presentation of vCJD differs from that of sCJD, often presenting first with psychiatric symptoms including behavioural changes, anxiety and depression (Zeidler *et al.*, 1997c). This is followed within weeks or months by a cerebellar ataxia and myoclonus. Memory disturbances and severe cognitive impairment appear later in the course of disease, resulting finally in akinetic mutism (Zeidler *et al.*, 1997b). Pathologically vCJD also differs from sCJD with characteristic amyloid plaques surrounded by vacuoles being detected throughout the cerebrum and cerebellum (Will *et al.*, 1996; Ironside & Bell, 1997; Ironside, 1998). These plaques have been termed 'florid plaques' and similar plaques have been detected in experimental passage of vCJD to mice (M. Bishop, unpublished results).

To date there have been 150 clinical cases of vCJD, all in individuals homozygous for the 129M polymorphism in PrP. Gene-targeted transgenic mice homozygous for either human 129M or 129V *PRNP* genes infected with vCJD have demonstrated that 129VV individuals may also be susceptible to vCJD (M. Bishop, unpublished results). Since many more individuals are likely to have been exposed to BSE than have succumbed to clinical disease there is also concern that there may be a carrier status, where individuals carry infectivity without showing clinical signs of disease, and that such individuals may be able to pass the disease on to others. Carrier status is an important issue not only for vCJD but throughout TSEs, particularly with sheep scrapie. The National Sheep Scrapie plan aims to eradicate scrapie by eliminating the most susceptible genotypes of sheep from the national herd, but it is not at present known whether scrapie carriers will remain and have the potential to transmit disease.

Whether such carrier states exist and their ability to pass disease on are therefore important questions yet to be answered.

More recently, two cases of possible human-to-human transmission of vCJD by blood transfusion have been reported (Llewelyn *et al.*, 2004; Peden *et al.*, 2004). While blood transmission of vCJD had been predicted from experimental studies in sheep, which had demonstrated that blood is an efficient route of transmission of both BSE and sheep scrapie (Houston *et al.*, 2000; Hunter *et al.*, 2002), it was unexpected that the second of these human cases appeared in an individual carrying 129MV PrP. From previous studies of iatrogenic CJD cases, 129MV heterozygotes might have been expected to be more resistant or have considerably longer incubation times of disease than either of the homozygous 129MM or 129VV individuals. However, clinical disease was not present in the 129MV case, suggesting that longer incubation times may indeed occur in these individuals, raising further concerns over the extent of sub-clinical disease in the population and the ability of such individuals to transmit disease.

CWD

CWD was first identified as a fatal wasting disease of captive mule deer in the late 1960s in Colorado and was classified as a TSE in 1978 (Williams & Young, 1980). More recently, CWD has become a serious problem in both game farms and wild deer and elk populations in North America, being detected at >10 % incidence in certain wild deer populations in Colorado and Wyoming. The mechanism of transmission is not known; however, evidence supports lateral transmission through direct animal-to-animal contact or by indirect exposure to infectivity through contaminated feed or water sources (Miller & Williams, 2003). New foci are thought to have arisen through commerce in live farmed deer and elk. Wild ruminants with CWD are found on the same ranges as cattle and sheep, raising concerns for transmission of CWD to these species. Moreover, there is also the concern that CWD like BSE could cross the human species barrier. *In vitro* studies have indicated an inefficient conversion of human PrP to a protease-resistant form by CWD-infected brain homogenate, raising the possibility of low-level transmission to humans (Raymond *et al.*, 2000). Epidemiological studies to date have not indicated any link between CJD and CWD; however, such studies are limited. Studies involving transmission of CWD to transgenic mice expressing human and cervid PrP are in progress to assess the potential for transmission of CWD to humans.

DIAGNOSING TSEs

One of the major problems still surrounding TSEs is the absence of a pre-mortem diagnostic test. While clinical signs of disease can give strong evidence of not only the presence of TSE but also the strain of TSE in the case of human disease, final confirmation of disease is only obtained following post-mortem examination of brain tissue. Macroscopic examination of the brain reveals spongiform change, neuronal loss and gliosis as major hallmarks of disease, and the deposition of PrP in the brain as amyloid plaques or diffuse deposits provides additional confirmation of TSE disease.

Although macroscopic examination can be used as a major diagnostic tool, the definitive confirmation of TSE disease is reliant on the direct measurement of infectivity by bioassay of tissue in mice (Bruce *et al.*, 1994, 1997, 2002). However, these assays take many months to perform, and are of limited use for large-scale screening programmes. Considerable effort has therefore gone into developing more rapid tests for detecting infectivity. Transgenic mice overexpressing a bovine PrP gene have allowed somewhat more rapid testing for vCJD and BSE by bioassay, but incubation times of disease are still over 200 days (Scott *et al.*, 1999; Buschmann *et al.*, 2000). A more rapid system for assaying infectivity has been developed using a tissue culture system in which the production of PrP^{Sc} is measured in cells infected with brain homogenate (Klohn *et al.*, 2003). While in this assay PrP^{Sc} production appears to correlate with infectivity, as measured by the mouse bioassay system, the cell line used in the assay can only be infected with a single strain of mouse scrapie. Thus at present this assay is unsuitable as a more general diagnostic tool, and it remains to be established whether a close correlation between PrP^{Sc} production and infectivity will be obtained using other TSE strains. If, however, other cell lines could be identified which can be infected with other TSE agents, such assay systems may become more widely applicable.

The limited ability of such cell lines to become infected with different TSE agents may be a useful research tool in understanding the susceptibility of particular cell populations to infectivity *in vivo*. While some cell lines such as the SMB line can be infected with a number of TSE strains (Birkett *et al.*, 2001), other cell lines appear to be very limited in their ability to take up and replicate different strains of infectivity (Klohn *et al.*, 2003). A study of the differences between such cell lines may indicate cellular factors important in sustaining a TSE infection and mechanisms that may be used to block the uptake or replication of infectivity.

The presence of PrP-res in brain tissue is widely accepted as a marker for the presence of TSE infectivity, and detection of disease-associated PrP-res has become the basis of all current diagnostic tests for TSE disease. Polyacrylamide-gel-based systems (e.g. Prionics Check WB) are designed to detect PrP-res in diseased tissue by immunoblot. The intensity of each of the three glycoforms (glycoform profile) and the molecular size of PrP-res detected by immunoblot analysis with specific anti-PrP antibodies has proved useful for the diagnosis of BSE and vCJD, since this particular strain of agent appears to retain its characteristic banding pattern on passage through different hosts.

Glycoform profile and molecular size of PrP-res using two different monoclonal antibodies to PrP may also be useful in distinguishing between sheep scrapie and sheep BSE and may thus provide an important diagnostic marker in screening for BSE in sheep (Thuring *et al.*, 2004). The glycoform pattern and molecular size of PrP in the case of other TSE strains, however, does not appear to provide as clear a diagnostic marker for the TSE strain, as some apparently unrelated strains of agent have similar glycoform patterns (Hope *et al.*, 1999). Indeed, it has been shown that several different PrP glycoform patterns can be present in a single brain (Somerville, 1999; Head *et al.*, 2004). There has been in recent years an increasing use of the analysis of PrP glycoform and molecular size by immunoblot analysis to define TSE strains (Collinge *et al.*, 1996; Parchi *et al.*, 1996, 1997, 1999), but while there is a lack of understanding of the relationship between PrP glycoform and the TSE strain, the method cannot be used extensively to define strains.

The need for surveillance for BSE and scrapie in Europe has led to the development of a number of rapid diagnostic tests which are based on either immunoblot or ELISA assay systems (Buschmann *et al.*, 2004b). Four rapid tests have passed the EU evaluation for BSE testing of cattle and are also currently recommended by the EU authorities for the testing of small ruminants. However, while all the rapid assay systems appear to be consistent for testing BSE in cattle, inconsistencies between the different assay systems have been observed in testing for sheep scrapie. The surveillance of sheep scrapie has suggested that it is more prevalent than previously estimated, and has also indicated that more strains of scrapie are present than have been previously recognized (Benestad *et al.*, 2003). In addition, a significant proportion of samples were found to be positive by one test and negative by another (Buschmann *et al.*, 2004b). It is not clear what the basis of this discrepancy is, but it is important to establish whether these tissues indeed harbour infectivity by bioassay in mice.

Understanding the exact relationship between disease-associated forms of PrP and infectivity is one of the major problems in assessing the reliability of PrP-res-based diagnostic assay systems. This issue has been further compounded by the identification of TSE disease in animals which do not appear to have a protease-resistant form of PrP (Lasmezas *et al.*, 1997; Manson *et al.*, 1999). Indeed, a model of disease has been shown to have high levels of infectivity in brain with undetectable levels of PrP-res by either immunoblot or ELISA-based assays, suggesting that PrP-res is not always a reliable diagnostic marker of disease (R. Barron, unpublished results).

Pre-mortem diagnosis of TSE disease would require the identification of infectivity in a readily accessible tissue or body fluid, and considerable effort has been put into assaying for disease-associated PrP in blood and serum. It has been assumed, based on comparison of PrP^{Sc} in the lymphoreticular system and brain tissue, that the levels of infectivity in blood are likely to be considerably lower than those seen in brain (Wadsworth *et al.*, 2001), and methods have therefore been developed to enrich for PrP^{Sc}, and to produce high-specificity antibodies for capture of PrP^{Sc} (Korth *et al.*, 1997; Wadsworth *et al.*, 2001; Paramithiotis *et al.*, 2003). One assay system that claims to display a level of sensitivity similar to bioassay and measure low levels of disease-associated PrP is the conformation-dependent immunoassay. This assay does not rely on the detection of PrP-res, but instead identifies disease-associated PrP by antibody binding to native and denatured forms of PrP (Safar *et al.*, 2002). The sensitivity of this assay system has not yet been validated in blood, and it will also be important to establish the relationship between the disease-associated PrP detected in this assay system and TSE infectivity.

Experimental transmission of TSEs by blood transfusion in sheep (Houston *et al.*, 2000; Hunter *et al.*, 2002) and the apparent transmission of vCJD by blood transfusion (Llewelyn *et al.*, 2004; Peden *et al.*, 2004) suggest that the levels of infectivity in blood may not be as low as previously estimated. The infectivity in blood may not therefore correlate with levels of PrP^{Sc} but may rather be associated with either a different molecular form of PrP or indeed molecular species other than PrP. Defining the exact molecular nature of the infectious agent in blood is central to producing a reliable pre-mortem diagnostic test.

In the absence of a PrP-based pre-mortem test, a number of approaches have been taken to identify other surrogate markers of disease. Erythroid-associated factor (ERAF) was identified as a candidate blood marker for preclinical diagnosis of TSE based on a reduced level of ERAF mRNA in haematopoietic tissue of rodent TSE, sheep scrapie and BSE (Miele *et al.*, 2001). Studies are currently in progress to assess its usefulness as a diagnostic marker in a variety of TSEs. Other potential surrogate markers are being sought by microarray and proteomic approaches which identify specific disease markers by comparing infected and uninfected tissues. However, the markers which have been identified to date by these systems have yet to be validated. It may be that a 'fingerprint' of a number of these surrogate markers could prove a useful diagnostic tool for TSEs.

DEFINING TSE STRAINS

Primary transmission of non-murine TSE isolates (e.g. scrapie, CJD) to mice usually results in long incubation times and low transmission rates. During subsequent serial mouse-to-mouse transmissions the incubation time generally shortens and stabilizes, and higher rates of transmission are observed (usually 100 %). The distribution of vacuoles in the brain (lesion profile) and other neuropathological features also generally stabilize, and these features become stable indefinitely on further passage in the same host. A TSE strain is defined on this set of characteristic properties (Fraser & Dickinson, 1973; Bruce *et al.*, 1991). Distinct TSE strains have been identified in mice by serially passaging infectious material from sheep, goats, cattle and humans. By using this method of strain typing, it has been clearly established that vCJD and BSE are the same TSE agent (Bruce *et al.*, 1997; Brown *et al.*, 2003) and vCJD was distinct from the cases of sCJD which were examined (Bruce *et al.*, 1997).

The prion or protein-only hypothesis defines the TSE infectious agent as a protease-resistant form of PrP which can self-replicate (Prusiner, 1982). The existence of multiple strains of TSE agents with different incubation times, clinical features and neuro-pathology has proved a challenge to the prion hypothesis. In other infectious diseases, strains arise from mutations or polymorphisms in the infectious agent's nucleic acid genome (DNA or RNA) which encodes the heritable strain-specific information of each agent. However, in order to accommodate strains in an infectious agent which may be composed entirely of protein, it has been proposed that each TSE strain represents a different stable conformation of abnormally folded PrP that can faithfully replicate and produce the diversity observed in disease (Telling *et al.*, 1996a; Safar *et al.*, 1998). This concept has been given some support by *in vitro* conversion assay systems in which the characteristics of PrP-res can be conferred to PrP^C, demonstrating that pathological forms of PrP have some capacity to propagate themselves (Bessen *et al.*, 1995). However, the failure of this cell-free conversion system to produce infectious material (Hill *et al.*, 1999) has hampered the ability to fully establish whether there is a link between TSE strain and PrP conformation. While TSE strains have been shown to be associated with differences in conformation (Safar *et al.*, 1998), degree of protease resistance (Tremblay *et al.*, 2004) and glycoform ratios (Collinge *et al.*, 1996; Parchi *et al.*, 1996), the exact relationship between these characteristics and the TSE strain is still to be defined.

The role of glycosylation of PrP as a basis for TSE strain determination and strain targeting is being addressed in experiments using transgenic mice (produced by gene targeting) which have altered patterns of PrP glycosylation (Fig. 2). Initial results indicate that the glycosylation pattern of host PrP can influence the incubation time of disease and the targeting of pathology and can modify the TSE strain as it is passaged through the host (N. Tuzi, unpublished results). Further analysis of these models will allow the precise relationship between TSE strain and PrP glycosylation to be established.

NEW TSE STRAINS

Extensive surveillance for BSE and scrapie in Europe and elsewhere has revealed the existence of a number of previously unrecognized TSE strains. Based on molecular

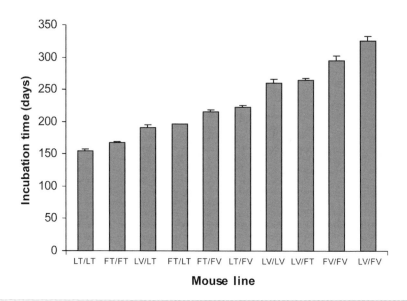

Fig. 3. Control of scrapie incubation time by codon 108 and 189 polymorphisms. Lines of gene-targeted transgenic mice were produced which expressed murine PrP-A containing either or both L108F and T189V polymorphisms. These mice (FT/FT, LV/LV and FV/FV) were crossed with each other, and with wild-type (LT/LT) mice to produce the full repertoire of genetic combinations at codons 108 and 189 on the same genetic background. Mice were challenged with ME7 scrapie. Disease incubation times are shown ± SEM.

strain typing and pathological examination of the brain, a new form of BSE has been identified in two older cattle in an Italian surveillance system (Casalone *et al.*, 2004). It has not yet been established whether this represents a newly diversified strain of TSE or whether this strain has existed for some time and been identified through the increased surveillance. The infectivity of the brains carrying this new form of PrP and the strain characteristics associated with it have still to be established by mouse bioassay. The ability of this strain to transmit to humans is also under investigation in transgenic mouse studies. New strains of cattle TSE have also been reported by groups in France and Japan (Yamakawa *et al.*, 2003; Biacabe *et al.*, 2004).

During the surveillance of sheep at slaughter a number of atypical cases of scrapie have been identified in Germany, France (Buschmann *et al.*, 2004b) and Norway (Benestad *et al.*, 2003). These cases demonstrate a molecular profile and distribution of PrP^Sc not previously recognized in sheep scrapie. Atypical cases have also appeared in sheep carrying a PrP genotype which was previously thought to have been resistant to scrapie (Buschmann *et al.*, 2004a), although confirmation that these cases carry infectivity remains to be established by mouse bioassay. However, BSE has been experimentally transmitted to sheep of this resistant genotype (Houston *et al.*, 2003), raising concern that no sheep are truly resistant to disease and that new scrapie strains may propagate in sheep previously thought to be resistant to disease.

HOST SUSCEPTIBILITY TO DISEASE

It has been recognized for many years that polymorphisms and mutations in host PrP can influence the incubation time of disease or the susceptibility of the host to disease. The L108T and/or F189V polymorphisms in murine PrP were shown to be the major factor influencing incubation time of scrapie in mice (Moore *et al.*, 1998), and more recently it has been demonstrated that while the 189 polymorphism has the major influence over incubation time, homozygosity at codon 108 leads to shorter incubation times than heterozygosity (Barron *et al.*, 2005) (Fig. 3). The influence of homozygosity in the N-terminus of PrP can also be demonstrated in CJD, where the majority of cases occur in individuals homozygous for methionine or valine at codon 129 (Palmer *et al.*, 1991; Zeidler *et al.*, 1997a; Alperovitch *et al.*, 1999). In sheep, the V136A, R154H and R171QH polymorphisms in sheep PrP have been used to establish a scale of susceptibility to scrapie, where VRQ/VRQ animals are most at risk of developing scrapie and ARR/ARR are thought to be resistant to disease.

The mechanism by which the PrP gene influences host susceptibility is not understood. The prion hypothesis proposed that identity between host PrP and the PrP sequence in the donor of infectivity leads to short incubation times and high susceptibility of the host to infection. Indeed, this was demonstrated in early transgenic models where transgenic mice expressing multiple copies of the hamster PrP gene were shown to be more susceptible to a hamster strain of scrapie than wild-type mice (Scott *et al.*, 1989). However, subsequent transgenic experiments have demonstrated that the mechanism is very much more complex. While overexpression of a bovine PrP gene in transgenic mice leads to a model which develops disease rapidly when inoculated with BSE (Scott *et al.*, 1999), replacement of the murine PrP gene with a bovine PrP gene by gene targeting leads to longer incubation times in the transgenic mice than in wild-type (P. Hart, unpublished results) (Fig. 4). This increase in incubation time is also observed on inoculation of gene-targeted transgenic mice expressing human PrP with BSE and vCJD, despite the apparent sequence compatibility with vCJD (Fig. 4). However, isolates of sCJD have been shown to transmit efficiently to the human PrP transgenic mice (Fig. 4), suggesting that the observed long incubation periods with vCJD are not due to incompatibility between some host factors and the transgene, but rather to a specific effect of the individual TSE strain. Thus increasing the sequence homology between host and donor PrP can increase or decrease incubation time but not in a predictable way. These experiments have also demonstrated a pronounced influence of the M129V polymorphism on transmission, with compatibility between host and donor PrP at codon 129 always leading to the shortest incubation times. Moreover, the

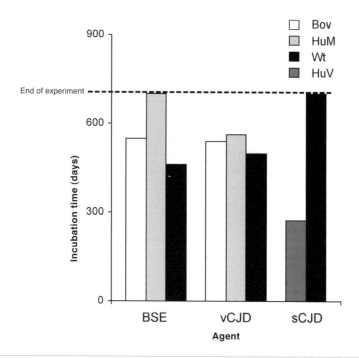

Fig. 4. Transmission of CJD and BSE to gene-targeted transgenic mice expressing human 129M (HuM), human 129V (HuV), bovine (Bov) and wild-type mouse (Wt) PrP genes. BSE transmitted to mice expressing wild-type mouse PrP-A with shorter incubation times than mice expressing bovine PrP. No transmission occurred in mice expressing human 129M PrP (HuM) 700 days post-inoculation (dotted line). Shorter incubation times were observed in wild-type mice than in both human 129M and bovine transgenic mice inoculated with vCJD. Type VV2A sCJD was transmitted efficiently to all mice expressing human 129V PrP (HuV), but no transmission occurred in wild-type 129/Ola mice 700 days post-inoculation. All lines were maintained on the same genetic background.

results have shown that M129V heterozygosity does not always protect the host from disease, as some CJD isolates produce the same susceptibility and incubation times in transgenic human PrP mice heterozygous and homozygous at codon 129 (M. Bishop, unpublished results).

The results of these experiments suggest that single amino acid alterations in PrP have a more pronounced effect on the susceptibility of a host to a particular agent than overall identity between host and donor PrP. Indeed, it has been known for some time that point mutations in the human PrP gene are associated with familial TSE, where the mutation itself is apparently sufficient to cause a spontaneous TSE to develop (Hsiao *et al.*, 1989; Petersen *et al.*, 1992; Kong *et al.*, 2004). It has been hypothesized that the mutations lead to an instability in PrP and thus make it more likely to misfold into the disease-associated form of PrP. This hypothesis was given considerable weight when it was demonstrated that overexpression of a murine PrP gene carrying a proline to

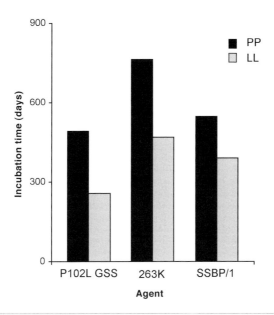

Fig. 5. Transmission of non-murine TSE agents to 101LL gene-targeted transgenic mice. Transmission of human P102L GSS, hamster 263K scrapie and sheep SSBP/1 all produced shortened incubation times in 101LL mice compared to 101PP (wild-type mice). Only P102L GSS contains the corresponding mutation in PrP. 263K and SSBP/1 are wild-type at this position.

leucine mutation at codon 101 [equivalent to the P102L mutation in human PrP associated with Gerstmann–Sträussler–Scheinker syndrome (GSS)] could produce a spontaneous neurodegenerative disease in mice, and that this disease could be passed on to other transgenic mice expressing a lower level of the same transgene (Hsiao *et al.*, 1990, 1994; Telling *et al.*, 1996b). While these experiments suggest that the mutation may be sufficient to cause TSE disease, there are some caveats to this, namely that the spontaneous disease was not associated with a protease-resistant form of PrP in the brain, and subsequent passage of the 'infectivity' was only possible in mice carrying the same transgene. Moreover, while spontaneous disease has been reported with the overexpression of a number of other PrP transgenes, including wild-type PrP (Westaway *et al.*, 1994; Perrier *et al.*, 2002; Tremblay *et al.*, 2004), none of these spontaneous diseases have transmitted to other mice.

However, when the same 101L mutation is introduced into the endogenous murine PrP gene by gene targeting, a somewhat different outcome is obtained. Mice heterozygous (101PL) or homozygous (101LL) for the transgene do not develop a spontaneous disease, nor can they transmit disease to either wild-type or 101LL transgenic mice (Manson *et al.*, 1999). While this result in itself may not be that surprising, since it could be anticipated that high expression levels of PrP are required to produce a spontaneous disease, what is remarkable about this mutation in the gene-targeted mice is the way in which it alters susceptibility to disease. With the exception of vCJD, most human TSE isolates are difficult to transmit to wild-type mice. However, 101LL mice show 100 % susceptibility and short incubation times when inoculated with infected brain homogenate from P102L GSS patients (Manson *et al.*, 1999) (Fig. 5). While this may demonstrate that donor and host identity at 101/102 increases the efficiency of transmission, a more surprising effect is seen when these mice are inoculated with SSBP/1 experimental sheep scrapie and 263K hamster scrapie (Barron *et al.*, 2001) (Fig. 5). These two agents, derived from hosts with proline at the equivalent position, have shorter incubation times in the 101LL mice than in the wild-type mice (101PP). The 101L mutation has therefore altered the susceptibility of the host to TSE agents from three different species, and the mechanism by which this is achieved is not dependent on overall PrP sequence compatibility between host and donor, or specifically at codon 101/102.

We are thus still some way from defining the mechanisms by which host PrP sequence influences susceptibility to disease, but it is hoped that *in vivo* studies using these well-defined PrP gene targeted models and *in vitro* studies using cell lines derived from these models will allow the mechanisms by which host PrP controls susceptibility to disease to be unravelled.

THE NATURE OF THE INFECTIOUS AGENT

The idea that a protein alone may be the source of infectivity in the TSE diseases was first put forward in 1967 (Gibbs, 1967; Griffith, 1967) and later developed as the prion hypothesis by Stanley Prusiner (Prusiner, 1982). Over three decades this hypothesis has gained much support, but the exact nature of what this elusive infectious agent is remains to be defined. While infectivity is generally associated with protease-resistant PrP in purified preparations from infected brain, there are some reports which claim to separate PrP and infectivity (Manuelidis *et al.*, 1987; Somerville & Dunn, 1996). The resistance of PrP knockout or null mice to TSE infectivity clearly demonstrates that PrP is required for the disease process but does not provide proof that PrP itself is the infectious agent (Bueler *et al.*, 1993; Manson *et al.*, 1994).

While the genetic linkage studies in human TSEs provide strong support for the prion hypothesis, the 102L mutation originally thought to be 100 % penetrant does not now appear to be so, with a recent report of monozygotic twins carrying the 102L mutation where one twin developed GSS and the other remained unaffected over 15 years later (Hamasaki *et al.*, 1998). Moreover, the expression of this mutation in murine PrP has shown a dramatic alteration in the susceptibility of 101LL mice to TSE infection, raising the possibility that the mutations are susceptibility factors rather than the primary cause of disease (Manson *et al.*, 1999; Barron *et al.*, 2001, 2003). Indeed, sheep scrapie was originally considered to be a genetic disease, as certain PrP genotypes were always associated with the development of scrapie. However, the existence of such sheep in scrapie-free countries such as New Zealand has shown scrapie to be an infectious rather than a genetic disease (Hunter *et al.*, 1997), and embryo transfer experiments have shown that sheep which would normally be expected to develop disease can survive if conditions of high sterility are maintained around time of birth and thereafter (J. D. Foster, personal communication).

Cell-free conversion assays have demonstrated that PrP-res can induce transformation of PrP^C into a protease-resistant form (Kocisko *et al.*, 1994). Moreover, many aspects of the disease process such as the species barrier and faithful replication of the abnormal form of PrP could be demonstrated in this system (Kocisko *et al.*, 1995; Raymond *et al.*, 2000). However, a final link between PrP-res and infectivity has not been achieved since it has not been possible to demonstrate the generation of infectivity in this system (Hill *et al.*, 1999). *In vitro* conversion assays based on cyclic amplification of the misfolded protein have also demonstrated increased production of PrP-res but not to date an increase in infectivity (Saborio *et al.*, 2001; Soto *et al.*, 2002).

Much support for the prion hypothesis has been taken from the study of yeast prions first put forward in 1994 (Wickner, 1994) to explain non-Mendelian transmission of two yeast genetic elements, [URE3] and [PSI+], which were proposed to be abnormally folded forms of Ure2 and Sup35, respectively. Subsequently, other prion proteins have been described in yeast, and misfolded proteins HET-s (Maddelein *et al.*, 2002) and Sup35 have been demonstrated to show self-propagation in an uninfected cell by introduction of a specific amyloid form of the protein. Moreover, the converted protein can propagate the characteristics of different strains used in the reaction, adding strength to the hypothesis that prion strains represent different abnormally folded forms of the protein (King & Diaz-Avalos, 2004; Tanaka *et al.*, 2004). However, while these experiments elegantly demonstrated that the principle of a protein acting as an infectious agent is possible in yeast, they do not prove that this phenomenon is occurring in mammalian TSEs.

Recently it has been reported that an abnormally folded fragment of recombinant PrP (amino acids 89–230) produced disease when inoculated into mice expressing a transgene of the same truncated protein at 16 times the level of wild-type PrP (Legname *et al.*, 2004). Subsequent passage of brain tissue from these mice produced disease in both wild-type and transgenic mice. While this experiment continues to add evidence to the prion hypothesis, the absence of soluble PrP and other aggregates of PrP as controls in this study leaves the association of PrP and infectivity still unresolved. Moreover, the abnormally folded protein used in these experiments did not demonstrate the normal protease resistance associated with disease-specific PrP. In addition, as mentioned previously, there are a number of descriptions of TSE disease occurring in the absence of a protease-resistant form of PrP (Lasmezas *et al.*, 1997; Manson *et al.*, 1999). If, therefore, PrP is the TSE infectious agent, these models suggest it is a form other than PrP-res. Laboratory models which contain high levels of infectivity in the absence of PrP^{Sc} are now being used to reinvestigate the nature of the infectious agent. These studies will determine whether infectivity is associated solely with PrP, or if other factors/molecules are involved. If PrP is shown to be the agent, these models may allow the identification of the specific form of PrP which is associated with infectivity.

While considerable progress has been made in the study of TSEs since the emergence of BSE, there are still major gaps in our knowledge of the basic mechanisms of these diseases. The species barrier and mechanisms of host resistance and susceptibility are not understood. Moreover, we have yet to establish exactly what defines a TSE strain and what determines how that strain targets specific areas of the brain. Importantly, the exact nature of the infectious agent still remains to be determined. Our ability to diagnose, control, eradicate or treat these diseases is still dependent on providing answers to these fundamental questions.

ACKNOWLEDGEMENTS

The work described in this chapter has been supported by BBSRC, MRC, DEFRA and DH. We would like to thank Angie Chong for Fig. 1 and Jane Shaw for Fig. 2. We would also like to thank Nora Hunter for reviewing the manuscript.

REFERENCES

Alperovitch, A., Zerr, I., Pocchiari, M. & 7 other authors (1999). Codon 129 prion protein genotype and sporadic Creutzfeldt–Jakob disease. *Lancet* **353**, 1673–1674.

Anderson, R. M., Donnelly, C. A., Ferguson, N. M. & 12 other authors (1996). Transmission dynamics and epidemiology of BSE in British cattle. *Nature* **382**, 779–788.

Barron, R. M., Thomson, V., Jamieson, E., Melton, D. W., Ironside, J., Will, R. & Manson, J. C. (2001). Changing a single amino acid in the N-terminus of murine PrP alters TSE incubation time across three species barriers. *EMBO J* **20**, 5070–5078.

Barron, R. M., Thomson, V., King, D., Shaw, J., Melton, D. W. & Manson, J. C. (2003). Transmission of murine scrapie to P101L transgenic mice. *J Gen Virol* **84**, 3165–3172.

Barron, R. M., Baybutt, H. N., Tuzi, N. L., McCormack, J., King, D., Moore, R. C., Melton, D. W. & Manson, J. C. (2005). Polymorphisms at codons 108 and 189 in murine PrP play distinct roles in the control of scrapie incubation time. *J Gen Virol* **86**, 859–868.

Baylis, M. (2002). The BSE-susceptible proportion of UK sheep. *Vet Rec* **151**, 481–482.

Benestad, S. L., Sarradin, P., Thu, B., Schonheit, J., Tranulis, M. A. & Bratberg, B. (2003). Cases of scrapie with unusual features in Norway and designation of a new type, Nor98. *Vet Rec* **153**, 202–208.

Bessen, R. A., Kocisko, D. A., Raymond, G. J., Nandan, S., Lansbury, P. T. & Caughey, B. (1995). Non-genetic propagation of strain-specific properties of scrapie prion protein. *Nature* **375**, 698–700.

Biacabe, A. G., Laplanche, J. L., Ryder, S. & Baron, T. (2004). Distinct molecular phenotypes in bovine prion diseases. *EMBO Rep* **5**, 110–114.

Birkett, C. R., Hennion, R. M., Bembridge, D. A., Clarke, M. C., Chree, A., Bruce, M. E. & Bostock, C. J. (2001). Scrapie strains maintain biological phenotypes on propagation in a cell line in culture. *EMBO J* **20**, 3351–3358.

Brown, D. A., Bruce, M. E. & Fraser, J. R. (2003). Comparison of the neuropathological characteristics of bovine spongiform encephalopathy (BSE) and variant Creutzfeldt–Jakob disease (vCJD) in mice. *Neuropathol Appl Neurobiol* **29**, 262–272.

Bruce, M. E., McConnell, I., Fraser, H. & Dickinson, A. G. (1991). The disease characteristics of different strains of scrapie in Sinc congenic mouse lines: implications for the nature of the agent and host control of pathogenesis. *J Gen Virol* **72**, 595–603.

Bruce, M. E., McBride, P. A., Jeffrey, M. & Scott, J. R. (1994). PrP in pathology and pathogenesis in scrapie-infected mice. *Mol Neurobiol* **8**, 105–112.

Bruce, M. E., Will, R. G., Ironside, J. W. & 10 other authors (1997). Transmissions to mice indicate that 'new variant' CJD is caused by the BSE agent. *Nature* **389**, 498–501.

Bruce, M. E., Boyle, A., Cousens, S., McConnell, I., Foster, J., Goldmann, W. & Fraser, H. (2002). Strain characterization of natural sheep scrapie and comparison with BSE. *J Gen Virol* **83**, 695–704.

Bueler, H., Aguzzi, A., Sailer, A., Greiner, R. A., Autenried, P., Aguet, M. & Weissmann, C. (1993). Mice devoid of PrP are resistant to scrapie. *Cell* **73**, 1339–1347.

Buschmann, A., Pfaff, E., Reifenberg, K., Muller, H. M. & Groschup, M. H. (2000). Detection of cattle-derived BSE prions using transgenic mice overexpressing bovine PrP(C). *Arch Virol Suppl* 75–86.

Buschmann, A., Luhken, G., Schultz, J., Erhardt, G. & Groschup, M. H. (2004a). Neuronal accumulation of abnormal prion protein in sheep carrying a scrapie-resistant genotype (PrPARR/ARR). *J Gen Virol* **85**, 2727–2733.

Buschmann, A., Biacabe, A.-G., Ziegler, U., Bencsik, A., Madec, J.-Y., Erhardt, G., Luhken, G., Baron, T. & Groschup, M. H. (2004b). Atypical scrapie cases in Germany and France are identified by discrepant reaction patterns in BSE rapid tests. *J Virol Methods* **117**, 27–36.

Casalone, C., Zanusso, G., Acutis, P., Ferrari, S., Capucci, L., Tagliavini, F., Monaco, S. & Caramelli, M. (2004). Identification of a second bovine amyloidotic spongiform encephalopathy: molecular similarities with sporadic Creutzfeldt–Jakob disease. *Proc Natl Acad Sci U S A* **101**, 3065–3070.

Collinge, J., Sidle, K., Meads, J., Ironside, J. & Hill, A. (1996). Molecular analysis of prion strain variation and the etiology of new variant CJD. *Nature* **383**, 685–690.

Donnelly, C. A., Ferguson, N. M., Ghani, A. C. & Anderson, R. M. (2002). Implications of BSE infection screening data for the scale of the British BSE epidemic and current European infection levels. *Proc R Soc Lond Ser B Biol Sci* **269**, 2179–2190.

Ferguson, N. M., Ghani, A. C., Donnelly, C. A., Hagenaars, T. J. & Anderson, R. M. (2002). Estimating the human health risk from possible BSE infection of the British sheep flock. *Nature* **415**, 420–424.

Foster, J. D., Parnham, D. W., Hunter, N. & Bruce, M. (2001). Distribution of the prion protein in sheep terminally affected with BSE following experimental oral transmission. *J Gen Virol* **82**, 2319–2326.

Fraser, H. & Dickinson, A. G. (1973). Scrapie in mice. Agent-strain differences in the distribution and intensity of grey matter vacuolation. *J Comp Pathol* **83**, 29–40.

Gibbs, C. J., Jr (1967). Search for infectious etiology in chronic and subacute degenerative diseases of the central nervous system. *Curr Top Microbiol Immunol* **40**, 44–58.

Griffith, J. S. (1967). Self replication and scrapie. *Nature* **215**, 1043–1044.

Hamasaki, S., Shirabe, S., Tsuda, R., Yoshimura, T., Nakamura, T. & Eguchi, K. (1998). Discordant Gerstmann-Straussler-Scheinker disease in monozygotic twins. *Lancet* **352**, 1358–1359.

Head, M. W., Bunn, T. J. R., Bishop, M. T. & 9 other authors (2004). Prion protein heterogeneity in sporadic but not variant Creutzfeldt–Jakob disease: UK cases 1991–2002. *Ann Neurol* **55**, 851–859.

Hill, A. F., Desbruslais, M., Joiner, S., Sidle, K. C. L., Gowland, I., Collinge, J., Doey, L. J. & Lantos, P. (1997). The same prion strain causes vCJD and BSE. *Nature* **389**, 448–450.

Hill, A. F., Antoniou, M. & Collinge, J. (1999). Protease-resistant prion protein produced in vitro lacks detectable infectivity. *J Gen Virol* **80**, 11–14.

Hilton, D. A., Ghani, A. C., Conyers, L., Edwards, P., McCardle, L., Penney, M., Ritchie, D. & Ironside, J. W. (2002). Accumulation of prion proteins in tonsil and appendix: review of tissue samples. *Br Med J* **325**, 633–634.

Hope, J., Wood, S. C. E. R., Birkett, C. R., Chong, A., Bruce, M. E., Cairns, D., Goldmann, W., Hunter, N. & Bostock, C. J. (1999). Molecular analysis of ovine prion protein identifies similarities between BSE and an experimental isolate of natural scrapie, CH1641. *J Gen Virol* **80**, 1–4.

Houston, F., Foster, J. D., Chong, A., Hunter, N. & Bostock, C. J. (2000). Transmission of BSE by blood transfusion in sheep. *Lancet* **356**, 999–1000.

Houston, F., Goldmann, W., Chong, A., Jeffrey, M., Gonzalez, L., Foster, J., Parnham, D. & Hunter, N. (2003). BSE in sheep bred for resistance to infection. *Nature* **423**, 498.

Hsiao, K., Baker, H. F., Crow, T. J., Poulter, M., Owen, F., Terwilliger, J. D., Westaway, D., Ott, J. & Prusiner, S. B. (1989). Linkage of a prion protein missense variant to Gerstmann-Straussler Syndrome. *Nature* **338**, 342–345.

Hsiao, K. K., Scott, M., Foster, D., Groth, D. F., DeArmond, S. J. & Prusiner, S. B. (1990). Spontaneous neurodegeneration in transgenic mice with mutant prion protein. *Science* **250**, 1587–1590.

Hsiao, K. K., Groth, D., Scott, M. & 7 other authors (1994). Serial transmission in rodents of neurodegeneration from transgenic mice expressing mutant prion protein. *Proc Natl Acad Sci U S A* **91**, 9126–9130.

Hunter, N., Cairns, D., Foster, J. D., Smith, G., Goldmann, W. & Donnelly, K. (1997). Is scrapie solely a genetic disease? *Nature* **386**, 137.

Hunter, N., Foster, J., Chong, A., McCutcheon, S., Parnham, D., Eaton, S., MacKenzie, C. & Houston, F. (2002). Transmission of prion diseases by blood transfusion. *J Gen Virol* **83**, 2897–2905.

Ironside, J. W. (1998). Neuropathological findings in new variant CJD and experimental transmission of BSE. *FEMS Immunol Med Microbiol* **21**, 91–95.

Ironside, J. & Bell, J. (1997). Florid plaques and new variant CJD. *Lancet* **350**, 1475.

Jeffrey, M., Martin, S., Barr, J., Chong, A. & Fraser, J. R. (2001). Onset of accumulation of PrPres in murine ME7 scrapie in relation to pathological and PrP immunohistochemical changes. *J Comp Pathol* **124**, 20–28.

Kao, R. R., Gravenor, M. B., Baylis, M., Bostock, C. J., Chihota, C. M., Evans, J. C., Goldmann, W., Smith, A. J. A. & McLean, A. R. (2002). The potential size and duration of an epidemic of bovine spongiform encephalopathy in British sheep. *Science* **295**, 332–335.

King, C. Y. & Diaz-Avalos, R. (2004). Protein-only transmission of three yeast prion strains. *Nature* **428**, 319–323.

Klohn, P. C., Stoltze, L., Flechsig, E., Enari, M. & Weissmann, C. (2003). A quantitative, highly sensitive cell-based infectivity assay for mouse scrapie prions. *Proc Natl Acad Sci U S A* **100**, 11666–11671.

Kocisko, D. A., Come, J. H., Priola, S. A., Chesebro, B., Raymond, G. J., Lansbury, P. T. & Caughey, B. (1994). Cell-free formation of protease-resistant prion protein. *Nature* **370**, 471–474.

Kocisko, D. A., Priola, S. A., Raymond, G. J., Chesebro, B., Lansbury, P. T. & Caughey, B. (1995). Species-specificity in the cell-free conversion of prion protein to protease-resistant forms – a model for the scrapie species barrier. *Proc Natl Acad Sci U S A* **92**, 3923–3927.

Kong, Q., Surewicz, K., Petersen, R. B. & 9 other authors (2004). Inherited prion diseases. In *Prion Biology and Diseases*, 2nd edn, pp. 673–775. Edited by S. B. Pruisner. Cold Spring Harbor, NY: Cold Spring Harbor Laboratory.

Korth, C., Stierli, B., Streit, P. & 14 other authors (1997). Prion (PrPSc)-specific epitope defined by a monoclonal antibody. *Nature* **390**, 74–77.

Lasmezas, C. I., Deslys, J., Robain, O. & 7 other authors (1997). Transmission of the BSE agent to mice in the absence of detectable abnormal prion protein. *Science* **275**, 402–405.

Legname, G., Baskakov, I. V., Nguyen, H. O. B., Riesner, D., Cohen, F. E., DeArmond, S. J. & Prusiner, S. B. (2004). Synthetic mammalian prions. *Science* **305**, 673–676.

Llewelyn, C. A., Hewitt, P. E., Knight, R. S. G., Amar, K., Cousens, S., Mackenzie, J. & Will, R. G. (2004). Possible transmission of variant Creutzfeldt–Jakob disease by blood transfusion. *Lancet* **363**, 417–421.

Maddelein, M. L., Reis, S. D., Duvezin-Caubet, S., Coulary-Salin, B. & Saupe, S. J. (2002). Amyloid aggregates of the HET-s prion protein are infectious. *Proc Natl Acad Sci U S A* **99**, 7402–7407.

Manson, J. C., Clarke, A. R., McBride, P. A., McConnell, I. & Hope, J. (1994). PrP gene dosage determines the timing but not the final intensity or distribution of lesions in scrapie pathology. *Neurodegeneration* **3**, 331–340.

Manson, J. C., Jamieson, E., Baybutt, H. & 10 other authors (1999). A single amino acid alteration (101L) introduced into murine PrP dramatically alters incubation time of transmissible spongiform encephalopathy. *EMBO J* **18**, 6855–6864.

Manuelidis, L., Sklaviadis, T. & Manuelidis, E. E. (1987). Evidence suggesting that PrP is not the infectious agent in Creutzfeldt–Jakob disease. *EMBO J* **6**, 341–347.

Miele, G., Manson, J. & Clinton, M. (2001). A novel erythroid-specific marker of transmissible spongiform encephalopathies. *Nat Med* **7**, 361–364.

Miller, M. W. & Williams, E. S. (2003). Horizontal prion transmission in mule deer: the gathering of deer during winter may foster the spread of chronic wasting disease. *Nature* **425**, 35–36.

Moore, R. C., Hope, J., McBride, P. A., McConnell, I., Selfridge, J., Melton, D. W. & Manson, J. C. (1998). Mice with gene targetted prion protein alterations show that Prnp, Sinc and Prni are congruent. *Nat Genet* **18**, 118–125.

Outram, G. W., Dickinson, A. G. & Fraser, H. (1973). Developmental maturation of susceptibility to scrapie in mice. *Nature* **241**, 536–537.

Palmer, M. S., Dryden, A. J., Hughes, J. T. & Collinge, J. (1991). Homozygous prion protein genotype predisposes to sporadic Creutzfeldt–Jakob disease. *Nature* **352**, 340–342.

Paramithiotis, E., Pinard, M., Lawton, T. & 19 other authors (2003). A prion protein epitope selective for the pathologically misfolded conformation. *Nat Med* **9**, 893–899.

Parchi, P., Castellani, R., Capellari, S. & 9 other authors (1996). Molecular basis of phenotypic variability in sporadic Creutzfeldt–Jakob disease. *Ann Neurol* **39**, 767–778.

Parchi, P., Capellari, S., Chen, S. G. & 8 other authors (1997). Typing prion isoforms. *Nature* **386**, 232–233.

Parchi, P., Giese, A., Capellari, S. & 15 other authors (1999). Classification of sporadic Creutzfeldt–Jakob disease based on molecular and phenotypic analysis of 300 subjects. *Ann Neurol* **46**, 224–233.

Peden, A. H., Head, M. W., Ritchie, D. L., Bell, J. E. & Ironside, J. W. (2004). Preclinical vCJD after blood transfusion in a PRNP codon 129 heterozygous patient. *Lancet* **364**, 527–529.

Perrier, V., Kaneko, K., Safar, J., Vergara, J., Tremblay, P., DeArmond, S. J., Cohen, F. E., Prusiner, S. B. & Wallace, A. C. (2002). Dominant-negative inhibition of prion replication in transgenic mice. *Proc Natl Acad Sci U S A* **99**, 13079–13084.

Petersen, R. B., Tabaton, M., Berg, L. & 16 other authors (1992). Analysis of the prion protein gene in thalamic dementia. *Neurology* **42**, 1859–1863.

Prusiner, S. B. (1982). Novel proteinaceous infectious particles cause scrapie. *Science* **216**, 136–144.

Prusiner, S. B. (1996). Molecular biology and pathogenesis of prion diseases. *Trends Biochem Sci* **21**, 482–487.

Raymond, G. J., Bossers, A., Raymond, L. D. & 7 other authors (2000). Evidence of a molecular barrier limiting susceptibility of humans, cattle and sheep to chronic wasting disease. *EMBO J* **19**, 4425–4430.

Saborio, G. P., Permanne, B. & Soto, C. (2001). Sensitive detection of pathological prion protein by cyclic amplification of protein misfolding. *Nature* **411**, 810–813.

Safar, J., Wille, H., Itrri, V., Groth, D., Serban, H., Torchia, M., Cohen, F. E. & Prusiner, S. B. (1998). Eight prion strains have PrPSc molecules with different conformations. *Nat Med* **4**, 1157–1165.

Safar, J. G., Scott, M., Monaghan, J. & 12 other authors (2002). Measuring prions causing bovine spongiform encephalopathy or chronic wasting disease by immuno-assays and transgenic mice. *Nat Biotechnol* **20**, 1147–1150.

Scott, M., Foster, D., Mirenda, C. & 9 other authors (1989). Transgenic mice expressing hamster prion protein produce species-specific scrapie infectivity and amyloid plaques. *Cell* **59**, 847–857.

Scott, M. R., Will, R., Ironside, J., Nguyen, H. O. B., Tremblay, P., DeArmond, S. J. & Prusiner, S. B. (1999). Compelling transgenetic evidence for transmission of bovine spongiform encephalopathy prions to humans. *Proc Natl Acad Sci U S A* **96**, 15137–15142.

Smith, P. & Bradley, R. (2003). Bovine spongiform encephalopathy (BSE) and its epidemiology. In *Prions for Physicians*, pp. 185–198. Edited by A. A. C.Weissmann, D. Dormont & N. Hunter. Oxford: Oxford University Press.

Somerville, R. A. (1999). Host and transmissible spongiform encephalopathy agent strain control glycosylation of PrP. *J Gen Virol* **80**, 1865–1872.

Somerville, R. A. & Dunn, A. J. (1996). The association between PrP and infectivity in scrapie and BSE infected mouse brain. *Arch Virol* **141**, 275–289.

Soto, C., Saborio, G. P. & Anderes, L. (2002). Cyclic amplification of protein misfolding: application to prion-related disorders and beyond. *Trends Neurosci* **25**, 390–394.

Stahl, N., Baldwin, M. A., Burlingame, A. L. & Prusiner, S. B. (1990). Identification of glycoinositol phospholipid linked and truncated forms of the scrapie prion protein. *Biochemistry (Mosc)* **29**, 8879–8884.

Tanaka, M., Chien, P., Naber, N., Cooke, R. & Weissman, J. S. (2004). Conformational variations in an infectious protein determine prion strain differences. *Nature* **428**, 323–328.

Telling, G., Parchi, P., DeArmond, S. & 7 other authors (1996a). Evidence for the conformation of the pathological isoform of the prion protein enciphering and propagating prion diversity. *Science* **274**, 2079–2082.

Telling, G. C., Haga, T., Torchia, M., Tremblay, P., Dearmond, S. J. & Prusiner, S. B. (1996b). Interactions between wild-type and mutant prion proteins modulate neurodegeneration in transgenic mice. *Genes Dev* **10**, 1736–1750.

Thuring, C. M. A., Erkens, J. H. F., Jacobs, J. G. & 8 other authors (2004). Discrimination between scrapie and bovine spongiform encephalopathy in sheep by molecular size, immunoreactivity, and glycoprofile of prion protein. *J Clin Microbiol* **42**, 972–980.

Tremblay, P., Ball, H. L., Kaneko, K., Groth, D., Hegde, R. S., Cohen, F. E., DeArmond, S. J., Prusiner, S. B. & Safar, J. G. (2004). Mutant PrPSc conformers induced by a synthetic peptide and several prion strains. *J Virol* **78**, 2088–2099.

van Keulen, L. J. M., Schreuder, B. E. C., Meloen, R. H., Mooij-Harkes, G., Vromans, M. E. W. & Langeveld, J. P. M. (1996). Immunohistochemical detection of prion protein in lymphoid tissues of sheep with natural scrapie. *J Clin Microbiol* **34**, 1228–1231.

Wadsworth, J. D. F., Joiner, S., Hill, A. F., Campbell, T. A., Desbruslais, M., Luthert, P. J. & Collinge, J. (2001). Tissue distribution of protease resistant prion protein in variant Creutzfeldt–Jakob disease using a highly sensitive immunoblotting assay. *Lancet* **358**, 171–180.

Westaway, D., Dearmond, S. J., Cayetanocanlas, J., Groth, D., Foster, D., Yang, S. L., Torchia, M., Carlson, G. A. & Prusiner, S. B. (1994). Degeneration of skeletal-muscle, peripheral-nerves, and the central-nervous-system in transgenic mice overexpressing wild-type prion proteins. *Cell* **76**, 117–129.

Wickner, R. B. (1994). [URE3] as an altered URE2 protein – evidence for a prion analog in *Saccharomyces cerevisiae*. *Science* **264**, 566–569.

Wilesmith, J. W. (2002). Preliminary epidemiological analyses of the first 16 cases of BSE born after July 31, 1996, in Great Britain. *Vet Rec* **151**, 451–452.

Wilesmith, J. W., Ryan, J. B. M. & Atkinson, M. J. (1991). Bovine spongiform encephalopathy – epidemiologic studies on the origin. *Vet Rec* **128**, 199–203.

Will, R. G., Ironside, J. W., Zeidler, M. & 7 other authors (1996). A new variant of Creutzfeldt–Jakob disease in the UK. *Lancet* **347**, 921–925.

Williams, E. S. & Young, S. (1980). Chronic wasting disease of captive mule deer: a spongiform encephalopathy. *J Wildl Dis* **16**, 89–98.

Yamakawa, Y., Hagiwara, K., Nohtomi, K., Nakamura, Y., Nishijima, M., Higuchi, Y., Sato, Y. & Sata, T. (2003). Atypical proteinase K-resistant prion protein (PrPres) observed in an apparently healthy 23-month-old Holstein steer. *Jpn J Infect Dis* **56**, 221–222.

Zeidler, M., Stewart, G., Cousens, S., Estibeiro, K. & Will, R. (1997a). Codon 129 genotype and new variant CJD. *Lancet* **350**, 668.

Zeidler, M., Stewart, G. E., Barraclough, C. R. & 9 other authors (1997b). New variant Creutzfeldt–Jakob disease: neurological features and diagnostic tests. *Lancet* **350**, 903–907.

Zeidler, M., Johnstone, E. C., Bamber, R. W. K. & 13 other authors (1997c). New variant Creutzfeldt–Jakob disease: psychiatric features. *Lancet* **350**, 908–910.

Influenza virus pathogenicity

Robert G. Webster, Aleksandr S. Lipatov and Erich Hoffmann

Division of Virology, Department of Infectious Diseases, St. Jude Children's Research Hospital, 332 North Lauderdale St, Memphis, TN 38105, USA

INTRODUCTION

Influenza A viruses are members of the *Orthomyxoviridae* and can be divided into 16 haemagglutinin (HA) and 9 neuraminidase (NA) subtypes based on antigenic and sequence differences in these surface glycoproteins (Wright & Webster, 2001). All of the subtypes occur in the aquatic birds of the world, where they co-exist harmoniously with their wild bird hosts (Webster *et al.*, 1992). It is only after transmission to another host (e.g. domestic chickens, humans, etc.) that the viruses evolve rapidly and cause severe disease. Thus host range transmission, adaptation to that host and successful transmission host-to-host are all prerequisites for an influenza virus to evolve into a strain that has established a stable lineage in that host and may or may not be highly pathogenic.

Influenza virus strains differ in pathogenicity: some viruses cause substantial mortality in their hosts, whereas others do not. The age, sex and immune status of the host population and the extent of antigenic differences between strains are of key importance. The serious outbreaks of disease in humans have corresponded to the introduction of 'new' pandemic strains into a totally susceptible host population; additionally, interpandemic variants can also be associated with high mortality rates. The disease resulting from influenza viral infection is a complex event involving both the virus and the host. The viral genes responsible for high degrees of pathogenicity are better resolved for influenza viruses in chickens and animal models than in humans. The HA is of critical importance but a combination of genes are necessary and include in addition to the HA one of the polymerase genes (PB2) and the nonstructural gene product (NS1).

SGM symposium 64: Molecular pathogenesis of virus infections.
Editors P. Digard, A. A. Nash & R. E. Randall. Cambridge University Press. ISBN 0 521 83248 9 ©SGM 2005

These genes are probably the minimal requirements, for the understanding of pathogenesis is really only beginning and the role of the host is largely unresolved. The field is in its infancy. Here we will consider what is known about pathogenesis, in poultry and animal models and humans, the viral gene segments involved and the strategies for future advancement in this area.

INFLUENZA VIRUSES WITH PATHOGENIC POTENTIAL

Of the 16 HA subtypes of influenza A viruses that occur naturally in aquatic birds, only 3 (H1, H2, H3) are known to have caused pandemics and epidemics in humans. However, we must keep in mind that historical records based on descriptions of disease signs indicate that influenza is an ancient disease dating back to ancient Greece, and the possibility is that other subtypes may have infected people before the mid-1800s when the first sero-archaeology records were available indicating earlier infections with H1, H2 and H3 viruses. Two subtypes (H1, H3) continue to circulate in pigs in the world and one subtype in horses (H3) (a second subtype in horses, H7 – equine 1, has probably disappeared).

Several subtypes of influenza A viruses have caused transitory infection of humans, most notably H5, H7 and H9. The World Health Organization (WHO) has ranked the influenza subtypes and their human pandemic potential: H1, H2 and H3 are in the highest category; H5, H6, H7 and H9 are in the next highest group; H4, H10 and H13 are of less concern; and H8, H11, H12, H14, H15 and H16 are of least concern (Table 1). These categories are based on subtypes that have caused transitory infections of humans or other mammals and those that have transferred between species. If we were to do a similar ranking for influenza viruses with pathogenic potential in domestic poultry, H5 and H7 would be in the highest category, H1, H3, H4, H6 and H9 in a lower category, and the remainder would fall in the category of being rarely isolated from land-based domestic poultry (Table 1). Regardless, all domestic poultry are not the same. Chickens are susceptible to a limited number of influenza subtypes while quail are susceptible to the majority of subtypes (Perez *et al.*, 2003; Makarova *et al.*, 2003) and domestic ducks are susceptible to all subtypes and usually show no disease signs with low or highly pathogenic H5 and H7 strains (Alexander, 2000).

HUMAN PANDEMICS AND EPIDEMICS

The pandemics of influenza in humans in the 20th century included 1918 H1N1 Spanish influenza, 1957 H2N2 Asian influenza, 1968 H3N2 Hong Kong influenza and 1977 H1N1 Russian influenza. These pandemics differed markedly in severity. The 1918 Spanish influenza pandemic killed between 50 and 100 million persons world-

Table 1. Influenza A virus subtypes ordered according to their possible pathogenic potential in humans or domestic poultry

Category	Subtypes	Comments
(a) Humans		
Highest potential	H1, H2, H3	Have caused pandemics and epidemics in humans
High potential	H5, H6, H7, H9	Infection of humans but failure to spread (H5, H7, H9) Infection of domestic poultry (H6)
Lower potential	H4, H10, H13	Transitory infection of mammals including pigs and/or domestic chickens
Lowest potential	H8, H11, H12, H14, H15, H16	Rarely found in land-based domestic poultry (H8, H11, H12) Rarely found even in aquatic birds (H8, H14, H15, H16)
(b) Domestic poultry (gallinaceous)		
Highest priority	H5, H7	Highly pathogenic strains cause lethal infections and possess multiple basic amino acids at the cleavage site of the HA
High priority	H1, H3, H4, H6, H9	Respiratory infection and drop in egg production in chickens (H9) Mixed infection with bacteria or other viral agents can cause lethality
Lowest priority	H2, H8, H10–H16	Rarely isolated from land-based gallinaceous birds

wide (Taubenberger *et al.*, 2001; Barry, 2004), the 1957 Asian approximately 1 million, and the 1968 Hong Kong approximately half a million. The 1977 Russian virus was generally mild, causing serious disease only in the younger populations, for the majority of the world population over 35 years had already been exposed to this virus. The known factors affecting the severity of these pandemics will be discussed later.

AVIAN INFLUENZA

Multiple outbreaks of highly pathogenic avian influenza (HPAI) caused by H5 and H7 influenza viruses have occurred in poultry throughout the world in 2003–2004 (for review, see Lipatov *et al.*, 2004; Stephenson *et al.*, 2004). The most recent outbreaks of HPAI include H5N1 in Asia, H7N3 in Canada (Hirst *et al.*, 2004) and H7N7 in the Netherlands (Fouchier *et al.*, 2004) – each of these outbreaks has been extremely pathogenic in gallinaceous poultry and each caused transitory infections, some with lethality in humans. The most devastating of these outbreaks has been H5N1 in Asia and we will focus on that later. The remaining subtypes of influenza A viruses cause inapparent disease in domestic poultry unless they are exacerbated by co-infecting organisms or the birds are stressed by environmental factors. To illustrate the genesis of a recent HPAI we will consider H5N1 in Asia.

H5N1 IN ASIA

The 1997 H5N1 virus

The precursor of the H5N1 influenza viruses that emerged in humans in Hong Kong in 1997 was A/goose/Guangdong/1/96 (H5N1). This highly pathogenic avian influenza virus killed 40 % of geese on a farm in Guangdong, China, in the summer of 1996 (Tang *et al.*, 1998) and was the source of the HA gene in the A/Hong Kong/156/97 (H5N1) virus that transmitted to a child in Hong Kong in 1997 (de Jong *et al.*, 1997; Xu *et al.*, 1999). It is unusual for influenza viruses to kill aquatic birds such as ducks (Alexander *et al.*, 1986), but little is known about their lethality to geese. The A/goose/Guangdong/1/96 (H5N1) virus reassorted with H9N2 and H6N1 influenza viruses and the reassortant became dominant in the live-poultry markets in Hong Kong in 1997.

The transmission of H5N1 virus to humans in Hong Kong in 1997 was the first confirmed avian-to-human transmission of influenza virus. Virological and epidemiological studies revealed that the live-poultry markets in Hong Kong were the source of the virus for humans (Shortridge *et al.*, 1998; Bridges *et al.*, 2002). The detection of highly pathogenic avian H5N1 virus in each of the live-poultry markets tested in Hong Kong in 1997 was surprising, because no poultry were dying in any of the markets. The probable explanation for this lack of deaths is that H9N2 viruses were co-circulating in the poultry and providing cross-reactive cell-mediated immunity. *In vivo* and *in vitro* studies by Seo & Webster (2001) demonstrated that cell-mediated immunity induced by H9N2 influenza viruses that share the six internal genes of H5N1/97 provided protection from overt H5N1-mediated disease but permitted shedding of H5N1 virus by these birds. Thus cell-mediated immunity can prevent clinical disease and permit the shedding of a sufficient amount of H5N1 virus to infect humans in the markets.

The H5N1/97 epidemic ceased when the entire poultry population of Hong Kong was culled in December 1997. The epidemic curve fell to zero after culling of the poultry. This event dramatically illustrates the role of poultry and poultry markets in the spread of H5N1 influenza viruses to humans. These viruses had not acquired the capacity for human-to-human transmission. No additional cases of human H5N1 infections were detected, and this genotype of H5N1 has not been isolated since.

The importance of the A/goose/Guangdong/1/96-like H5N1 viruses as a continuing source of virus in southeastern China has been very much underestimated in the re-emergence of H5N1 viruses in Hong Kong poultry markets in 2001, 2002 and 2003 (Fig. 1). The A/goose/Guangdong/1/96-like H5N1 viruses continued to circulate in southern China after the eradication, in 1998, of the human and poultry H5N1 viruses (Cauthen *et al.*, 2000; Webster *et al.*, 2002). Additionally, this H5N1 goose virus

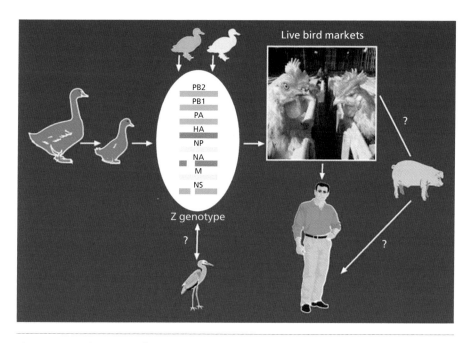

Fig. 1. Genesis of H5N1/04 influenza virus. Diagrammatic representation of the evolution of the Z genotype of H5N1 influenza virus that became dominant in Asia in 2004. The HA and NA genes were derived from viruses related to A/goose/Guangdong/1/96 (H5N1) that had become established in domestic ducks in coastal China (Chen *et al.*, 2004). The NA has a deletion of 19 amino acids in the stalk. The A/goose/Guangdong/1/96 (H5N1)-like virus reassorted with an unknown virus from avian sources (green duck) to obtain the polymerase gene segments (PB2, PB1, PA) and the nonstructure (NS) gene segments. The NS gene product, NS1, had a deletion of five amino acids. Additional reassortment events with an unknown avian virus (yellow duck) provided the nucleoprotein (NP) and matrix genes. The resulting Z genotype (Li *et al.*, 2004) became dominant in Vietnam, Thailand, Laos, Cambodia and Malaysia with high lethality for domestic poultry. This virus probably spread through the live-poultry markets and poultry farms to humans. To date there is no evidence for continued human-to-human transmission. Although H5N1/04 viruses have been detected in pigs in Vietnam (Y.-K. Choi and others, personal communication), there is no convincing evidence of pig-to-pig spread. The roles of wild migrating birds remain controversial; crows probably spread the H5N1 viruses locally in Japan. H5N1 viruses were isolated from dead egrets, falcons and other wild birds in Asia [Ellis *et al.*, 2004; OIE website (http://www.oie.int/)]; however, it is unresolved whether these birds contracted the virus locally as scavengers. There are unconfirmed reports of H5N1 in wild ducks in Siberia (Russian news agencies).

reassorted with influenza viruses in domestic ducks and transmitted to ducks in the region (Chen *et al.*, 2004). These goose H5N1 viruses were the source of both the HA and the NA genes of the H5N1 viruses that became dominant throughout eastern Asia in 2004.

Emergence of H5N1 viruses with novel properties

Although from 1997 through early 2002 the H5N1 influenza viruses in Asia acquired different constellations of internal genes, their HA and NA remained antigenically conserved (Guan *et al.*, 2002; Chen *et al.*, 2004). However, in 2002 the HA of the H5N1 virus underwent marked antigenic drift (Guan *et al.*, 2004; Sturm-Ramirez *et al.*, 2004). The resulting viruses were highly pathogenic in chickens but nonpathogenic in ducks. This situation changed dramatically in November 2002 when H5N1 viruses were isolated from dead wild birds in Hong Kong. The most remarkable property of the H5N1 genotype from late 2002 was its high pathogenicity for ducks and other aquatic birds, a property rarely found in nature. The previous event of significance to aquatic birds had occurred in 1961, when A/tern/South Africa/61 (H5N3) killed terns (Becker, 1966).

In early February 2003, H5N1 virus genetically similar to the virus killing aquatic birds re-emerged in a family in Fujian, China. The daughter died of a respiratory infection of undiagnosed cause while visiting Fujian; the father and son developed severe respiratory illness after their return to Hong Kong. The father died and the son recovered. Infection with H5N1 influenza virus was confirmed in father and son.

In 2001 and 2002 there was a multiplicity of different H5N1 genotypes co-circulating in poultry in southeastern China. At least nine genotypes were circulating in southern China in 2002 (Li *et al.*, 2004). The H5N1 virus that infected humans in 2003 [A/HK/212/03 (H5N1)] was the forerunner of the H5N1 virus that would become dominant throughout Asia in 2004.

The avian and human H5N1 influenza epidemic of 2004

During 2003, the Z genotype became dominant in southern China (Li *et al.*, 2004). The unprecedented magnitude of the bird flu epidemic in Asian countries in 2004 – when H5N1 virus was infecting birds in China, Japan, South Korea, Thailand, Vietnam, Indonesia, Cambodia and Laos – resulted in the destruction of hundreds of millions of poultry, mainly chickens. In most of these countries, outbreaks of highly lethal H5N1 avian influenza were confined to poultry, but in at least two countries the virus transmitted to humans, and most of the persons infected died (27/20 deaths in Vietnam and 17/12 deaths in Thailand).

Additionally, the H5N1 has expanded its host range. Tigers and leopards fed diseased chicken carcasses died and domestic cats can be infected and naturally transmitted the virus to contact cats (Kuiken *et al.*, 2004).

By June 2004, each of the countries that had had H5N1 influenza virus infections in poultry considered that their domestic poultry were free of H5N1 virus. However, the H5N1 virus resurged in July 2004 in poultry in Vietnam, Thailand and China, and the first cases of H5N1 infection in poultry appeared in Malaysia in August 2004. The resurgence of H5N1 virus in the summer months was of particular concern, and the possibility was considered that the clean-up and disinfection done in January

2004 had been incomplete or that the H5N1 virus had become established in birds (perhaps ducks) in the region.

Is H5N1 influenza endemic in Asia?

The resurgence of highly pathogenic H5N1 influenza viruses on poultry farms in Thailand, Vietnam and China in July 2004 and the subsequent infection and death of humans in August 2004 raise the possibility that highly pathogenic avian H5N1 influenza viruses are now endemic in poultry in Asia. The available evidence indicates that H5N1 is endemic in domestic ducks in southern China (Chen *et al.*, 2004), and the above information supports the likelihood that H5N1 virus is now endemic in domestic ducks throughout southern Asia. Although the Z genotype of H5N1 influenza virus has been dominant in Asia, the virus is heterogeneous in its pathogenicity in ducks. Examination of multiple H5N1 isolates of the Z genotype for pathogenicity in ducks reveals that these isolates can be divided into three broad groups: those that cause neurological signs and kill all inoculated ducks, those that kill a percentage of ducks without causing neurological signs, and those that replicate but cause no overt disease (D. J. Hulse and others, personal communication). All of these H5N1 viruses are highly pathogenic in gallinaceous poultry.

REVERSE GENETICS

The first reverse-genetics system to be developed for influenza A virus was the ribonucleoprotein (RNP)-transfection method (Enami *et al.*, 1990; Luytjes *et al.*, 1989). After *in vitro* transcription of virus-like vRNA by the T7 RNA polymerase and reconstitution of viral RNP molecules, genetically altered RNP segments were introduced into eukaryotic cells by transfection. Infection with influenza helper virus resulted in the generation of viruses possessing a gene derived from cloned cDNA. The establishment of the RNA polymerase I (pol I)-driven synthesis of vRNA molecules *in vivo* allowed the intracellular production of RNP complexes (Neumann *et al.*, 1994). In this system, virus-like cDNA was inserted between the pol I promoter and terminator sequences (Zobel *et al.*, 1993). Unlike the mRNA transcripts synthesized by RNA polymerase II (pol II), pol-I-generated RNAs lack both a 5′ cap and a 3′ poly(A) tail. Functional vRNP molecules could be generated either by infection with helper virus or by cotransfection of protein expression plasmids encoding PB1, PB2, PA or nucleoprotein (NP) (Neumann *et al.*, 1994; Pleschka *et al.*, 1996). However, the need for helper virus limits the practical value of these methods since a strong selection system is required to eliminate helper virus.

To overcome the technical limitations, helper virus free, plasmid-only systems were designed to allow the generation of virus *de novo*. Four pol II plasmids representing PB1, PB2, PA and NP protein and eight pol I plasmids representing the eight vRNAs

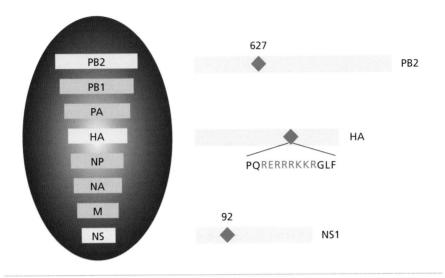

Fig. 2. Pathogenicity of influenza viruses: a polygenic trait. Diagrammatic representation of influenza virus gene segments associated with pathogenicity. Those gene segments considered to be associated with high pathogenicity are highlighted and the amino acid residues that have been linked to high pathogenicity of avian or human influenza viruses are shown.

were cotransfected, resulting in the generation of infectious virus (Fodor *et al.*, 1999; Neumann *et al.*, 1999). The eight-plasmid pol I pol II expression system avoided the use of separate plasmids for protein expression, thus simplifying the method of generation of influenza A virus entirely from cloned cDNA (Hoffmann *et al.*, 2000). In this system, viral cDNA is inserted between the human pol I promoter and a terminator sequence. This entire pol I transcription unit is flanked by a pol II promoter and a poly(A) site. The orientation of the two transcription units allows the synthesis of negative-sense viral RNA and positive-sense mRNA from one viral cDNA template.

The high efficiency of *de novo* generation of influenza virus now allows the routine use of the technique for generating any desired mutation in any of the eight segments. The ability to mutate each gene now allows the testing of hypotheses about the significance of residues associated with pathogenicity (Hatta *et al.*, 2001) (Fig. 2). Using the DNA transfection method it has been demonstrated that seed virus for inactivated and attenuated vaccines can be generated rapidly (Hoffmann *et al.*, 2002; Jin *et al.*, 2003; Subbarao *et al.*, 2003; Webby *et al.*, 2004).

The generation of recombinant viruses has the potential to create and release either intentionally or accidentally viruses into the environment with changed cell tropism and increased pathogenic potential. Thus it is important to assess the risk of each experiment and conduct the research under appropriate biosafety level conditions.

PATHOGENICITY

Pathogenesis is the disease caused by an infectious agent (virus) in a particular host. The severity or virulence is the capacity of that virus when compared with similar agents to produce disease in that host. Molecular determinants underlying the pathogenicity of avian influenza viruses for mammalian hosts remain unclear. The available data implicate the HA, PB2 and NS genes in the pathogenicity of influenza virus and are considered below (Fig. 2).

Role of the HA

Although the pathogenicity of influenza viruses is a polygenic trait, HA plays a pivotal role in determining whether the pathogenicity is high or low (Webster & Rott, 1987). It was shown that the broad tissue tropism and systemic replication in infected chickens are determined by the structure of the cleavage site of the HA that is cleaved by host cell proteases (Klenk & Garten, 1994; Chen *et al.*, 1998; Steinhauer, 1999). The cleavage site of the HA of nonpathogenic avian H5 and H7 influenza viruses in the aquatic bird reservoirs usually contains few basic amino acids, but during rapid evolution of these influenza viruses, they can acquire multiple basic amino acids. These basic amino acids are thought to be acquired through a 'stuttering' action by the polymerase during replication (Perdue *et al.*, 1996). The HAs of the highly pathogenic influenza viruses, all of which possess multiple basic amino acids at their cleavage site, are cleaved by the ubiquitous subtilisin-like proteases furin and PC6 (Stieneke-Grober *et al.*, 1992; Horimoto *et al.*, 1994). The ease of cleavage conferred by the presence of multiple basic amino acids results, after infection, in systemic spread of the virus, its replication in all organs, and death of chickens, turkeys and most gallinaceous birds. The cleavability of HA is also affected by carbohydrate-bearing residues in the vicinity of the HA cleavage site, the presence of which blocks access by activating enzymes (Kawaoka *et al.*, 1984), and by mutations that result in the loss of carbohydrate-bearing residues near the cleavage site or in the lengthening of the cleavage-site sequence, either of which events increases the cleavability of the HA (Horimoto *et al.*, 1994). The HA of nonpathogenic influenza viruses in domestic poultry, in contrast, is typically less easily cleavable or less accessible to the activating enzymes. These nonpathogenic viruses are therefore unable to enter cells and effect systemic infection, and they replicate predominantly in the respiratory and intestinal tracts. Signs of disease resulting from infection with a virus of low pathogenicity can vary from imperceptible to severe, depending on the presence of co-infecting bacteria or viruses (or both).

An alternative strategy by which a nonpathogenic avian influenza virus can become highly pathogenic is recombination. Recombination, which involves the intragenic insertion of genetic material from another virus or host gene, has been detected in H7 but not in H5 influenza viruses. High pathogenicity was recently acquired by recombination in, for example, A/chicken/Chile/4322/02 (H7N3), in which 30 nucleotides were inserted into the region of the HA gene encoding the cleavage site. The inserted sequence was similar to a portion (positions 1268–1297) of the NP gene of A/gull/Maryland/704/77 (H13N6) (Suarez *et al.*, 2004). The resulting highly pathogenic H7N3 Chilean variant was derived from a virus of low pathogenicity and was remarkable in having only three basic amino acids at the HA cleavage site.

High pathogenicity has also been acquired in the laboratory by recombination occurring during *in vitro* passaging of A/turkey/Oregon/71 (H7N3) and A/seal/Massachusetts/1/80 (H7N7) (Khatchikian *et al.*, 1989; Orlich *et al.*, 1994). Preliminary reports suggest that the highly pathogenic Canadian H7N3/04 virus that originated in British Columbia became highly pathogenic as a result of a recombination event that inserted 21 nucleotides from its own matrix gene segment (Hirst *et al.*, 2004).

The highly pathogenic avian H5 viruses that have infected humans in 1997, 2003 and 2004 and the highly pathogenic avian H7 viruses that transmitted to humans in the Netherlands in 2003 and Canada in 2004 all possessed HAs with multiple basic amino acids at the cleavage site (Table 2).

The 1918 Spanish influenza was the most devastating influenza pandemic in recorded history, killing between 50 and 100 million persons worldwide (Taubenberger *et al.*, 2001). The complete nucleotide sequence of seven of the eight gene segments of the 1918 Spanish influenza virus has been established by the 'tour de force' studies of Jeffrey Taubenberger and his group (Taubenberger *et al.*, 2001; Reid *et al.*, 2004). To date, these sequence analyses have not provided insight into the extreme pathogenesis caused by these viruses; however, the sequence of the PB2 gene is not yet available. In order to attempt to resolve the question of extreme pathogenesis, hybrid viruses have been constructed containing 1918 gene segments reconstructed from the primary nucleotide sequence and using reverse-genetics technology to prepare influenza viruses. Constructs containing 1918 HA and 1918 NA gene segments with six 'internal' genes from either an A/WSN/33 (H1N1) (WSN) or A/PR/8/34 (H1N1) (PR8) were as pathogenic in mice as the WSN or PR8 parental viruses but not more so (Tumpey *et al.*, 2002, 2004). The 1918 HA/NA-WSN recombinant virus showed increased up-regulation of genes associated with activated T cells and macrophages, as well as genes involved in apoptosis, tissue injury and oxidative damage that were not observed in mice infected with a reassortment containing A/New Caledonia/1/99 (H1N1) HA/NA-WSN (Kash *et al.*, 2004). Thus the high pathogenicity of these constructs was probably being provided by the 1918 HA/NA and not by the WSN internal gene segments. However, both WSN and PR8 are highly laboratory passaged strains and it is possible that the 'internal' gene segments required to complement high pathogenicity have been modified. To resolve

Table 2. Critical residues for host range and pathogenicity in the following gene products

Virus	Proteins and residues		
	PB2	**HA**	**NS1**
H1N1 1918 'Spanish' influenza	Lys_{627}??*	Monobasic cleavage site	Asp_{92}
H5N1/97 (Hong Kong bird flu)	Lys/Glu_{627}†	Multibasic cleavage site	Glu_{92}
H5N1/04 (Vietnam bird flu)	Lys_{627}	Multibasic cleavage site	Deletion of 5 aa
H7N7 Netherlands/03	Lys_{627}	Multibasic cleavage site	Asp_{92}
H7N3 Canada/04	Glu_{627}	Multibasic cleavage site	Asp_{92}

*The sequence of the PB2 gene of 1918 'Spanish' influenza has not yet been reported. Analysis of PB2 sequences available at the Influenza Sequence Database (www.flu.lanl.gov) revealed that all known H1N1 influenza viruses of human and swine origin isolated in the period 1930–1947 contain Lys at position 627. Therefore, it is probable that the PB2 protein of 1918 'Spanish' influenza will also contain Lys_{627}.

†One group of human H5N1/97 isolates contain Lys_{627} in the PB2, another Glu_{627}.

this possibility studies by the Kawaoka group (Kobasa *et al.*, 2004) with constructs containing 1918 HA and NA and the internal genes from A/Kawasaki/173/2001 (H1N1) virus was shown to be more pathogenic than the A/Kawasaki/173/2001 (H1N1) parent. Additional constructs containing only the HA of 1918 and the other seven genes from A/Kawasaki/173/2001 (H1N1) established that only the HA was required for high pathogenicity. The constructs containing the 1918 HA and the remaining seven gene segments from A/Kawasaki/173/2001 (H1N1) spread throughout the entire lungs of mice and caused extensive pneumonia; this construct induced high levels of macrophage-derived chemokines and cytokines in the mouse as has been described for the 1918 HA/NA-WSN recombinant virus (Kash *et al.*, 2004). Resolution of the actual residues in the 1918 HA remains to be done and the possible additional contribution of the 1918 PB2 gene combinations remains to be determined (Table 2).

Role of the PB2

The first evidence for the role of polymerase protein PB2 and particularly the amino acid at position 627 in the determination of host range of influenza viruses was provided by the work of Subbarao *et al.* (1993). They showed that a reassortant virus containing a PB2 gene of avian origin and the remaining genes from a human influenza virus is restricted in growth in the respiratory tract of squirrel monkeys and humans, and fails to produce plaques in MDCK cells. However, these features can be restored by the adaptation of this virus to MDCK cells, and a single amino acid substitution, Glu→Lys, at position 627 in the PB2 was responsible for this restoration (Subbarao *et al.*, 1993). Subsequently, PB2 has been implicated in the pathogenicity of H5N1/97 influenza viruses in the mouse model. Using reverse genetics, Hatta *et al.* (2001) showed that Lys_{627} in the polymerase protein PB2 is crucial for high virulence and systemic replication of A/Hong Kong/483/97 (H5N1) virus in mice. The amino acid at position 627 of PB2 determines the efficiency of virus replication in mouse (not avian) cells, but this amino acid does not determine viral tropism toward different organs in the mouse (Shinya *et al.*, 2004). The analysis of amino acid sequences available in the Influenza Sequence Database shows that PB2 of H5N1/97, H5N1/04 and H7N7/03 influenza viruses transmitted to humans contains Lys at position 627 (Table 2). The sequence of the PB2 gene of 1918 'Spanish' flu has not yet been reported. However, analysis of PB2 sequences shows that all 'old' influenza viruses of human and swine origin contain Lys at position 627. Therefore, it is likely that the PB2 protein of 1918 'Spanish' influenza will also contain Lys_{627}.

Role of the NS gene/NS1 protein

The NS gene segment of influenza A viruses encodes two proteins, NS1 and the nuclear export protein NS2. The NS1 protein contributes to viral pathogenesis by allowing the virus to disarm the interferon (IFN)-based defence system of the host cell (Garcia-Sastre, 2001). It was shown that the NS gene of H5N1/97 virus dramatically increased the pathogenicity of PR8 virus in pigs (Seo *et al.*, 2002). The authors hypothesized that the NS gene of H5N1/97 viruses confers resistance to the antiviral effects of IFNs and tumour necrosis factor alpha (TNF-α) and that the unique properties of the H5N1/97 NS1 protein required Glu_{92} (Seo *et al.*, 2002). Additional studies showed that the NS gene of the H5N1/97 virus supports the high pathogenicity of PR8 virus in mice (Lipatov & Webster, 2004). Reassortant virus bearing the NS gene of H5N1/97 induced increases in inflammatory cytokines and chemokines in infected mouse lungs; in contrast, the concentration of the anti-inflammatory cytokines decreased. As in the pig model, the capacity of the H5N1/97 NS gene to support high pathogenicity in mice is related to Glu_{92} of the NS1 protein. This amino acid at position 92 of the NS1 protein is unique to H5N1/97 viruses (Table 2).

The NS gene encoding the NS1 protein with unique deletion of five amino acids was first identified in H5N1 virus isolated from poultry in Hong Kong in 2001 (Guan *et al.*, 2002). Human and avian H5N1 viruses isolated of 2003 and 2004 contained similar NS genes (Guan *et al.*, 2004; Govorkova *et al.*, 2005) (Table 2). Studies *in vitro* in pig lung epithelial cells showed that PR8 reassortant viruses carrying the NS gene from avian H5N1/01 virus are resistant to the antiviral effects of IFNs and TNF-α (Seo *et al.*, 2002). However, our preliminary data on pathogenicity of PR8 reassortant viruses with the NS genes of H5N1/01, H5N1/03 and H5N1/04 showed that these viruses are attenuated. It was determined that the human H5N1/03 isolate induces high levels of TNF-α and IP-10 in primary human macrophages (Guan *et al.*, 2004), but the contribution of the NS gene to this phenomenon is questionable.

WSN reassortant viruses containing the NS gene from 1918 'Spanish' influenza were studied in a mouse model and in human lung epithelial cells. The results of these studies are conflicting. In mice, WSN mouse-adapted virus reassortants with either the whole NS gene or only the NS1 gene of the 1918 pandemic influenza virus were less pathogenic than the original WSN virus (Basler *et al.*, 2001). The human origin of the NS gene may have significantly reduced the pathogenicity of the reassortant in mice. This explanation is supported by work showing that a virus containing the NS1 gene of the 1918 pandemic strain blocks the expression of IFN-regulated genes in human lung cells more efficiently than its parental WSN virus (Geiss *et al.*, 2002).

DISCUSSION AND CONCLUDING REMARKS

As indicated in the Introduction, our understanding of the pathogenicity of influenza is in its infancy. The previous sections have outlined what is known about the pathogenicity of human and avian influenza viruses. In their natural environment, influenza viruses are completely nonpathogenic and live in harmony with their wild aquatic bird hosts. Highly pathogenic strains of influenza evolve after spread to other hosts. Thus transmission to other hosts is an important aspect of emergence of highly pathogenic strains of influenza and may be an immediate rather than fully adapted state. The evolution of a parasite that kills its host is not very successful from the perspective of the parasite. Historically, highly pathogenic strains of influenza do not circulate for long. The 1918 Spanish influenza virus caused peak mortality in November 1918, by 1919 the virus was less pathogenic and by 1920 the virus had lost its extreme virulence. In recent years, highly pathogenic H5 and H7 avian influenza viruses have been eradicated or controlled by culling but prior to 1900 and the development of a domestic poultry industry this was not done. Regardless, highly pathogenic avian influenza viruses have not persisted in nature – all H5 and H7 outbreaks evolve from nonpathogenic strains after interspecies transmission (Röhm *et al.*, 1995). Thus transmissibility and generation of highly pathogenic influenza viruses are interrelated. The currently circulating H5N1/04 influenza viruses in Vietnam and Thailand are extremely pathogenic in chickens and transmit successfully, while in humans these viruses are extremely pathogenic (killing 34 of 43 infected persons) but have to date shown limited ability to transmit person to person. In aquatic birds, the H5N1/04 currently circulating in Asia is evolving rapidly towards a less pathogenic state. In December 2002, the H5N1 influenza virus circulating in Hong Kong killed the majority of exotic waterfowl in Kowloon Park (ducks, flamingos, swans, geese, etc.) (Sturm-Ramirez *et al.*, 2004; Ellis *et al.*, 2004). By mid-July 2004 the H5N1 influenza viruses circulating in poultry and people in Vietnam and Thailand could be divided into three groups in terms of duck pathogenicity: those that caused 100 % lethality, those that killed a few ducks and the majority that cause no apparent disease signs (D. J. Hulse and others, personal communication). This latter group were benign in ducks but still lethal for chickens. Additionally, these viruses were shed for prolonged periods by domestic ducks, promoting the selection of antigenic variants (D. J. Hulse and others, personal communication). Thus overall in nature with influenza viruses there is evolution towards harmony with the host. In the case of the H5N1/04 virus this can mask and increase the spread of highly pathogenic H5N1 influenza viruses for gallinaceous poultry and for humans.

Reverse genetics for influenza viruses now permits resolution of the gene segments and amino acid residues involved in both transmissibility and pathogenicity of influenza viruses. This technology (reverse genetics) permits the preparation of an influenza virus with designated gene composition and amino acid substitutions (Fig. 3). Thus reverse genetics has been used to tailor-make a H5N1/04 vaccine strain to A/Vietnam/1203/04 (H5N1) that is nonpathogenic for chickens, ferrets and embryonated eggs and produces high yields of virus. This vaccine contains the modified HA gene and the NA of A/Vietnam/1203/04 (H5N1) plus six 'internal' genes from classical A/PR/8/34 (H1N1) that provide high growth potential. Thus reverse genetics allows the rapid production of vaccine strains and also provides insight into pathogenesis.

One of the key questions regarding H5N1/04 influenza viruses is their enormous pathogenicity in poultry and humans. These H5N1/04 strains of influenza will kill chickens after intravenous inoculation in less than 20 h and kill ferrets, the mammalian model for influenza (Govorkova et al., 2005). How is this possible and what is the mechanism of rapid killing? It seems unlikely to be the virus load and more likely to be a 'cytokine storm'. The data are not yet available but the indications are that different gene segments and combinations of segments are associated with pro-inflammatory cytokines: NS1 with A/Hong Kong/156/97 (Seo et al., 2002); NS1 and NP with H5N1/03 (Cheung et al., 2002; Guan et al., 2004); HA with Spanish 1918 (Kobasa et al., 2004).

To obtain insight into the molecular mechanisms involved in high pathogenicity it is probable that in the process influenza viruses with high pathogenic potential will have to be generated. These studies are controversial from the perspective of biosecurity and must only be done in approved high containment facilities with highly trained staff. Much attention is now being given to updating the guidelines for facilities that must be used for studying such influenza viruses. However, the real biosecurity issues will come down to the institutional oversight committees and their understanding of the guidelines proposed and the experimental studies proposed. A key issue is training of personnel; the most sophisticated containment facilities do not provide high-level biosecurity unless the personnel are highly trained. It seems unlikely that we will

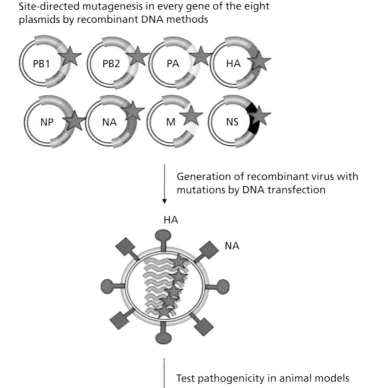

Site-directed mutagenesis in every gene of the eight plasmids by recombinant DNA methods

Generation of recombinant virus with mutations by DNA transfection

Test pathogenicity in animal models

Fig. 3. Plasmid-only system for analysis of the molecular basis of pathogenicity of influenza virus. The eight viral RNA gene segments are amplified by RT-PCR and cloned into plasmids. Each of the viral gene segments can be manipulated by standard genetic engineering methods. After generation of a set of recombinant viruses with different mutations, viral growth and pathogenic potential are tested in cell culture or animal models.

understand the molecular basis of pathogenicity without constructing influenza viruses with high pathogenic potential – the knowledge to be gained will have enormous potential for good in terms of understanding intervention strategies, drug development and future control measures. At this time the key question regarding H5N1/04 in Asia is whether continuing human-to-human transmission can occur. There has been at least one case of human-to-human transmission of H5N1 in a family in Thailand but to date further transmission has not been detected.

Human-to-human transmission could be acquired by continuing rapid accumulation of mutations, for the possibility is that the 1918 Spanish influenza virus transmitted the entire virus from some yet unresolved avian source directly to humans and pigs. Whether the virus first went through pigs is similarly unresolved. The second mechanism for acquiring human-to-human transmission was utilized by the Asian/57 and Hong Kong/68 pandemics; these viruses acquired novel antigens from avian sources and transmissibility from the previous human viruses. Which mechanism will the H5N1 utilize to achieve transmissibility in humans and will it be highly pathogenic? There does not appear to be any restrictions on the ability of the H5N1/04 viruses to reassort with human influenza viruses, for the H5N1/04 vaccine strain containing six 'internal' genes from A/PR/8/34 (H1N1) and modified HA and intact NA genes from A/Vietnam/1203/04 (H5N1) was readily constructed for vaccine purposes. The wide distribution of highly pathogenic H5N1/04 in domestic poultry, particularly in ducks in Southeast Asia, raises the possibility that this H5N1 virus will achieve human-to-human transmissibility.

In addition to the above considerations is the 'intent to harm'. If nature does succeed in completing the ongoing evolution of a highly pathogenic virus, its consequence for humanity could be disastrous. Similarly, man-made constructs could be disastrous for humanity. Despite these concerns, the understanding of pathogenicity will be the key to future intervention strategies for controlling the emergence of pandemic influenza.

ACKNOWLEDGEMENTS

We thank Janet R. Davies for editorial assistance, and Julie Groff for illustrations. Influenza research at St. Jude Children's Research Hospital is supported by NIAID contract AI95357 and Cancer Center Support (CORE) grant CA-21765 from the National Institutes of Health and by the American Lebanese Syrian Associated Charities (ALSAC). At the University of Hong Kong influenza studies were supported by contract AI95357 from the National Institutes of Health (USA), from the Wellcome Trust Grant 067072/D/02/Z and by the Ellison Medical Foundation. We thank Carol Walsh for manuscript preparation and administrative assistance.

REFERENCES

Alexander, D. J. (2000). A review of avian influenza in different bird species. *Vet Microbiol* **77**, 3–13.

Alexander, D. J., Parsons, G. & Manvell, R. J. (1986). Experimental assessment of the pathogenicity of eight avian influenza A viruses of H5 subtype for chickens, turkeys, ducks and quail. *Avian Pathol* **15**, 647–662.

Barry, J. M. (2004). *The Great Influenza. The Epic Story of the Deadliest Plague in History.* New York: Viking.

Basler, C. F., Reid, A. H., Dybing, J. K. & 9 other authors (2001). Sequence of the 1918 pandemic influenza virus nonstructural gene (NS) segment and characterization of recombinant viruses bearing the 1918 NS genes. *Proc Natl Acad Sci U S A* **98**, 2746–2751.

Becker, W. B. (1966). The isolation and classification of Tern virus: influenza A-Tern South Africa – 1961. *J Hyg (London)* **64**, 309–320.

Bridges, C. B., Lim, W., Hu-Primmer, J. & 9 other authors (2002). Risk of influenza A (H5N1) infection among poultry workers, Hong Kong, 1997-1998. *J Infect Dis* **185**, 1005–1010.

Cauthen, A. N., Swayne, D. E., Schultz-Cherry, S., Perdue, M. L. & Suarez, D. L. (2000). Continued circulation in China of highly pathogenic avian influenza viruses encoding the hemagglutinin gene associated with the 1997 H5N1 outbreak in poultry and humans. *J Virol* **74**, 6592–6599.

Chen, H., Deng, G., Li, Z. & 7 other authors (2004). The evolution of H5N1 influenza viruses in ducks in southern China. *Proc Natl Acad Sci U S A* **101**, 10452–10457.

Chen, J., Lee, K. H., Steinhauer, D. A., Stevens, D. J., Skehel, J. J. & Wiley, D. C. (1998). Structure of the hemagglutinin precursor cleavage site, a determinant of influenza pathogenicity and the origin of the labile conformation. *Cell* **95**, 409–417.

Cheung, C. Y., Poon, L. L., Lau, A. S., Luk, W., Lau, Y. L., Shortridge, K. F., Gordon, S., Guan, Y. & Peiris, J. S. (2002). Induction of proinflammatory cytokines in human macrophages by influenza A (H5N1) viruses: a mechanism for the unusual severity of human disease? *Lancet* **360**, 1831–1837.

de Jong, J. C., Claas, E. C., Osterhaus, A. D., Webster, R. G. & Lim, W. L. (1997). A pandemic warning? *Nature* **389**, 554.

Ellis, T. M., Bousfield, B. R., Bissett, L. A. & 7 other authors (2004). Investigation of outbreaks of highly pathogenic H5N1 avian influenza in waterfowl and wild birds in Hong Kong in late 2002. *Avian Pathol* **33**, 492–505.

Enami, M., Luytjes, W., Krystal, M. & Palese, P. (1990). Introduction of site-specific mutations into the genome of influenza virus. *Proc Natl Acad Sci U S A* **87**, 3802–3805.

Fodor, E., Devenish, L., Engelhardt, O. G., Palese, P., Brownlee, G. G. & Garcia-Sastre, A. (1999). Rescue of influenza A virus from recombinant DNA. *J Virol* **73**, 9679–9682.

Fouchier, R. A., Schneeberger, P. M., Rozendaal, F. W. & 11 other authors (2004). Avian influenza A virus (H7N7) associated with human conjunctivitis and a fatal case of acute respiratory distress syndrome. *Proc Natl Acad Sci U S A* **101**, 1356–1361.

Garcia-Sastre, A. (2001). Inhibition of interferon-mediated antiviral responses by influenza A viruses and other negative-strand RNA viruses. *Virology* **279**, 375–384.

Geiss, G. K., Salvatore, M., Tumpey, T. M. & 8 other authors (2002). Cellular transcriptional profiling in influenza A virus-infected lung epithelial cells: the role of the nonstructural NS1 protein in the evasion of the host innate defense and its potential contribution for pandemic influenza. *Proc Natl Acad Sci U S A* **99**, 10736–10741.

Govorkova, E. A., Rehg, J. E., Krauss, S. & 11 other authors (2005). Lethality to ferrets of H5N1 influenza viruses isolated from humans and poultry in 2004. *J Virol* **79**, 2191–2198

Guan, Y., Peiris, J. S., Lipatov, A. S., Ellis, T. M., Dyrting, K. C., Krauss, S., Zhang, L. J., Webster, R. G. & Shortridge, K. F. (2002). Emergence of multiple genotypes of H5N1 avian influenza viruses in Hong Kong SAR. *Proc Natl Acad Sci U S A* **99**, 8950–8955.

Guan, Y., Poon, L. L., Cheung, C. Y. & 10 other authors (2004). H5N1 influenza: a protean pandemic threat. *Proc Natl Acad Sci U S A* **101**, 8156–8161.

Hatta, M., Gao, P., Halfmann, P. & Kawaoka, Y. (2001). Molecular basis for high virulence of Hong Kong H5N1 influenza A viruses. *Science* **293**, 1840–1842.

Hirst, M., Astell, C. R., Griffith, M. & 24 other authors (2004). A novel avian influenza H7N3 strain associated with an avian influenza outbreak British Columbia. *Emerg Infect Dis* **10**, 2192–2195.

Hoffmann, E., Neumann, G., Kawaoka, Y., Hobom, G. & Webster, R. G. (2000). A DNA transfection system for generation of influenza A virus from eight plasmids. *Proc Natl Acad Sci U S A* **97**, 6108–6113.

Hoffmann, E., Krauss, S., Perez, D., Webby, R. & Webster, R. G. (2002). Eight-plasmid system for rapid generation of influenza virus vaccines. *Vaccine* **20**, 3165–3170.

Horimoto, T., Nakayama, K., Smeekens, S. P. & Kawaoka, Y. (1994). Proprotein-processing endoproteases PC6 and furin both activate hemagglutinin of virulent avian influenza viruses. *J Virol* **68**, 6074–6078.

Jin, H., Lu, B., Zhou, H., Ma, C., Zhao, J., Yang, C. F., Kemble, G. & Greenberg, H. (2003). Multiple amino acid residues confer temperature sensitivity to human influenza virus vaccine strains (FluMist) derived from cold-adapted A/Ann Arbor/6/60. *Virology* **306**, 18–24.

Kash, J. C., Basler, C. F., Garcia-Sastre, A. & 7 other authors (2004). Global host immune response: pathogenesis and transcriptional profiling of type A influenza viruses expressing the hemagglutinin and neuraminidase genes from the 1918 pandemic virus. *J Virol* **78**, 9499–9511.

Kawaoka, Y., Naeve, C. W. & Webster, R. G. (1984). Is virulence of H5N2 influenza viruses in chickens associated with loss of carbohydrate from the hemagglutinin? *Virology* **139**, 303–316.

Khatchikian, D., Orlich, M. & Rott, R. (1989). Increased viral pathogenicity after insertion of a 28S ribosomal RNA sequence into the haemagglutinin gene of an influenza virus. *Nature* **340**, 156–157.

Klenk, H.-D. & Garten, W. (1994). Host cell proteases controlling virus pathogenicity. *Trends Microbiol* **2**, 39–43.

Kobasa, D., Takada, A., Shinya, K. & 16 other authors (2004). Enhanced virulence of influenza A viruses with the haemagglutinin of the 1918 pandemic virus. *Nature* **431**, 703–707.

Kuiken, T., Rimmelzwaan, G., van Riel, D., van Amerongen, G., Baars, M., Fouchier, R. & Osterhaus, A. (2004). Avian H5N1 influenza in cats. *Science* **306**, 241.

Li, K. S., Guan, Y., Wang, J. & 19 other authors (2004). Genesis of a highly pathogenic and potentially pandemic H5N1 influenza virus in eastern Asia. *Nature* **430**, 209–213.

Lipatov, A. S. & Webster, R. G. (2004). Factors determining high pathogenicity of Hong Kong H5N1/97 influenza viruses in mammals. *Int Congr Ser* **1263**, 59–62.

Lipatov, A. S., Govorkova, E. A., Webby, R. J., Ozaki, H., Peiris, M., Guan, Y., Poon, L. & Webster, R. G. (2004). Influenza: emergence and control. *J Virol* **78**, 8951–8959.

Luytjes, W., Krystal, M., Enami, M., Pavin, J. D. & Palese, P. (1989). Amplification, expression, and packaging of foreign gene by influenza virus. *Cell* **59**, 1107–1113.

Makarova, N. V., Ozaki, H., Kida, H., Webster, R. G. & Perez, D. R. (2003). Replication and transmission of influenza viruses in Japanese quail. *Virology* **310**, 8–15.

Neumann, G., Zobel, A. & Hobom, G. (1994). RNA polymerase I-mediated expression of influenza viral RNA molecules. *Virology* **202**, 477–479.

Neumann, G., Watanabe, T., Ito, H. & 9 other authors (1999). Generation of influenza A viruses entirely from cloned cDNAs. *Proc Natl Acad Sci U S A* **96**, 9345–9350.

Orlich, M., Gottwald, H. & Rott, R. (1994). Nonhomologous recombination between the hemagglutinin gene and the nucleoprotein gene of an influenza virus. *Virology* **204**, 462–465.

Perdue, M. L., Garcia, M., Beck, J., Brugh, M. & Swayne, D. E. (1996). An Arg-Lys insertion at the hemagglutinin cleavage site of an H5N2 avian influenza isolate. *Virus Genes* **12**, 77–84.

Perez, D. R., Lim, W., Seiler, J. P., Guan, Y., Peiris, M., Shortridge, K. F. & Webster, R. G. (2003). Role of quail in the interspecies transmission of H9 influenza A viruses: molecular changes on HA that correspond to adaptation from ducks to chickens. *J Virol* **77**, 3148–3156.

Pleschka, S., Jaskunas, R., Engelhardt, O. G., Zurcher, T., Palese, P. & Garcia-Sastre, A. (1996). A plasmid-based reverse genetics system for influenza A virus. *J Virol* **70**, 4188–4192.

Reid, A. H., Taubenberger, J. K. & Fanning, T. G. (2004). Evidence of an absence: the genetic origins of the 1918 pandemic influenza virus. *Nat Rev Microbiol* **2**, 909–914.

Röhm, C., Horimoto, T., Kawaoka, Y., Suss, J. & Webster, R. G. (1995). Do hemagglutinin genes of highly pathogenic avian influenza viruses constitute unique phylogenetic lineages? *Virology* **209**, 664–670.

Seo, S. H. & Webster, R. G. (2001). Cross-reactive, cell-mediated immunity and protection of chickens from lethal H5N1 influenza virus infection in Hong Kong poultry markets. *J Virol* **75**, 2516–2525.

Seo, S. H., Hoffmann, E. & Webster, R. G. (2002). Lethal H5N1 influenza viruses escape host anti-viral cytokine responses. *Nat Med* **8**, 950–954.

Shinya, K., Hamm, S., Hatta, M., Ito, H., Ito, T. & Kawaoka, Y. (2004). PB2 amino acid at position 627 affects replicative efficiency, but not cell tropism, of Hong Kong H5N1 influenza A viruses in mice. *Virology* **320**, 258–266.

Shortridge, K. F., Zhou, N. N., Guan, Y. & 12 other authors (1998). Characterization of avian H5N1 influenza viruses from poultry in Hong Kong. *Virology* **252**, 331–342.

Steinhauer, D. A. (1999). Role of hemagglutinin cleavage for the pathogenicity of influenza virus. *Virology* **258**, 1–20.

Stephenson, I., Nicholson, K. G., Wood, J. M., Zambon, M. C. & Katz, J. M. (2004). Confronting the avian influenza threat: vaccine development for a potential pandemic. *Lancet Infect Dis* **4**, 499–509.

Stieneke-Grober, A., Vey, M., Angliker, H., Shaw, E., Thomas, G., Roberts, C., Klenk, H.-D. & Garten, W. (1992). Influenza virus hemagglutinin with multibasic cleavage site is activated by furin, a subtilisin-like endoprotease. *EMBO J* **11**, 2407–2414.

Sturm-Ramirez, K. M., Ellis, T., Bousfield, B. & 7 other authors (2004). Reemerging H5N1 influenza viruses in Hong Kong in 2002 are highly pathogenic to ducks. *J Virol* **78**, 4892–4901.

Suarez, D. L., Senne, D. A., Banks, J. & 11 other authors (2004). Recombination resulting in virulence shift in avian influenza outbreak, Chile. *Emerg Infect Dis* **10**, 693–699.

Subbarao, E. K., London, W. & Murphy, B. R. (1993). A single amino acid in the PB2 gene of influenza A virus is a determinant of host range. *J Virol* **67**, 1761–1764.

Subbarao, K., Chen, H., Swayne, D. & 8 other authors (2003). Evaluation of a genetically modified reassortant H5N1 influenza A virus vaccine candidate generated by plasmid-based reverse genetics. *Virology* **305**, 192–200.

Tang, X., Tian, G., Zhao, J. & Zhou, K. Y. (1998). Isolation and characterization of prevalent strains of avian influenza viruses in China. *Chin J Anim Poult Infect Dis* **20**, 1–5 (in Chinese).

Taubenberger, J. K., Reid, A. H., Janczewski, T. A. & Fanning, T. G. (2001). Integrating historical, clinical and molecular genetic data in order to explain the origin and virulence of the 1918 Spanish influenza virus. *Philos Trans R Soc Lond B Biol Sci* **356**, 1829–1839.

Tumpey, T. M., Garcia-Sastre, A., Mikulasova, A., Taubenberger, J. K., Swayne, D. E., Palese, P. & Basler, C. F. (2002). Existing antivirals are effective against influenza viruses with genes from the 1918 pandemic virus. *Proc Natl Acad Sci U S A* **99**, 13849–13854.

Tumpey, T. M., Garcia-Sastre, A., Taubenberger, J. K., Palese, P., Swayne, D. E. & Basler, C. F. (2004). Pathogenicity and immunogenicity of influenza viruses with genes from the 1918 pandemic virus. *Proc Natl Acad Sci U S A* **101**, 3166–3171.

Webby, R. J., Perez, D. R., Coleman, J. S. & 8 other authors (2004). Responsiveness to a pandemic alert: use of reverse genetics for rapid development of influenza vaccines. *Lancet* **363**, 1099–1103.

Webster, R. G. & Rott, R. (1987). Influenza virus A pathogenicity: the pivotal role of hemagglutinin. *Cell* **50**, 665–666.

Webster, R. G., Bean, W. J., Gorman, O. T., Chambers, T. M. & Kawaoka, Y. (1992). Evolution and ecology of influenza A viruses. *Microbiol Rev* **56**, 152–179.

Webster, R. G., Guan, Y., Peiris, M. & 9 other authors (2002). Characterization of H5N1 influenza viruses that continue to circulate in geese in southeastern China. *J Virol* **76**, 118–126.

Wright, P. F. & Webster, R. G. (2001). Orthomyxoviruses. In *Fields Virology*, 4th edn, pp. 1533–1579. Edited by D. M. Knipe & P. M. Howley. Philadelphia: Lippincott Williams & Wilkins.

Xu, X., Subbarao, K., Cox, N. J. & Guo, Y. (1999). Genetic characterization of the pathogenic influenza A/Goose/Guangdong/1/96 (H5N1) virus: similarity of its hemagglutinin gene to those of H5N1 viruses from the 1997 outbreaks in Hong Kong. *Virology* **261**, 15–19.

Zobel, A., Neumann, G. & Hobom, G. (1993). RNA polymerase I catalysed transcription of insert viral cDNA. *Nucleic Acids Res* **21**, 3607–3614.

RNAi as an antiviral mechanism and therapeutic approach

Ronald P. van Rij and Raul Andino

Department of Microbiology and Immunology, University of California, San Francisco, CA 94143-2280, USA

INTRODUCTION

The observation that small amounts of double-stranded (ds) RNA can effectively trigger gene silencing in the nematode *Caenorhabditis elegans* (Fire *et al.*, 1998) had a major impact on modern molecular biology. In the last few years, similar dsRNA-initiated mechanisms have been identified and studied in different model organisms. Originally coined for single-stranded RNA-based gene silencing in *C. elegans* (Rocheleau *et al.*, 1997), the term RNA interference (RNAi) or RNA silencing now defines a group of gene-silencing mechanisms triggered by dsRNA that is conserved in virtually all eukaryotes (with the notable exception of *Saccharomyces cerevisiae*). In addition to its use as an experimental tool for sequence-specific down-regulation of genes, RNAi has emerged as an important mechanism for the natural control of gene expression.

RNAi has influenced the field of molecular virology in two aspects. First, RNAi can be used to suppress virus replication, either by directly targeting the viral genome or by targeting essential host factors (Gitlin & Andino, 2003). These observations hold tremendous potential for the development of RNAi-based therapeutics. RNAi also provides novel tools to study virus–host interactions and to alter gene expression in viruses that are difficult to manipulate genetically. Second, RNAi appears to provide a natural defence mechanism directed against parasitic nucleic acids, such as transposons and viruses (Baulcombe, 2004; Gitlin & Andino, 2003; Plasterk, 2002; Voinnet, 2001). In this chapter, the therapeutic and natural antiviral potential of RNAi will be discussed with an emphasis on animal viruses. However, since the role of RNAi as an antiviral

SGM symposium 64: Molecular pathogenesis of virus infections.
Editors P. Digard, A. A. Nash & R. E. Randall. Cambridge University Press. ISBN 0 521 83248 9 ©SGM 2005

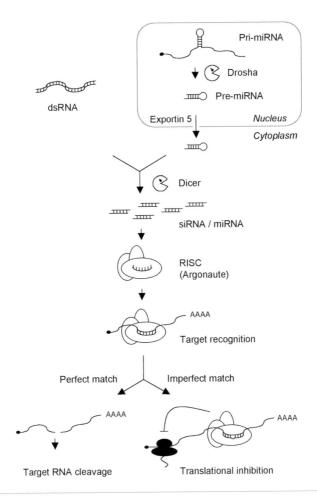

Fig. 1. Overview of the RNAi pathway. Cytoplasmic double-stranded RNA is cleaved by the RNase-III-like protein Dicer into small interfering RNAs (siRNA). The double-stranded siRNAs are unwound and incorporated in the RNA-induced silencing complex (RISC). RISC scans the pool of cytoplasmic RNA and, upon recognition of perfect complementarity, cleaves the target RNA. The microRNA pathway is initiated by recognition of stem–loop structures in endogenous transcripts (primary miRNA). The nuclear localized RNase-III-like protein Drosha cleaves these into precursor miRNAs (pre-miRNAs), which are exported to the cytoplasm by Exportin 5. In the cytoplasm, these pre-miRNAs are cleaved by Dicer and the mature miRNAs are incorporated into RISC. Upon recognition of imperfect target sites, translational repression is induced via an unknown mechanism.

mechanism has been firmly established in plants (Baulcombe, 2004; Voinnet, 2001), we will briefly review this supporting evidence as well.

THE MECHANISM OF RNAi

The core of the RNAi machinery is functionally and biochemically highly conserved in organisms as diverse as plants, nematodes and mammals. However, despite the conservation, variation occurs between different model organisms (for more detailed species-specific information see reviews by Baulcombe, 2004; Hannon, 2003; Lippman & Martienssen, 2004; Meister & Tuschl, 2004).

The RNAi response is triggered by dsRNA, which is cleaved by the RNase-III-like enzyme Dicer into small interfering RNA (siRNA) (Fig. 1) (Bernstein *et al.*, 2001). These siRNAs are short stretches of dsRNA ~21–25 bp long, each strand containing a 2-nucleotide 3′ overhang (Elbashir *et al.*, 2001a). The siRNAs are unwound and one of the two strands (the guide strand) is loaded into the multi-protein complex RISC (RNA-induced silencing complex) (Hammond *et al.*, 2000). RISC scans the endogenous pool of cytoplasmic RNA and, upon recognition of a perfectly matched complementary sequence, cleaves the target RNA, resulting in its eventual degradation. Within RISC, an Argonaute family member (Ago2 in *Drosophila melanogaster* and mammals) is in direct contact with the guide RNA (Tomari *et al.*, 2004) and is thought to be responsible for the 'slicing' activity – the cleavage of the target RNA – via its RNase-H-like domain (Liu *et al.*, 2004; Song *et al.*, 2004).

In addition to this 'RNA cleavage mode', the RNAi machinery is involved in translational inhibition via an undefined mechanism. This pathway is initiated by micro-RNAs (miRNAs), a class of small, non-coding RNAs encoded in the genome (Fig. 1) (Ambros, 2004; Bartel, 2004; He & Hannon, 2004). Primary miRNA transcripts (pri-miRNAs), ranging from several hundred to several thousand base pairs in size, are cleaved by the nuclear RNase-III-like enzyme Drosha into ~70-nt stem–loop structures, the precursor miRNAs (pre-miRNA) (Denli *et al.*, 2004; Gregory *et al.*, 2004; Y. Lee *et al.*, 2003). These pre-miRNAs are exported to the cytoplasm by the nuclear export factor Exportin 5 (Bohnsack *et al.*, 2004; Lund *et al.*, 2004; Yi *et al.*, 2003), where they are cleaved into their mature ~22 nt form by Dicer. Mature miRNAs are loaded into RISC where they guide the inhibition of translation of their target mRNAs without destroying them. In general, miRNAs exhibit only near-complete complementarity to their target site. However, miRNAs in plants and some miRNAs in mammals exhibit complete complementarity to their target mRNA, directing mRNA cleavage (Mansfield *et al.*, 2004; Yekta *et al.*, 2004).

miRNAs have been detected in association with polyribosomes, the sites of active translation (Kim *et al.*, 2004). Furthermore, biochemical evidence suggests a physical association of ribosomes with active RISC and RISC components (Caudy *et al.*, 2002; Hammond *et al.*, 2000; Pham *et al.*, 2004). These results suggest that miRNA/siRNA-mediated post-transcriptional gene silencing occurs during active translation, and that cytoplasmic mRNAs are scanned by RISC during this process. Formal proof for such a scanning mechanism, however, is still lacking.

In addition to the post-transcriptional modes of gene silencing, the RNAi machinery has been implicated in transcriptional gene silencing, *de novo* methylation of DNA and changes in chromatin structure in several organisms (Lippman & Martienssen, 2004), and in programmed DNA elimination in the binucleated protozoan *Tetrahymena thermophila* (Mochizuki & Gorovsky, 2004). In this chapter, the term RNAi is used to indicate its RNA cleavage activity, unless noted otherwise.

RNAi AS AN EXPERIMENTAL TOOL

The natural mechanism of RNAi has been developed into versatile tools to inhibit gene expression experimentally (Dykxhoorn *et al.*, 2003; Engelke, 2003; Hannon & Rossi, 2004). In insects and nematodes, long segments of dsRNA can be introduced into cells, mimicking the natural initiation of the RNAi pathway. In mammals, the RNAi pathway must be triggered by dsRNA shorter than ~30 bp to prevent the activation of the interferon pathway. Synthetic siRNAs with the same structural characteristics as the siRNA intermediate in the natural RNAi pathway are potent inducers of gene silencing (Elbashir *et al.*, 2001c). The efficiency of siRNAs to mediate gene silencing depends on several, not fully understood, sequence requirements. One of the major determinants for a functional siRNA is the thermodynamic stability of the 5′ end of the antisense strand (Khvorova *et al.*, 2003; Schwarz *et al.*, 2003). Still, with even the most recent design rules for siRNAs, the level of knockdown needs to be verified experimentally, and several siRNAs may need to be tested to obtain efficient gene silencing. Alternatively, siRNAs can be generated *in vitro* by enzymic cleavage of long dsRNA by bacterial RNase III or human Dicer (Myers *et al.*, 2003; Yang *et al.*, 2002), an approach that is independent of a single predicted siRNA/target sequence. To obtain gene silencing for prolonged periods of time, several systems for intracellular expression of siRNA have been developed. Short hairpin RNAs (shRNAs) can be expressed as a short sequence in sense and antisense orientation, separated by a non-complementary loop. Dicer activity is required for processing of these shRNAs, which channels these artificial RNAs into the RNAi pathway. Generally pol III promoters are used, such as the H1 and U6 promoters that normally drive the expression of small nuclear RNAs (Brummelkamp *et al.*, 2002; Paddison *et al.*, 2002), but the use of pol I and II promoters has also been successful (McCown *et al.*, 2003; Xia *et al.*, 2002). shRNA expression cassettes may be incorporated in lentiviral and adenovirus vectors (Barton & Medzhitov, 2002; Xia *et al.*, 2002), allowing for the prolonged and stable expression of the construct in cells that are difficult to transfect.

In general, when using well-designed siRNAs or shRNA plasmids, a potent and sequence-specific knockdown of gene expression is achieved. However, several groups have reported off-target effects on gene expression in mammalian cells (for review see Hannon & Rossi, 2004). One source of these non-specific effects is the activation of the protein kinase R (PKR)-mediated interferon response by dsRNA, which results in sequence-independent destruction of RNA and a generalized repression of protein synthesis (Goodbourn *et al.*, 2000; Katze *et al.*, 2002). Although it is thought that PKR is activated by dsRNAs that are longer than 30 bp, some studies have reported the activation of the interferon response by siRNAs or shRNAs under some conditions and in certain cell types (Bridge *et al.*, 2003; Persengiev *et al.*, 2004; Sledz *et al.*, 2003). Therefore, it is crucial to control for activation of the interferon response and verify sequence specificity when using RNAi as an experimental tool.

RNAi AS A THERAPEUTIC ANTIVIRAL APPROACH

Since the specific down-regulation of gene expression by siRNAs was initially described in mammals (Elbashir *et al.*, 2001c), an ever-expanding number of studies have shown that virus replication can be suppressed by RNAi in tissue culture (Gitlin & Andino, 2003; Lopez & Arias, 2004; Stevenson, 2003). Suppression of virus replication can be achieved using two approaches: targeting the viral genome directly, or targeting host factors that are essential for virus replication. Although clinical applications are perhaps several years away, these studies provide proof-of-principle for a novel class of antiviral agents based on RNAi technology. In addition, these studies may provide tools to study the interactions between virus and host in the pathogenesis of virus infections.

Targeting the viral genome by RNAi

Table 1 summarizes the panel of viruses that has been suppressed by targeting the viral genome by RNAi. These studies differ greatly in experimental approach and results; however, several general conclusions can be drawn. Although differences in the level of suppression occur, it seems that in principle any virus can be targeted by RNAi. Different methods to induce RNAi were successful, including the use of synthetic siRNA, shRNA expressed from a plasmid or delivered by a viral vector, as well as enzymatically generated siRNA (Gitlin *et al.*, 2005; Kronke *et al.*, 2004). Few studies provide side-by-side comparisons of these different delivery methods; however, the method of choice will ultimately be dictated by the clinical feasibility, rather than optimal performance in tissue culture.

Accessibility of viral RNA to RISC. Even though genomic and replicating viral RNA are usually protected by proteinaceous or membranous structures, it seems that single-stranded viral RNAs (such as translating RNAs) are sensitive to the RNAi machinery. Indeed, this notion is supported by RNA-silencing studies of rotavirus, a virus with a segmented dsRNA genome. Targeting two structural genes that are not involved in genome replication (VP4 and VP7) resulted in reduced levels of the corresponding proteins (spike and capsid) and associated abnormalities in virion structure. However,

Table 1. Examples of RNAi-mediated suppression of mammalian virus replication by directly targeting the viral genome

Virus	Virus family	Genome*	Reference
Human immunodeficiency virus 1	*Retroviridae*	ssRNA > cDNA	Stevenson (2003) (review) Hu *et al.* (2004) (review)
Hepatitis C virus	*Flaviviridae*	(+) ssRNA	Randall & Rice (2004) (review)
West Nile virus	*Flaviviridae*	(+) ssRNA	McCown *et al.* (2003)
SARS-associated coronavirus	*Coronaviridae*	(+) ssRNA	He *et al.* (2003) Wang *et al.* (2004)
Poliovirus	*Picornaviridae*	(+) ssRNA	Gitlin *et al.* (2002)
Human rhinovirus 16	*Picornavividae*	(+) ssRNA	Phipps *et al.* (2004)
Respiratory syncytial virus	*Paramyxoviridae*	(−) ssRNA	Bitko & Barik (2001)
Influenza A	*Orthomyxoviridae*	Segmented; (−) ssRNA	Ge *et al.* (2003) McCown *et al.* (2003)
Rotavirus	*Reoviridae*	Segmented; dsRNA	Dector *et al.* (2002) Silvestri *et al.* (2004)
Hepatitis delta virus	*Deltavirus*	Circular; ssRNA	Chang & Taylor (2003)
Hepatitis B virus	*Hepadnaviridae*	Circular; partially dsDNA	McCaffrey *et al.* (2003) Giladi *et al.* (2003)
Herpes simplex virus 1	*Herpesviridae*	dsDNA	Bhuyan *et al.* (2004)
Human cytomegalovirus	*Herpesviridae*	dsDNA	Wiebusch *et al.* (2004)
Epstein–Barr virus	*Herpesviridae*	dsDNA	Chang *et al.* (2004)
Murine herpesvirus 68	*Herpesviridae*	dsDNA	Jia & Sun (2003)
Human herpesvirus 6B	*Herpesviridae*	dsDNA	Yoon *et al.* (2004)
Human papillomavirus 16	*Papovaviridae*	Circular; dsDNA	Hall & Alexander (2003)
JC virus	*Papovaviridae*	Circular; dsDNA	Radhakrishnan *et al.* (2004) Orba *et al.* (2004)

*(+), Positive-stranded; (−), negative-stranded.

the 11 RNA segments were present in equimolar ratio in the virion (Arias *et al.*, 2004; Silvestri *et al.*, 2004). This result may imply that there are two pools of RNA during rotavirus replication: translating RNA that is susceptible to RNAi, and replicating RNA that is shielded from RISC by association within the viroplasm or with RNA-binding protein.

In positive-stranded RNA viruses, the viral genomic RNA is used both as a messenger for translation and as a template for negative-strand synthesis. Thus an alternative approach was used to address the question of whether replicating viral RNAs are susceptible to RNAi-induced degradation. The miRNA *let-7* sequence was inserted in the 5′ untranslated region (UTR) of the poliovirus (PV) genome in a sense [PV *let-7*(+)]

and antisense [PV *let-7*(–)] orientation. Transfection of viral RNA into HeLa cells, which express high levels of *let-7*, resulted in virus production of PV *let-7*(+), but not of PV *let-7*(–) (Gitlin *et al.*, 2005). This result indicated that the endogenous *let-7*-loaded RISC can cleave the positive strand of PV *let-7*(–) upon recognition of the perfect complementary site. However, the negative strand of PV *let-7*(+), which contains the perfect complementary site for *let-7*-loaded RISC, cannot be targeted.

In hepatitis delta virus infection, only the viral mRNA, not the circular genome or its exact complement, could be targeted by RNAi (Chang & Taylor, 2003). It was suggested that the nuclear localization of these RNAs is probably responsible for the resistance to RNAi. With respect to human immunodeficiency virus (HIV) as a target for RNAi, it is clear that spliced and unspliced RNAs are suitable targets for RNAi; however, there remains some controversy about the accessibility of the incoming genomic viral RNA for RISC (Stevenson, 2003).

Another important issue with respect to viruses as targets of RNAi is the susceptibility of UTRs versus coding sequences. Although 3′ UTRs of cellular mRNA are routinely used as targets for RNAi, the UTRs of RNA viruses contain complex RNA structures and binding sites for cellular proteins. These domains and their interactions with other proteins or RNA could shield the RNA from RISC-mediated recognition and cleavage. Indeed, it has been observed that certain regions of the 5′ UTR in hepatitis C virus can be targeted by RNAi, whereas other regions seem to be less accessible (Kronke *et al.*, 2004; Yokota *et al.*, 2003).

Virus escape from RNAi. Viruses are notorious for their ability to develop resistance to antiviral drugs. In that respect, it bears no surprise that virus escape from RNAi suppression has been described. Although the RNAi machinery may tolerate a certain degree of mismatches between the siRNA and its target, single mutations in the central region can abolish target RNA cleavage (Elbashir *et al.*, 2001b). Indeed, single mismatches within the targeted region or its entire deletion have been described in escape from RNAi by PV and HIV-1 (Boden *et al.*, 2003; Das *et al.*, 2004; Gitlin *et al.*, 2005). Initiating the RNAi response using long dsRNA (in cells lacking an interferon response) or using a pool of siRNAs generated *in vitro* by RNase III cleavage of long dsRNA prevented virus escape of PV (Gitlin *et al.*, 2005). Treatment with a combination of different siRNAs, providing an RNAi equivalent of a multidrug regimen, may thus be a feasible approach to avoid virus escape.

Targeting cellular factors by RNAi

Viruses can also be suppressed by targeting host factors required for virus replication. The first application of this approach was the demonstration that the vacuolar protein sorting machinery is involved in HIV-1 budding. In that study, RNAi-mediated down-regulation of the TSG101 protein in this pathway reduced virus replication due to a defect in viral budding. The observed defect was reverted by expression of an 'RNAi resistant' TSG101 construct that contains seven silent mutations in the siRNA target region (Garrus et al., 2001). Recently, a number of studies have reported the use of RNAi to study the role of host factors in virus infections (for examples see Chiu et al., 2004; Gao et al., 2004; Surabhi & Gaynor, 2002). These studies are likely to enhance understanding of the complex interactions between virus and host. Furthermore, these studies may identify potential targets for therapeutic intervention. For example, HIV-1 replication can be suppressed by knockdown of its two major entry coreceptors, chemokine receptors CCR5 and CXCR4 (M. T. Lee et al., 2003; Qin et al., 2003). Of note, a small percentage of healthy individuals lack CCR5 expression without obvious abnormalities, due to homozygosity for a 32 bp deletion in their CCR5 gene (Liu et al., 1996). This makes CCR5 an excellent target for an RNAi-based therapeutic approach for HIV-1 infection.

In vivo studies of RNAi as an antiviral approach

Tissue culture experiments have established that viruses can be targeted by RNAi. Several groups have successfully extended these observations to in vivo models of infections with hepatitis B, influenza and foot-and-mouth disease virus (Chen et al., 2004; Ge et al., 2004; Giladi et al., 2003; McCaffrey et al., 2003; Tompkins et al., 2004). In these studies, siRNA or plasmids expressing shRNA were introduced into mice using hydrodynamic intravenous administration (Giladi et al., 2003; McCaffrey et al., 2003; Tompkins et al., 2004), subcutaneous injection in the neck (Chen et al., 2004) or intravenous injection of siRNAs in complex with the cationic polymer poly-ethyleneimine (PEI) (Ge et al., 2004). Recently, it was shown that intranasal delivery of siRNAs in complex with a transfection agent or with the natural polymer chitosan could inhibit replication of two respiratory viruses, parainfluenza virus and respiratory syncytial virus (Zhang et al., 2005; Bitko et al., 2005). Surprisingly, intranasal siRNAs were also effective when delivered without a carrier (Bitko et al., 2005). Furthermore, siRNAs could inhibit virus replication when delivered up to 3 days after inoculation, albeit that this approach was less effective than when siRNAs were administrated before or at the day of inoculation (Bitko et al., 2005). This result and similar findings in other studies (Zhang et al., 2005; Ge et al., 2004) indicate that siRNAs may be used therapeutically to treat virus infections.

Together, these studies provide proof-of-principle for the potential of RNAi-based antiviral strategies. However, delivery of suppressive amounts of siRNAs to the sites of replication will be the major obstacle for clinical applications (Soutschek et al., 2004). The use of naked siRNA or siRNAs in complex with PEI or chitosan (Bitko et al., 2005;

Ge *et al.*, 2004; Zhang *et al.*, 2005) may be such approaches that are compatible with clinical use in humans.

RNAi to study virus replication and pathogenesis
In addition to the potential development of RNAi-based therapeutics, RNAi may influence the field of virology on another level. As described earlier, interactions between virus and host during the virus life cycle are numerous. Thus far, essential host factors have been identified using classical cloning-expression and biochemical approaches. RNAi may provide a more rapid and less labour-intensive approach for the identification of essential host factors in virus replication.

In addition, RNAi may provide an opportunity to perform reverse genetics for segmented viruses. Theoretically, it should be possible to decrease expression of a viral gene by RNAi and provide a mutated viral gene *in trans* by expression from plasmid vectors. However, as illustrated earlier, it may be difficult to target replicating genomes for degradation, thus limiting the possibility of generating replication-competent mutant virus. As an alternative, one could generate viruses that contain wild-type viral RNA segments, but are pseudotyped for mutated structural proteins.

RNAi AS AN INNATE ANTIVIRAL MECHANISM
One of the major remaining questions in the RNAi field is whether the RNA-silencing machinery has a natural antiviral role in mammals. The observed susceptibility of virus to synthetic siRNAs does not prove that an RNAi response is initiated in a natural infection. The critical role of RNAi in the defence against viruses has been firmly established in plants. Although viruses that infect *C. elegans* have not been identified, transposon mobilization provides a model for parasitic nucleic acids in this organism. These model organisms give important information on potential interactions between RNAi and virus. However, major differences in the RNAi pathway between different organisms exist, and therefore findings in one model organism may not apply to others.

An active uptake mechanism for dsRNA and systemic spread of a silencing signal (the molecular nature of which is unknown) occur in both plants and nematodes (Voinnet *et al.*, 1998; Winston *et al.*, 2002). Also, in both these model organisms, an RNA-dependent RNA polymerase activity is responsible for the generation of secondary siRNAs, generated to regions flanking the primary trigger dsRNA (Himber *et al.*, 2003; Sijen *et al.*, 2001; Vaistij *et al.*, 2002). These features seem to be absent in mammals. Furthermore, the number of Argonaute and Dicer-like genes varies among different model organisms. For example, the *Arabidopsis thaliana* genome contains ten Argonaute homologues and four Dicer-like genes, whereas mammals have four Argonaute homologues and only a single Dicer gene. The different Dicer-like proteins

Table 2. Examples of interactions between viruses and the RNAi machinery

Category	Virus	Genome	Activity/mechanism	Reference
Exploitation	Epstein–Barr virus	dsDNA	Virus-encoded miRNAs modulate host and virus gene expression	Pfeffer *et al.* (2004)
Targeted by RNAi	Plant viruses	RNA + DNA	Suppression of virus replication and spread	Voinnet (2001) (review)
Targeted by Dicer	Potato virus X	ssRNA	Double-stranded intermediates of replication	Hamilton & Baulcombe (1999)
	CymRSV*	ssRNA	Highly base-paired structure in single-stranded defective interfering RNA	Szittya *et al.* (2002)
	Geminiviruses	ssDNA	Overlapping transcripts from bidirectional transcription	Chellappan *et al.* (2004)
Inhibition of RNAi	Plant viruses	RNA + DNA	Suppression of RNAi via different mechanisms	Roth *et al.* (2004) (review)
	Influenza NS1	ssRNA	Interaction with ds and siRNA	Li *et al.* (2004)
	Nodamura virus B2	ssRNA	Interaction with dsRNA and suppression of Dicer activity	C. Sullivan & D. Ganem†
	Adenoviral VA1 RNA	dsDNA	Competitive binding to Exportin 5 and Dicer	Lu & Cullen (2004)

*CymRSV, Cymbidium ringspot tombusvirus.

†Unpublished observations.

process different sources of dsRNA, although functional redundancy seems to exist (Baulcombe, 2004). Two different classes of siRNA are generated in plants, which presumably have different functions (Hamilton *et al.*, 2002).

Here, a brief summary of the major findings on the role of RNAi in virus infections and transposon silencing in plants and worms is given (for more detailed reviews see Baulcombe, 2004; Plasterk, 2002; Voinnet, 2001).

RNAi versus transposons in *C. elegans*

The first evidence that transposon silencing is one of the functions of RNAi was provided by genetic experiments in *C. elegans*. In normal wild-type nematodes, transposition only occurs in somatic cells, whereas transposons are silenced in the germline. Worm mutants that show germline transposition were found to be defective in RNAi (Ketting *et al.*, 1999; Tabara *et al.*, 1999). Subsequent studies identified transposon-derived ds and siRNAs corresponding mainly to the terminal inverted repeat regions. These dsRNAs are likely derived from read-through transcription of flanking genes and subsequent intramolecular base-pairing between the terminal inverted repeat regions (Sijen & Plasterk, 2003).

RNAi versus viruses in plants

The role of RNAi (or post-transcriptional gene silencing in plant nomenclature) in virus infections has been extensively studied in plants. Small antisense RNA species of ~25 nt were first identified in virus infections and other forms of post-transcriptional gene silencing in plants (Hamilton & Baulcombe, 1999). Several findings underline the critical role of RNAi in antiviral defence.

(1) Production of virus-derived siRNA during virus infection (Hamilton & Baulcombe, 1999). One obvious source of these siRNAs is the cleavage of dsRNA intermediates in replication of ssRNA viruses by Dicer-like proteins (Angell & Baulcombe, 1997). However, highly base-paired structures in single-stranded defective interfering RNAs have also been reported as targets for Dicer (Szittya *et al.*, 2002). Single-stranded DNA viruses in plants also trigger an RNA-silencing response, even though there is no dsRNA replication intermediate. In this case, dsRNA is likely generated by bidirectional transcription from the dsDNA intermediate in replication (Chellappan *et al.*, 2004) (Table 2).

(2) Increased virus replication in the absence of RNAi. Mutants in which RNAi components are functionally inactivated display an enhanced susceptibility to virus infection (Mourrain *et al.*, 2000).

(3) Local and systemic spread of the RNA-silencing signal. Inoculation of lower leaves with a movement-defective mutant of potato virus X resulted in systemic spread and RNA silencing in upper leaves of tobacco plants (Voinnet *et al.*, 2000). It thus seems that a silencing signal travels ahead of the virus infection to uninfected tissues, thereby limiting virus infection when the virus reaches these cells.

(4) Virus suppressors of gene silencing (Roth *et al.*, 2004). To enable their replication and spread in the presence of a strong RNAi response, plant viruses have developed suppressors of silencing. In a screen of 15 plant viruses from 9 different virus families, 11 viruses were shown to suppress RNAi in plants (Voinnet *et al.*, 1999). Among these were viruses with DNA genomes, further lending support to the notion that the plant antiviral RNAi response can target DNA viruses. Suppressors have been identified that target different aspects of the silencing machinery. For example, potyvirus HC-pro blocks cleavage of dsRNA into siRNAs, the cucumber mosaic virus 2b protein suppresses the systemic spread of the silencing signal (Brigneti *et al.*, 1998), and p19 from tombusviruses binds siRNAs with high specificity, thus sequestering siRNA and preventing the spread of the silencing signal beyond the site of local infection (Vargason *et al.*, 2003; Ye *et al.*, 2003). This diversity of silencing suppressors was recently highlighted by the identification

of three distinct suppressors in a single viral genome (of citrus tristeza virus) (Lu *et al.*, 2004).

RNAi versus viruses in insects

It is becoming increasingly clear that RNAi provides a natural antiviral mechanism in insects. Flock house virus (FHV), a member of the *Nodaviridae* family that was isolated from naturally infected insects (grass grub, *Costelytra zealandica*), is targeted by the RNAi machinery in *Drosophila* S2 cells (Li *et al.*, 2002). Virus-derived siRNAs are produced over time, and Argonaute 2 depletion results in an increase in viral RNA levels. The FHV B2 protein seems to be essential for accumulation of viral RNA, whereas Argonaute 2 depletion relieves the dependence on the B2 protein in replication. The B2 can act as a suppressor of silencing in plants, and could functionally substitute the cucumber mosaic virus 2b protein (also a suppressor of RNA silencing) (Li *et al.*, 2002). Subsequent studies indicated that Nodamura virus (NoV) with a deletion of the B2 protein replicates in an Argonaute-2-depleted environment, but requires the B2 protein for replication in the presence of a functional RNAi machinery in *Drosophila* S2 cells and in a cell line from the mosquito *Anopheles gambiae* (Li *et al.*, 2004).

Further evidence for an innate antiviral RNAi response in insects comes from infections of mosquito cells with Sindbis virus (SINV) vectors expressing exogenous genes from an artificial second subgenomic promoter (sgSin). Infection of a cell line derived from the mosquito *Aedes albopictus* with sgSin expressing dengue virus genomic sequences resulted in resistance to subsequent challenges with dengue virus. In accordance with an RNAi antiviral response, 21–23-nt virus-derived siRNAs were observed in these infected cultures (Adelman *et al.*, 2001; Sanchez-Vargas *et al.*, 2004). The Sindbis-based expression system was also able to down-regulate endogenous gene expression *in vivo* in *Aedes aegyptii* mosquitos. Similar to the virus-induced gene-silencing phenomenon in plants (Voinnet, 2001), expression of (part of) the cellular trypsin gene by sgSin vectors resulted in degradation of the corresponding endogenous mRNA, silencing of trypsin protein expression, and production of siRNAs corresponding to this gene (Sanchez-Vargas *et al.*, 2004). Furthermore, SINV replication also induced an RNAi response *in vivo* in the silkmoth *Bombyx mori*, as virus-derived siRNA production was observed in infected animals (Uhlirova *et al.*, 2003). Further evidence for a role for the RNAi machinery was obtained by studies in an infection model of O'nyong-nyong virus (from the same family as SINV, i.e. *Togaviridae*, genus *Alphavirus*) in its natural vector, the mosquito *Anopheles gambiae*. Here, it was shown that co-injection of virus with dsRNA targeting two Argonaute proteins (Ago2 and 3) resulted in increased titres and virus spread (Keene *et al.*, 2004). In addition to these experimental infections, small RNAs derived from the endogenous Drosophila C virus

(DCV) were cloned from *Drosophila* embryos, while profiling the expression of small RNAs during *Drosophila* development (Aravin *et al.*, 2003). Together, these findings suggest that an antiviral RNAi response is initiated *in vitro* and *in vivo* in different virus infection models in insects, including virus infections of their natural hosts.

Thus far, RNAi in *Drosophila* seems to be a cell-autonomous process. Although secondary siRNAs have been described in embryonic extracts (Lipardi *et al.*, 2001), no evidence for transitive RNAi was obtained *in vivo* (Roignant *et al.*, 2003). In accordance, no RNA-dependent RNA polymerase (a requirement for the generation of secondary siRNAs) has been identified in the *Drosophila* genome. Furthermore, using tissue-specific expression of inverted repeats, it was shown that RNAi was confined to these tissues in the entire organism (Roignant *et al.*, 2003). Thus the amplification and systemic spread of the RNAi signal that generates systemic antiviral activity in plants (Himber *et al.*, 2003; Vaistij *et al.*, 2002) seems to be absent in insects. This implies that a potential antiviral RNAi response is unlikely to provide a systemic defence against viruses, but might restrict virus replication at the local level.

RNAi versus viruses: the case in mammals

Upon virus infection, a complex system of innate and adaptive immune responses is initiated in mammals. One of the innate responses to virus infection consists of the dsRNA-initiated PKR-mediated interferon response. Activation of this pathway leads to expression of type I interferons and interferon-inducible antiviral genes, non-specific inhibition of protein translation and apoptosis (Goodbourn *et al.*, 2000; Katze *et al.*, 2002). In light of this possibly competing dsRNA-initiated immune mechanism, it is possible that the antiviral role of RNAi has been lost during mammalian evolution. To establish whether RNAi represents an innate antiviral response in mammals, the same points as described earlier for plant viruses need to be addressed, i.e. production of virus-derived siRNA, enhanced virus replication in the absence of a functional RNAi response, and identification of virus-encoded suppressors of RNAi. Many of these issues are still unresolved. Thus far, the major emphasis has been placed on the identification of virus-encoded suppressors of RNAi (Table 2).

As described earlier, FHV lacking the B2 protein requires exogenous expression of the silencing suppressor B2 or knockdown of the RNAi machinery for efficient replication in *Drosophila* cells (Li *et al.*, 2002). Using this infection model, the dsRNA-binding NS1 protein from influenza virus and the E3L protein from vaccinia virus could substitute the B2 function (Li *et al.*, 2004). Furthermore, influenza NS1 was also able to bind siRNAs *in vitro* and suppress RNA silencing in plants (Bucher *et al.*, 2004; Lichner *et al.*, 2003). This indicates that NS1 can act as a suppressor of RNA silencing, at least in these models. In mammalian infection, however, the function of influenza NS1 as an interferon antagonist may explain the defect in replication observed upon deletion of a functional NS1 protein (Bergmann *et al.*, 2000; Wang *et al.*, 2000). To truly establish an antiviral function for RNAi in influenza infection requires therefore further investigation.

NoV replication is under control of an antiviral RNAi response, which is antagonized by B2 in insect cell culture (see earlier) (Li *et al.*, 2004). The virus infects and replicates in insects and mammalian hosts, providing an opportunity to study the role of B2 in mammalian cells as well. NoV B2 protein was able to inhibit shRNA and siRNA-initiated RNAi in mammalian cells by binding to pre-Dicer substrates and was able to inhibit Dicer cleavage of dsRNA *in vitro*. The single-stranded guide strand was mainly detected in immunoprecipitations of B2 protein, indicating that the protein may also interact with RNA loaded in RISC (C. Sullivan & D. Ganem, unpublished results). Defects in RNA replication of mutants without functional B2 protein varied from negligible to severe in different mammalian cell lines (Johnson *et al.*, 2004). These results could imply that an antiviral RNAi response limits replication of NoV, differing in efficacy among different mammalian cell lines.

Inhibition of the RNAi pathway mediated by non-coding viral RNA was observed for adenovirus. The 160-nt adenoviral RNA VA1 is expressed at high levels during infection and has been shown to be involved in blocking activation of PKR. In addition, VA1 inhibits RNAi induced by shRNA and miRNAs, but does not inhibit siRNA-initiated RNAi. Inhibition was mediated by competitive binding to the nuclear export factor Exportin 5 and to Dicer. Attempts to study RNAi in virus infection were complicated by the severe cytopathic effects of wild-type adenoviral infections. However, shRNA-based silencing of the luciferase reporter gene was modestly reversed by infection with a severely attenuated adenovirus that expressed low levels of VA1 (Lu & Cullen, 2004).

Together, these studies suggest that mammalian viruses may encode viral gene products that are able to inhibit RNAi in mammalian cells. By extension, this suggests that RNAi is indeed a functional antiviral response in mammals. However, many of the experiments in these studies were performed under conditions of overexpression of the protein of interest, and it remains unclear whether the timing and level of expression in the setting of virus infection are high enough to observe a suppressive effect on RNAi. Formal proof of an antiviral role of RNAi in mammals will depend on the identification of virus-derived siRNAs *in vivo* or *in vitro*. Thus far, no virus-derived siRNAs have been identified in tissue culture models for infection with hepatitis C virus, HIV and yellow fever virus (S. Pfeffer & T. Tuschl, unpublished observations) and in PV and Theiler's virus infections (M. C. Saleh & R. Andino, unpublished observations). Furthermore, it needs to be established whether the absence of a functional RNAi machinery, such as in genetic knockouts for the crucial RNAi components Dicer and Argonaute 2, results in an increase in virus replication. It will be of further interest to determine whether virus infection induces the up- or down-regulation of components of the RNAi machinery.

Similar to the situation in *Drosophila*, RNAi in mammals seem to be a cell-autonomous process, due to the absence of an uptake and amplification mechanism of RNAi. A mammalian antiviral RNAi response therefore may not be able to provide a systemic defence. However, RNAi may play a role in local restriction of virus infection, whereas other branches of the immune system may kick in after the initial phase of infection.

Finally, a remarkable interaction between viruses and the RNAi machinery was observed in Epstein–Barr virus (EBV) infection. While cloning small RNAs from a cell line latently infected with EBV, two different classes of virus-encoded miRNAs were identified (Pfeffer *et al.*, 2004). These miRNAs are encoded as fold-back precursors in the UTRs and introns of two viral genes. Predicted targets for these viral miRNAs include host cellular genes, such as regulators of cell proliferation and apoptosis, chemokines and cytokines, transcriptional regulators and components of signal transduction pathways. In addition, several potential virus targets were predicted. One of the viral miRNAs is perfectly complementary to the 3′ UTR of the viral DNA polymerase BALF5 and may target this gene for degradation. Indeed, cleavage products with a size that corresponds to the predicted miRNA-mediated cleavage product have been described (Furnari *et al.*, 1993). Thus EBV, and probably other DNA viruses, has found a way to exploit the miRNA pathway for regulation of its own viral genes and for interference with host gene expression. Rather than proving a point of RNAi as an antiviral pathway, it once again underlines the potential of this pathway for control of gene expression.

CONCLUSION

In recent years, RNAi has emerged as a major force in molecular and cell biology and virology. RNAi can easily be exploited as an experimental tool for functional knockdown of cellular genes. Furthermore, RNAi holds tremendous potential for the development of a novel class of drugs, with potential applications to different human diseases, including infections by important viral pathogens. Although proof-of-principle has been provided in tissue culture and mouse models, the major challenge will be the development of a delivery system for siRNA that is compatible with clinical use in humans.

In plants and insects, a strong case has been made for the function of RNAi as a natural antiviral mechanism. A parallel between RNAi and the immune system has been previously drawn (Gitlin & Andino, 2003; McManus, 2004; Plasterk, 2002; Zamore, 2004). However, whether RNAi indeed represents a novel arm of the antiviral immune response in mammals is still a wide-open question.

ACKNOWLEDGEMENTS

This work was supported by an EMBO long-term fellowship to R. v. R. and by NIH grant AI40085 to R. A. We thank C. Sullivan & D. Ganem, and S. Pfeffer & T. Tuschl, for permission to cite their unpublished observations. We thank Ron Geller, Maria Carla Saleh, Marco Vignuzzi and Derek Wells for critically reading the manuscript.

REFERENCES

Adelman, Z. N., Blair, C. D., Carlson, J. O., Beaty, B. J. & Olson, K. E. (2001). Sindbis virus-induced silencing of dengue viruses in mosquitoes. *Insect Mol Biol* **10**, 265–273.

Ambros, V. (2004). The functions of animal microRNAs. *Nature* **431**, 350–355.

Angell, S. M. & Baulcombe, D. C. (1997). Consistent gene silencing in transgenic plants expressing a replicating potato virus X RNA. *EMBO J* **16**, 3675–3684.

Aravin, A. A., Lagos-Quintana, M., Yalcin, A., Zavolan, M., Marks, D., Snyder, B., Gaasterland, T., Meyer, J. & Tuschl, T. (2003). The small RNA profile during *Drosophila melanogaster* development. *Dev Cell* **5**, 337–350.

Arias, C. F., Dector, M. A., Segovia, L., Lopez, T., Camacho, M., Isa, P., Espinosa, R. & Lopez, S. (2004). RNA silencing of rotavirus gene expression. *Virus Res* **102**, 43–51.

Bartel, D. P. (2004). MicroRNAs: genomics, biogenesis, mechanism, and function. *Cell* **116**, 281–297.

Barton, G. M. & Medzhitov, R. (2002). Retroviral delivery of small interfering RNA into primary cells. *Proc Natl Acad Sci U S A* **99**, 14943–14945.

Baulcombe, D. (2004). RNA silencing in plants. *Nature* **431**, 356–363.

Bergmann, M., Garcia-Sastre, A., Carnero, E., Pehamberger, H., Wolff, K., Palese, P. & Muster, T. (2000). Influenza virus NS1 protein counteracts PKR-mediated inhibition of replication. *J Virol* **74**, 6203–6206.

Bernstein, E., Caudy, A. A., Hammond, S. M. & Hannon, G. J. (2001). Role for a bi-dentate ribonuclease in the initiation step of RNA interference. *Nature* **409**, 363–366.

Bhuyan, P. K., Kariko, K., Capodici, J., Lubinski, J., Hook, L. M., Friedman, H. M. & Weissman, D. (2004). Short interfering RNA-mediated inhibition of herpes simplex virus type 1 gene expression and function during infection of human keratinocytes. *J Virol* **78**, 10276–10281.

Bitko, V. & Barik, S. (2001). Phenotypic silencing of cytoplasmic genes using sequence-specific double-stranded short interfering RNA and its application in the reverse genetics of wild type negative-strand RNA viruses. *BMC Microbiol* **1**, 34.

Bitko, V., Musiyenko, A., Shulyayeva, O. & Barik, S. (2005). Inhibition of respiratory viruses by nasally administered siRNA. *Nat Med* **11**, 50–55.

Boden, D., Pusch, O., Lee, F., Tucker, L. & Ramratnam, B. (2003). Human immuno-deficiency virus type 1 escape from RNA interference. *J Virol* **77**, 11531–11535.

Bohnsack, M. T., Czaplinski, K. & Gorlich, D. (2004). Exportin 5 is a RanGTP-dependent dsRNA-binding protein that mediates nuclear export of pre-miRNAs. *RNA* **10**, 185–191.

Bridge, A. J., Pebernard, S., Ducraux, A., Nicoulaz, A. L. & Iggo, R. (2003). Induction of an interferon response by RNAi vectors in mammalian cells. *Nat Genet* **34**, 263–264.

Brigneti, G., Voinnet, O., Li, W. X., Ji, L. H., Ding, S. W. & Baulcombe, D. C. (1998). Viral pathogenicity determinants are suppressors of transgene silencing in *Nicotiana benthamiana*. *EMBO J* **17**, 6739–6746.

Brummelkamp, T. R., Bernards, R. & Agami, R. (2002). A system for stable expression of short interfering RNAs in mammalian cells. *Science* **296**, 550–553.

Bucher, E., Hemmes, H., de Haan, P., Goldbach, R. & Prins, M. (2004). The influenza A virus NS1 protein binds small interfering RNAs and suppresses RNA silencing in plants. *J Gen Virol* **85**, 983–991.

Caudy, A. A., Myers, M., Hannon, G. J. & Hammond, S. M. (2002). Fragile X-related protein and VIG associate with the RNA interference machinery. *Genes Dev* **16**, 2491–2496.

Chang, J. & Taylor, J. M. (2003). Susceptibility of human hepatitis delta virus RNAs to small interfering RNA action. *J Virol* **77**, 9728–9731.

Chang, Y., Chang, S. S., Lee, H. H., Doong, S. L., Takada, K. & Tsai, C. H. (2004). Inhibition of the Epstein-Barr virus lytic cycle by Zta-targeted RNA interference. *J Gen Virol* **85**, 1371–1379.

Chellappan, P., Vanitharani, R. & Fauquet, C. M. (2004). Short interfering RNA accumulation correlates with host recovery in DNA virus-infected hosts, and gene silencing targets specific viral sequences. *J Virol* **78**, 7465–7477.

Chen, W., Yan, W., Du, Q., Fei, L., Liu, M., Ni, Z., Sheng, Z. & Zheng, Z. (2004). RNA interference targeting VP1 inhibits foot-and-mouth disease virus replication in BHK-21 cells and suckling mice. *J Virol* **78**, 6900–6907.

Chiu, Y. L., Cao, H., Jacque, J. M., Stevenson, M. & Rana, T. M. (2004). Inhibition of human immunodeficiency virus type 1 replication by RNA interference directed against human transcription elongation factor P-TEFb (CDK9/CyclinT1). *J Virol* **78**, 2517–2529.

Das, A. T., Brummelkamp, T. R., Westerhout, E. M., Vink, M., Madiredjo, M., Bernards, R. & Berkhout, B. (2004). Human immunodeficiency virus type 1 escapes from RNA interference-mediated inhibition. *J Virol* **78**, 2601–2605.

Dector, M. A., Romero, P., Lopez, S. & Arias, C. F. (2002). Rotavirus gene silencing by small interfering RNAs. *EMBO Rep* **3**, 1175–1180.

Denli, A. M., Tops, B. B., Plasterk, R. H., Ketting, R. F. & Hannon, G. J. (2004). Processing of primary microRNAs by the Microprocessor complex. *Nature* **432**, 231–235.

Dykxhoorn, D. M., Novina, C. D. & Sharp, P. A. (2003). Killing the messenger: short RNAs that silence gene expression. *Nat Rev Mol Cell Biol* **4**, 457–467.

Elbashir, S. M., Lendeckel, W. & Tuschl, T. (2001a). RNA interference is mediated by 21- and 22-nucleotide RNAs. *Genes Dev* **15**, 188–200.

Elbashir, S. M., Martinez, J., Patkaniowska, A., Lendeckel, W. & Tuschl, T. (2001b). Functional anatomy of siRNAs for mediating efficient RNAi in *Drosophila melanogaster* embryo lysate. *EMBO J* **20**, 6877–6888.

Elbashir, S. M., Harborth, J., Lendeckel, W., Yalcin, A., Weber, K. & Tuschl, T. (2001c).

Duplexes of 21-nucleotide RNAs mediate RNA interference in cultured mammalian cells. *Nature* **411**, 494–498.

Engelke, D. R. (editor) (2003). *RNA Interference (RNAi) – Nuts and Bolts of RNAi Technology*. Eagleville, PA: DNA Press.

Fire, A., Xu, S., Montgomery, M. K., Kostas, S. A., Driver, S. E. & Mello, C. C. (1998). Potent and specific genetic interference by double-stranded RNA in *Caenorhabditis elegans*. *Nature* **391**, 806–811.

Furnari, F. B., Adams, M. D. & Pagano, J. S. (1993). Unconventional processing of the 3′ termini of the Epstein-Barr virus DNA polymerase mRNA. *Proc Natl Acad Sci U S A* **90**, 378–382.

Gao, X., Wang, H. & Sairenji, T. (2004). Inhibition of Epstein-Barr virus (EBV) reactivation by short interfering RNAs targeting p38 mitogen-activated protein kinase or c-myc in EBV-positive epithelial cells. *J Virol* **78**, 11798–11806.

Garrus, J. E., von Schwedler, U. K., Pornillos, O. W. & 9 other authors (2001). Tsg101 and the vacuolar protein sorting pathway are essential for HIV-1 budding. *Cell* **107**, 55–65.

Ge, Q., McManus, M. T., Nguyen, T., Shen, C. H., Sharp, P. A., Eisen, H. N. & Chen, J. (2003). RNA interference of influenza virus production by directly targeting mRNA for degradation and indirectly inhibiting all viral RNA transcription. *Proc Natl Acad Sci U S A* **100**, 2718–2723.

Ge, Q., Filip, L., Bai, A., Nguyen, T., Eisen, H. N. & Chen, J. (2004). Inhibition of influenza virus production in virus-infected mice by RNA interference. *Proc Natl Acad Sci U S A* **101**, 8676–8681.

Giladi, H., Ketzinel-Gilad, M., Rivkin, L., Felig, Y., Nussbaum, O. & Galun, E. (2003). Small interfering RNA inhibits hepatitis B virus replication in mice. *Mol Ther* **8**, 769–776.

Gitlin, L. & Andino, R. (2003). Nucleic acid-based immune system: the antiviral potential of mammalian RNA silencing. *J Virol* **77**, 7159–7165.

Gitlin, L., Karelsky, S. & Andino, R. (2002). Short interfering RNA confers intracellular antiviral immunity in human cells. *Nature* **418**, 430–434.

Gitlin, L., Stone, J. K. & Andino, R. (2005). Poliovirus escape from RNAi: siRNA-target recognition and implications for therapeutic approaches. *J Virol* **79**, 1027–1035.

Goodbourn, S., Didcock, L. & Randall, R. E. (2000). Interferons: cell signalling, immune modulation, antiviral response and virus countermeasures. *J Gen Virol* **81**, 2341–2364.

Gregory, R. I., Yan, K. P., Amuthan, G., Chendrimada, T., Doratotaj, B., Cooch, N. & Shiekhattar, R. (2004). The Microprocessor complex mediates the genesis of microRNAs. *Nature* **432**, 235–240.

Hall, A. H. & Alexander, K. A. (2003). RNA interference of human papillomavirus type 18 E6 and E7 induces senescence in HeLa cells. *J Virol* **77**, 6066–6069.

Hamilton, A. J. & Baulcombe, D. C. (1999). A species of small antisense RNA in posttranscriptional gene silencing in plants. *Science* **286**, 950–952.

Hamilton, A., Voinnet, O., Chappell, L. & Baulcombe, D. (2002). Two classes of short interfering RNA in RNA silencing. *EMBO J* **21**, 4671–4679.

Hammond, S. M., Bernstein, E., Beach, D. & Hannon, G. J. (2000). An RNA-directed nuclease mediates post-transcriptional gene silencing in *Drosophila* cells. *Nature* **404**, 293–296.

Hannon, G. J. (editor) (2003). *RNAi, a Guide to Gene Silencing*. Cold Spring Harbor, NY: Cold Spring Harbor Laboratory.

Hannon, G. J. & Rossi, J. J. (2004). Unlocking the potential of the human genome with RNA interference. *Nature* **431**, 371–378.

He, L. & Hannon, G. J. (2004). MicroRNAs: small RNAs with a big role in gene regulation. *Nat Rev Genet* **5**, 522–531.

He, M. L., Zheng, B., Peng, Y., Peiris, J. S., Poon, L. L., Yuen, K. Y., Lin, M. C., Kung, H. F. & Guan, Y. (2003). Inhibition of SARS-associated coronavirus infection and replication by RNA interference. *JAMA (J Am Med Assoc)* **290**, 2665–2666.

Himber, C., Dunoyer, P., Moissiard, G., Ritzenthaler, C. & Voinnet, O. (2003). Transitivity-dependent and -independent cell-to-cell movement of RNA silencing. *EMBO J* **22**, 4523–4533.

Hu, W. Y., Bushman, F. D. & Siva, A. C. (2004). RNA interference against retroviruses. *Virus Res* **102**, 59–64.

Jia, Q. & Sun, R. (2003). Inhibition of gammaherpesvirus replication by RNA interference. *J Virol* **77**, 3301–3306.

Johnson, K. L., Price, B. D., Eckerle, L. D. & Ball, L. A. (2004). Nodamura virus non-structural protein B2 can enhance viral RNA accumulation in both mammalian and insect cells. *J Virol* **78**, 6698–6704.

Katze, M. G., He, Y. & Gale, M., Jr (2002). Viruses and interferon: a fight for supremacy. *Nat Rev Immunol* **2**, 675–687.

Keene, K. M., Foy, B. D., Sanchez-Vargas, I., Beaty, B. J., Blair, C. D. & Olson, K. E. (2004). From the cover: RNA interference acts as a natural antiviral response to O'nyong-nyong virus (Alphavirus; Togaviridae) infection of *Anopheles gambiae*. *Proc Natl Acad Sci U S A* **101**, 17240–17245.

Ketting, R. F., Haverkamp, T. H., van Luenen, H. G. & Plasterk, R. H. (1999). Mut-7 of *C. elegans*, required for transposon silencing and RNA interference, is a homolog of Werner syndrome helicase and RNaseD. *Cell* **99**, 133–141.

Khvorova, A., Reynolds, A. & Jayasena, S. D. (2003). Functional siRNAs and miRNAs exhibit strand bias. *Cell* **115**, 209–216.

Kim, J., Krichevsky, A., Grad, Y., Hayes, G. D., Kosik, K. S., Church, G. M. & Ruvkun, G. (2004). Identification of many microRNAs that copurify with polyribosomes in mammalian neurons. *Proc Natl Acad Sci U S A* **101**, 360–365.

Kronke, J., Kittler, R., Buchholz, F., Windisch, M. P., Pietschmann, T., Bartenschlager, R. & Frese, M. (2004). Alternative approaches for efficient inhibition of hepatitis C virus RNA replication by small interfering RNAs. *J Virol* **78**, 3436–3446.

Lee, M. T., Coburn, G. A., McClure, M. O. & Cullen, B. R. (2003). Inhibition of human immunodeficiency virus type 1 replication in primary macrophages by using Tat- or CCR5-specific small interfering RNAs expressed from a lentivirus vector. *J Virol* **77**, 11964–11972.

Lee, Y., Ahn, C., Han, J. & 8 other authors (2003). The nuclear RNase III Drosha initiates microRNA processing. *Nature* **425**, 415–419.

Li, H. W., Li, W. X. & Ding, S. W. (2002). Induction and suppression of RNA silencing by an animal virus. *Science* **296**, 1319–1321.

Li, W. X., Li, H., Lu, R. & 9 other authors (2004). Interferon antagonist proteins of influenza and vaccinia viruses are suppressors of RNA silencing. *Proc Natl Acad Sci U S A* **101**, 1350–1355.

Lichner, Z., Silhavy, D. & Burgyan, J. (2003). Double-stranded RNA-binding proteins could suppress RNA interference-mediated antiviral defences. *J Gen Virol* **84**, 975–980.

Lipardi, C., Wei, Q. & Paterson, B. M. (2001). RNAi as random degradative PCR: siRNA primers convert mRNA into dsRNAs that are degraded to generate new siRNAs. *Cell* **107**, 297–307.

Lippman, Z. & Martienssen, R. (2004). The role of RNA interference in heterochromatic silencing. *Nature* **431**, 364–370.

Liu, J., Carmell, M. A., Rivas, F. V., Marsden, C. G., Thomson, J. M., Song, J. J., Hammond, S. M., Joshua-Tor, L. & Hannon, G. J. (2004). Argonaute2 is the catalytic engine of mammalian RNAi. *Science* **305**, 1437–1441.

Liu, R., Paxton, W. A., Choe, S. & 7 other authors (1996). Homozygous defect in HIV-1 coreceptor accounts for resistance of some multiply-exposed individuals to HIV-1 infection. *Cell* **86**, 367–377.

Lopez, S. & Arias, C. F. (editors) (2004). Special Issue: Viral gene silencing by RNA interference. *Virus Res* **102**, 1–124.

Lu, R., Folimonov, A., Shintaku, M., Li, W. X., Falk, B. W., Dawson, W. O. & Ding, S. W. (2004). Three distinct suppressors of RNA silencing encoded by a 20-kb viral RNA genome. *Proc Natl Acad Sci U S A* **101**, 15742–15747.

Lu, S. & Cullen, B. R. (2004). Adenovirus VA1 noncoding RNA can inhibit small interfering RNA and MicroRNA biogenesis. *J Virol* **78**, 12868–12876.

Lund, E., Guttinger, S., Calado, A., Dahlberg, J. E. & Kutay, U. (2004). Nuclear export of microRNA precursors. *Science* **303**, 95–98.

Mansfield, J. H., Harfe, B. D., Nissen, R. & 10 other authors (2004). Micro-RNA-responsive 'sensor' transgenes uncover Hox-like and other development-ally regulated patterns of vertebrate microRNA expression. *Nat Genet* **36**, 1079–1083.

McCaffrey, A. P., Nakai, H., Pandey, K., Huang, Z., Salazar, F. H., Xu, H., Wieland, S. F., Marion, P. L. & Kay, M. A. (2003). Inhibition of hepatitis B virus in mice by RNA interference. *Nat Biotechnol* **21**, 639–644.

McCown, M., Diamond, M. S. & Pekosz, A. (2003). The utility of siRNA transcripts produced by RNA polymerase i in down regulating viral gene expression and replication of negative- and positive-strand RNA viruses. *Virology* **313**, 514–524.

McManus, M. T. (2004). Small RNAs and immunity. *Immunity* **21**, 1–20.

Meister, G. & Tuschl, T. (2004). Mechanisms of gene silencing by double-stranded RNA. *Nature* **431**, 343–349.

Mochizuki, K. & Gorovsky, M. A. (2004). Small RNAs in genome rearrangement in *Tetrahymena*. *Curr Opin Genet Dev* **14**, 181–187.

Mourrain, P., Beclin, C., Elmayan, T. & 11 other authors (2000). *Arabidopsis* SGS2 and SGS3 genes are required for posttranscriptional gene silencing and natural virus resistance. *Cell* **101**, 533–542.

Myers, J. W., Jones, J. T., Meyer, T. & Ferrell, J. E., Jr (2003). Recombinant Dicer efficiently converts large dsRNAs into siRNAs suitable for gene silencing. *Nat Biotechnol* **21**, 324–328.

Orba, Y., Sawa, H., Iwata, H., Tanaka, S. & Nagashima, K. (2004). Inhibition of virus production in JC virus-infected cells by postinfection RNA interference. *J Virol* **78**, 7270–7273.

Paddison, P. J., Caudy, A. A., Bernstein, E., Hannon, G. J. & Conklin, D. S. (2002). Short hairpin RNAs (shRNAs) induce sequence-specific silencing in mammalian cells. *Genes Dev* **16**, 948–958.

Persengiev, S. P., Zhu, X. & Green, M. R. (2004). Nonspecific, concentration-dependent stimulation and repression of mammalian gene expression by small interfering RNAs (siRNAs). *RNA* **10**, 12–18.

Pfeffer, S., Zavolan, M., Grasser, F. A. & 8 other authors (2004). Identification of virus-encoded microRNAs. *Science* **304**, 734–736.

Pham, J. W., Pellino, J. L., Lee, Y. S., Carthew, R. W. & Sontheimer, E. J. (2004). A Dicer-2-dependent 80s complex cleaves targeted mRNAs during RNAi in *Drosophila*. *Cell* **117**, 83–94.

Phipps, K. M., Martinez, A., Lu, J., Heinz, B. A. & Zhao, G. (2004). Small interfering RNA molecules as potential anti-human rhinovirus agents: in vitro potency, specificity, and mechanism. *Antiviral Res* **61**, 49–55.

Plasterk, R. H. (2002). RNA silencing: the genome's immune system. *Science* **296**, 1263–1265.

Qin, X. F., An, D. S., Chen, I. S. & Baltimore, D. (2003). Inhibiting HIV-1 infection in human T cells by lentiviral-mediated delivery of small interfering RNA against CCR5. *Proc Natl Acad Sci U S A* **100**, 183–188.

Radhakrishnan, S., Gordon, J., Del Valle, L., Cui, J. & Khalili, K. (2004). Intracellular approach for blocking JC virus gene expression by using RNA interference during viral infection. *J Virol* **78**, 7264–7269.

Randall, G. & Rice, C. M. (2004). Interfering with hepatitis C virus RNA replication. *Virus Res* **102**, 19–25.

Rocheleau, C. E., Downs, W. D., Lin, R., Wittmann, C., Bei, Y., Cha, Y. H., Ali, M., Priess, J. R. & Mello, C. C. (1997). Wnt signaling and an APC-related gene specify endoderm in early *C. elegans* embryos. *Cell* **90**, 707–716.

Roignant, J. Y., Carre, C., Mugat, B., Szymczak, D., Lepesant, J. A. & Antoniewski, C. (2003). Absence of transitive and systemic pathways allows cell-specific and isoform-specific RNAi in *Drosophila*. *RNA* **9**, 299–308.

Roth, B. M., Pruss, G. J. & Vance, V. B. (2004). Plant viral suppressors of RNA silencing. *Virus Res* **102**, 97–108.

Sanchez-Vargas, I., Travanty, E. A., Keene, K. M., Franz, A. W., Beaty, B. J., Blair, C. D. & Olson, K. E. (2004). RNA interference, arthropod-borne viruses, and mosquitoes. *Virus Res* **102**, 65–74.

Schwarz, D. S., Hutvagner, G., Du, T., Xu, Z., Aronin, N. & Zamore, P. D. (2003). Asymmetry in the assembly of the RNAi enzyme complex. *Cell* **115**, 199–208.

Sijen, T. & Plasterk, R. H. (2003). Transposon silencing in the *Caenorhabditis elegans* germ line by natural RNAi. *Nature* **426**, 310–314.

Sijen, T., Fleenor, J., Simmer, F., Thijssen, K. L., Parrish, S., Timmons, L., Plasterk, R. H. & Fire, A. (2001). On the role of RNA amplification in dsRNA-triggered gene silencing. *Cell* **107**, 465–476.

Silvestri, L. S., Taraporewala, Z. F. & Patton, J. T. (2004). Rotavirus replication: plus-sense templates for double-stranded RNA synthesis are made in viroplasms. *J Virol* **78**, 7763–7774.

Sledz, C. A., Holko, M., de Veer, M. J., Silverman, R. H. & Williams, B. R. (2003). Activation of the interferon system by short-interfering RNAs. *Nat Cell Biol* **5**, 834–839.

Song, J. J., Smith, S. K., Hannon, G. J. & Joshua-Tor, L. (2004). Crystal structure of Argonaute and its implications for RISC slicer activity. *Science* **305**, 1434–1437.

Soutschek, J., Akinc, A., Bramlage, B. & 22 other authors (2004). Therapeutic silencing of an endogenous gene by systemic administration of modified siRNAs. *Nature* **432**, 173–178.

Stevenson, M. (2003). Dissecting HIV-1 through RNA interference. *Nat Rev Immunol* **3**, 851–858.

Surabhi, R. M. & Gaynor, R. B. (2002). RNA interference directed against viral and cellular targets inhibits human immunodeficiency virus Type 1 replication. *J Virol* **76**, 12963–12973.

Szittya, G., Molnar, A., Silhavy, D., Hornyik, C. & Burgyan, J. (2002). Short defective interfering RNAs of tombusviruses are not targeted but trigger post-transcriptional gene silencing against their helper virus. *Plant Cell* **14**, 359–372.

Tabara, H., Sarkissian, M., Kelly, W. G., Fleenor, J., Grishok, A., Timmons, L., Fire, A. & Mello, C. C. (1999). The rde-1 gene, RNA interference, and transposon silencing in *C. elegans*. *Cell* **99**, 123–132.

Tomari, Y., Matranga, C., Haley, B., Martinez, N. & Zamore, P. D. (2004). A protein sensor for siRNA asymmetry. *Science* **306**, 1377–1380.

Tompkins, S. M., Lo, C. Y., Tumpey, T. M. & Epstein, S. L. (2004). Protection against lethal influenza virus challenge by RNA interference in vivo. *Proc Natl Acad Sci U S A* **101**, 8682–8686.

Uhlirova, M., Foy, B. D., Beaty, B. J., Olson, K. E., Riddiford, L. M. & Jindra, M. (2003). Use of Sindbis virus-mediated RNA interference to demonstrate a conserved role of Broad-Complex in insect metamorphosis. *Proc Natl Acad Sci U S A* **100**, 15607–15612.

Vaistij, F. E., Jones, L. & Baulcombe, D. C. (2002). Spreading of RNA targeting and DNA methylation in RNA silencing requires transcription of the target gene and a putative RNA-dependent RNA polymerase. *Plant Cell* **14**, 857–867.

Vargason, J. M., Szittya, G., Burgyan, J. & Tanaka Hall, T. M. (2003). Size selective recognition of siRNA by an RNA silencing suppressor. *Cell* **115**, 799–811.

Voinnet, O. (2001). RNA silencing as a plant immune system against viruses. *Trends Genet* **17**, 449–459.

Voinnet, O., Vain, P., Angell, S. & Baulcombe, D. C. (1998). Systemic spread of sequence-specific transgene RNA degradation in plants is initiated by localized introduction of ectopic promoterless DNA. *Cell* **95**, 177–187.

Voinnet, O., Pinto, Y. M. & Baulcombe, D. C. (1999). Suppression of gene silencing: a general strategy used by diverse DNA and RNA viruses of plants. *Proc Natl Acad Sci U S A* **96**, 14147–14152.

Voinnet, O., Lederer, C. & Baulcombe, D. C. (2000). A viral movement protein prevents spread of the gene silencing signal in *Nicotiana benthamiana*. *Cell* **103**, 157–167.

Wang, X., Li, M., Zheng, H., Muster, T., Palese, P., Beg, A. A. & Garcia-Sastre, A. (2000). Influenza A virus NS1 protein prevents activation of NF-kappaB and induction of alpha/beta interferon. *J Virol* **74**, 11566–11573.

Wang, Z., Ren, L., Zhao, X., Hung, T., Meng, A., Wang, J. & Chen, Y. G. (2004). Inhibition of severe acute respiratory syndrome virus replication by small interfering RNAs in mammalian cells. *J Virol* **78**, 7523–7527.

Wiebusch, L., Truss, M. & Hagemeier, C. (2004). Inhibition of human cytomegalovirus replication by small interfering RNAs. *J Gen Virol* **85**, 179–184.

Winston, W. M., Molodowitch, C. & Hunter, C. P. (2002). Systemic RNAi in *C. elegans* requires the putative transmembrane protein SID-1. *Science* **295**, 2456–2459.

Xia, H., Mao, Q., Paulson, H. L. & Davidson, B. L. (2002). siRNA-mediated gene silencing in vitro and in vivo. *Nat Biotechnol* **16**, 16.

Yang, D., Buchholz, F., Huang, Z., Goga, A., Chen, C. Y., Brodsky, F. M. & Bishop, J. M. (2002). Short RNA duplexes produced by hydrolysis with *Escherichia coli* RNase III mediate effective RNA interference in mammalian cells. *Proc Natl Acad Sci U S A* **99**, 9942–9947.

Ye, K., Malinina, L. & Patel, D. J. (2003). Recognition of small interfering RNA by a viral suppressor of RNA silencing. *Nature* **426**, 874–878.

Yekta, S., Shih, I. H. & Bartel, D. P. (2004). MicroRNA-directed cleavage of HOXB8 mRNA. *Science* **304**, 594–596.

Yi, R., Qin, Y., Macara, I. G. & Cullen, B. R. (2003). Exportin-5 mediates the nuclear export of pre-microRNAs and short hairpin RNAs. *Genes Dev* **17**, 3011–3016.

Yokota, T., Sakamoto, N., Enomoto, N. & 8 other authors (2003). Inhibition of intracellular hepatitis C virus replication by synthetic and vector-derived small interfering RNAs. *EMBO Rep* **4**, 602–608.

Yoon, J. S., Kim, S. H., Shin, M. C., Hong, S. K., Jung, Y. T., Khang, I. G., Shin, W. S., Kim, C. C. & Paik, S. Y. (2004). Inhibition of herpesvirus-6B RNA replication by short interference RNAs. *J Biochem Mol Biol* **37**, 383–385.

Zamore, P. D. (2004). Plant RNAi: how a viral silencing suppressor inactivates siRNA. *Curr Biol* **14**, R198–200.

Zhang, W., Yang, H., Kong, X. & 7 other authors (2005). Inhibition of respiratory syncytial virus infection with intranasal siRNA nanoparticles targeting the viral NS1 gene. *Nat Med* **11**, 56–62.

Neurons and host immunity conspire to maintain herpes simplex virus in a latent state

Michael L. Freeman,[1,2] Vilma Decman[1,3] and Robert L. Hendricks[1,4,5]

Department of Ophthalmology[1], Graduate Programs in Molecular Virology and Microbiology[2] and Immunology[3], and Departments of Immunology[4] and Molecular Genetics and Biochemistry[5], University of Pittsburgh School of Medicine, Pittsburgh, PA 15213, USA

ESTABLISHMENT OF A LATENT INFECTION

Herpes simplex virus 1 (HSV-1) is a large DNA virus that encodes more than 80 proteins. Although humans are the only natural host of HSV-1, infections can be induced experimentally in a variety of animals, including mice, rabbits, rats and guinea pigs. Primary infection usually occurs on mucosal surfaces following exposure to infected secretions. A lytic infection of the epithelia ensues in which HSV-1 genes are expressed sequentially as immediate early (α) genes, whose products are primarily involved in transactivation of early and late genes and immune evasion; early (β) genes, whose products are primarily involved in regulating viral DNA synthesis; and late (γ) genes, which primarily encode viral structural proteins. The latter are subdivided into γ_1 genes, which are expressed early in the lytic cycle, but whose expression is enhanced after viral DNA synthesis; and γ_2 genes, which are only expressed after the initiation of DNA synthesis.

During a primary infection, HSV-1 invades the neurons that innervate the infected mucosal surface. The virus is transported by retrograde axonal transport to the neuronal nuclei that are housed in the sensory ganglia. In animal models, HSV-1 replicates briefly in neurons, and progressive infection of surrounding neurons is observed. The virus then establishes a latent infection in which functional viral genomes are retained in neuronal nuclei, but no virus replication occurs.

MAINTENANCE OF HSV-1 LATENCY

In humans and in some animal models (rabbit and guinea pig), HSV-1 sporadically reactivates from latency in sensory neurons, is transported by anterograde axonal

SGM symposium 64: Molecular pathogenesis of virus infections.
Editors P. Digard, A. A. Nash & R. E. Randall. Cambridge University Press. ISBN 0 521 83248 9 ©SGM 2005

transport, and is shed at peripheral sites, potentially leading to recurrent disease. HSV-1 establishes a latent infection in the sensory ganglia of mice, but does not spontaneously reactivate. However, exposure of humans, rabbits and mice to emotional or physical stress or immunosuppression can induce HSV-1 reactivation and shedding at the periphery (Padgett *et al.*, 1998; Pereira *et al.*, 2003; Sainz *et al.*, 2001). Recurrent infections of the cornea are accompanied by immunopathology that can lead to corneal scarring and permanent loss of vision (Hendricks, 1999). HSV-1 corneal disease is the most frequent infectious cause of blindness in the United States, and a leading indication for corneal transplantation (Liesegang, 2001). Since most HSV-1 disease is caused by reactivation of the virus from latency and transport to the periphery, it is important to understand the factors that are responsible for inhibiting viral gene expression during latency.

IS THERE A ROLE FOR THE HOST IMMUNE SYSTEM IN THE MAINTENANCE OF HSV-1 LATENCY?

Many of the known stimuli that induce HSV-1 reactivation from latency are associated with immunosuppression, including the immunosuppressive properties of stress and stress hormones, and iatrogenic immunosuppression. The association of HSV-1 reactivation with immunosuppression, combined with the recently observed, long-term retention of $CD8^+$ T cells in HSV-1 latently infected mouse and human ganglia, are suggestive of a role for $CD8^+$ T cells in maintaining the virus in a latent state (Stroop & Schaefer, 1987; Liu *et al.*, 1996; Halford *et al.*, 1996; Shimeld *et al.*, 1995). However, before considering such a role for $CD8^+$ T cells, it is important to clearly define the terms latency and reactivation. For the purpose of all subsequent discussion, latency will be defined as the retention of a functional HSV-1 genome in the extended absence of virus particles. From a pragmatic standpoint, HSV-1 reactivation should be defined in terms of virion formation, since disease can only occur when latency gives way to production of infectious virus. Moreover, defining latency as terminating at the point of virion formation will accommodate expression of a limited array of lytic genes during latency, as is seen with all other members of the herpesvirus family.

$CD8^+$ T cell function is regulated by cognate interaction with an epitope comprised of an antigenic peptide bound to an appropriate MHC class I molecule on the surface of a target cell. For $CD8^+$ T cells to regulate HSV-1 latency, the target cells would of necessity be the neurons that harbour latent virus. The theory that $CD8^+$ T cells recognize an HSV-1 epitope on latently infected neurons flies in the face of two widely held beliefs: that no HSV-1 proteins are produced during latency, and that neurons do not express MHC class I molecules. The ensuing sections will present evidence that some HSV lytic gene expression occurs in latently infected neurons, that infected neurons do express MHC class I molecules, that $CD8^+$ T cells do, indeed, detect viral epitopes on latently infected neurons, and that the interaction of CD8$^+$ T cells with latently infected neurons prevents full reactivation and virion formation.

DO NEURONS THAT HARBOUR LATENT HSV-1 EXPRESS ANY VIRAL LYTIC GENES?

The highly regulated expression of 80 HSV genes during lytic infection appears to be largely abrogated during latency. Latency-associated transcripts (LATs) are the only abundant viral transcripts present in latently infected neurons (Stevens *et al.*, 1987). However, no LAT translation product has been reproducibly observed. HSV transcripts other than LATs are not usually detected in latently infected neurons by *in situ* hybridization (Croen *et al.*, 1987; Stevens *et al.*, 1987). These findings lead to the general concept that the latent viral genome is both transcriptionally and translationally silent, allowing the latent virus to effectively hide from the host immune system. However, recent studies employing more sensitive detection methods show expression of HSV immediate early (α), early (β) and even late (γ_1) viral genes in latently infected mouse neurons (Feldman *et al.*, 2002; Chen *et al.*, 1997, 2002; Kramer & Coen, 1995). Moreover, viral antigens have been detected in latently infected trigeminal ganglia (TG) *in vivo* (Feldman *et al.*, 2002; Sawtell, 2003) and in latently infected neurons in *ex vivo* TG cultures (Liu *et al.*, 2000). Thus the concept that the virus is able to 'hide' from the immune system during latency now appears less tenable.

DO INFECTED NEURONS EXPRESS MHC CLASS I?

Neurons are among a very small number of cell types that do not normally express detectable MHC class I molecules. Yet the concept that CD8$^+$ T cells can directly monitor HSV-1 gene expression in latently infected neurons assumes expression of an MHC class I/viral peptide complex on the surface of the neuron. How then can one make these two concepts compatible? A clue comes from reports by Pereira and others (Pereira *et al.*, 1994; Pereira & Simmons, 1999) demonstrating that sensory neurons express readily detectable MHC class I during acute HSV-1 infection. These investigators documented MHC class I expression on sensory neurons during and briefly following the period of HSV-1 replication within the sensory ganglia. This was consistent with the observation that termination of HSV-1 replication in the peripheral nervous system was dependent on CD8$^+$ T cells (Simmons & Tscharke, 1992). Interestingly, HSV-1 replication and the associated MHC class I expression did not appear to result in loss of viability of the neurons. In contrast, once latency was established in the sensory ganglia, MHC class I expression was no longer detectable on the latently infected neurons. These findings invite the speculation that HSV-1 lytic genes and MHC class I genes might be concordantly regulated. Evidence presented earlier and that to follow suggests that a low level of viral gene expression is permitted within latently infected neurons, though the level of protein expression in most neurons might be too low to detect by conventional means. A similar argument might be made for MHC class I expression. Since CD8[+] T cell stimulation requires extremely low levels of epitope expression (as little as one MHC class I/peptide complex per cell), it is conceivable that viral epitopes are present on the surface of latently infected neurons at a very early stage in the reactivation process.

EVIDENCE FOR CD8[+] T CELL CONTROL OF HSV-1 LATENCY

Although the immune response at the primary site of HSV-1 infection has been well characterized, less is known about the immunological events that occur in sensory ganglia during latency. However, substantial evidence is now emerging in support of a role for T lymphocytes in controlling HSV-1 latency. Expression of chemokines such as RANTES that are essential for attracting T cells, as well as T-cell-derived antiviral cytokines (i.e. IFN-γ and TNF-α), is detectable in latently infected TG for at least 180 days after HSV-1 corneal infection (Liu *et al.*, 1996; Halford *et al.*, 1996, 1997; Cantin *et al.*, 1995; Chen *et al.*, 2000). Moreover, histological analysis of mouse TG after HSV-1 corneal infection revealed that both CD4[+] and CD8[+] T cells accumulate and are retained in the ganglion seemingly for the life of the animal (Shimeld *et al.*, 1995; Liu *et al.*, 1996). The CD8[+] T cells were found in close apposition to the neuron cell bodies within the ophthalmic branch of the TG. The relevance of these findings in HSV-1-infected humans was revealed in a recent report showing localization of CD8[+] T cells to neuron cell bodies in TG of humans with a history of recurrent HSV-1 infections (Theil *et al.*, 2003). Importantly, the CD8[+] T cells selectively associated with neurons that harboured latent HSV-1.

The majority of the CD8[+] T cells in latently infected TG of mice are specific for a single immunodominant epitope on the HSV-1 γ_1 gene product glycoprotein B (gB$_{498-505}$) (Khanna *et al.*, 2003). Interestingly, many of these gB$_{498-505}$-specific CD8[+] T cells acquired expression of the recent activation marker CD69 while present in the latently infected ganglia, and polarized their T-cell receptor to the junction with neuron cell bodies. The latter findings provide strong evidence for the concept that low levels of certain viral proteins (in particular gB) are produced in neurons, and recognized by CD8[+] T cells within the ganglion. It is reasonable to conclude, therefore, that CD8[+] T cells can respond to viral gene expression in latently infected neurons prior to full reactivation and virion formation.

A role for CD8[+] T cells in blocking HSV-1 reactivation from latency has received further support from studies incorporating *ex vivo* cultures of latently infected TG. In one such study, the CD8[+] T cells that were present in latently infected TG of mice 14 days after HSV-1 corneal infection were shown to completely inhibit reactivation in *ex vivo* TG cultures (Liu *et al.*, 2000). In addition, a $gB_{498-505}$-specific $CD8^+$ T cell clone was shown to block HSV-1 reactivation from latency in a dose-dependent, antigen-specific, and MHC-restricted fashion when added to $CD8^+$ T cell-depleted ganglion cultures (Khanna *et al.*, 2003). The protected cultures were shown by RT-PCR to lack transcripts for the viral γ_2 gene for glycoprotein H (gH). These studies demonstrated an exquisite $CD8^+$ T cell regulation of HSV-1 gene expression during latency in which recognition of a γ_1 gene product, gB, induced a response that terminated the viral life cycle prior to expression of the γ_2 gene gH.

Several studies support the notion that $CD8^+$ T cells prevent HSV-1 reactivation from latency in part through the production of IFN-γ. As noted previously, IFN-γ mRNA and proteins are consistently detected in latently infected ganglia (Cantin *et al.*, 1995; Halford *et al.*, 1996; Shimeld *et al.*, 1997; Liu *et al.*, 1996). Although HSV-1 does not spontaneously reactivate in IFN-$\gamma^{-/-}$ or IFN-$\gamma R^{-/-}$ mice, the incidence of stress-induced reactivation in these mice was significantly higher than that of wild-type mice (Cantin *et al.*, 1999; Minami *et al.*, 2002). While the results of this study suggested a role for IFN-γ in preventing HSV-1 reactivation from latency *in vivo*, the investigators did not rule out the alternative possibility that the increased susceptibility to reactivation resulted from a higher copy number of viral genomes in the sensory neurons of IFN-$\gamma^{-/-}$ and IFN-$\gamma R^{-/-}$ mice. However, subsequent *in vitro* studies demonstrated that IFN-γ could block HSV-1 reactivation from latency when added to cultures of latently infected TG early in the reactivation process (Liu *et al.*, 2001). There appeared to be a 24 h window of opportunity after which HSV-1 reactivation was refractory to IFN-γ inhibition. Recent data clearly demonstrate that IFN-γ can block multiple steps in the viral life cycle during reactivation from latency in some neurons, whereas other neurons appear to be refractory to the inhibitory effect of IFN-γ (unpublished data).

Emerging data support the theory that the net effect of cross-regulation between IFN-γ and the HSV-1 α gene product ICP0 might influence the balance between HSV-1 latent and lytic infection. ICP0 is a promiscuous transactivator that is required for efficient HSV-1 reactivation from latency (Halford & Schaffer, 2001; Halford *et al.*, 2001). Early after infection, HSV-1 genomes associate with cellular nuclear domain 10 (ND10) bodies (Maul, 1998; Burkham *et al.*, 2001). Viral gene transcription and DNA replication occur in globular replication compartments whose formation requires prior dissociation of ND10 bodies (Everett *et al.*, 1998a; Everett, 2000). ICP0 targets the PML component of ND10 bodies for destruction in proteasomes, augmenting HSV-1 gene expression and virion formation. It has been proposed that ICP0-regulated destruction of ND10 bodies might also facilitate HSV-1 reactivation from latency (Everett *et al.*, 1998b). In addition, ICP0 facilitates HSV-1 replication by targeting

IFN-induced antiviral proteins for destruction by proteasomes (Lin *et al.*, 2004; Chee *et al.*, 2003).

IFN-γ counters the effects of ICP0 in several ways. During lytic infection, IFN-γ inhibits production of ICP0 transcripts (Harle *et al.*, 2002; Taylor *et al.*, 1998) and strongly up-regulates expression of the various components of ND10 bodies (Chelbi-Alix *et al.*, 1995; Grotzinger *et al.*, 1996). IFN-γ also induces the production of the cyclin-dependent kinase (cdk) inhibitors p21$^{WAF1/CIP1}$ and p27^{Kip1}, resulting in decreased cdk2 and cdk4 activity (Mandal *et al.*, 1998). Work of Schaffer and colleagues demonstrated that cdk2 expression is up-regulated in neurons exhibiting HSV-1 reactivation from latency, and cdk2 is required for HSV-1 replication and α and β gene expression (Schang *et al.*, 1998, 2002). Furthermore, cdk2 is required for ICP0 post-translational modifi-cations necessary for its transactivating activity (Davido *et al.*, 2002), but not for its ability to degrade the components of ND10 bodies (Davido *et al.*, 2003). Together these findings suggest a model in which the presence of IFN-γ early in the reactivation process might block the functions of ICP0 that are required for HSV-1 reactivation from latency. A delay in the presence of IFN-γ would allow ICP0 protein to accumulate in the latently infected neuron, enhancing expression of viral lytic cycle genes, and inhibiting IFN-γ function.

As noted earlier, the addition of IFN-γ to *ex vivo* cultures of latently infected TG can block HSV-1 reactivation in some, but not all, neurons. In contrast, the addition of HSV-1-specific CD8$^+$ T cells to such cultures blocks reactivation in all neurons. These findings demonstrate that CD8$^+$ T cells employ more than one effector mechanism in blocking HSV-1 reactivation from latency, and that individual latently infected neurons differentially respond to these effector molecules. The identity of the other CD8$^+$ T cell effector molecule(s) that inhibits HSV-1 reactivation from latency remains to be determined.

IF HSV-1-SPECIFIC CD8$^+$ T CELLS ARE SO EFFECTIVE AT BLOCKING HSV-1 REACTIVATION FROM LATENCY, WHY DO PEOPLE GET RECURRENT HERPETIC DISEASE?

As noted earlier, CD8$^+$ T cells in both mice and humans are retained in the TG in close apposition to latently infected neurons, and murine CD8$^+$ T cells can completely block HSV-1 reactivation from latency in *ex vivo* TG cultures. Why then is recurrent herpetic disease so prevalent? At least two explanations present themselves. One possible explanation lies in the fact that the HSV-1 α gene product ICP47 can block the transport of peptides into the endoplasmic reticulum for loading on MHC class I (Goldsmith *et al.*, 1998; Hill *et al.*, 1995). The effect of ICP47 on HSV-1 epitope recognition on neurons might be especially profound given that these cells normally express very low to undetectable levels of MHC class I. In that regard, it is interesting to note that: (a) ICP47 blocks TAP transport of peptides far more effectively in human cells than in mouse cells; and (b) HSV-1 spontaneously reactivates from latency in humans but not or at a very low frequency in mice. It is conceivable that the propensity of HSV-1 to reactivate in humans is related to an inhibitory effect of ICP47 on epitope expression on human neurons.

Stress might be another factor that influences the susceptibility of some people to recurrent herpetic disease. Glucocorticoids are effector molecules of the hypothalamic-pituitary axis (HPA); in mice, corticosterone is the major HPA-derived hormone. Corticosterone influences T-cell function in a variety of ways. It regulates production of a variety of cytokines that influence T-cell migration and proliferation (McEwen *et al.*, 1997; Noisakran *et al.*, 1998; Bonneau, 1996); regulates IL-2 receptor α and β and IL-6 receptor expression (Ashwell *et al.*, 2000); induces T-cell apoptosis; and blocks the T-cell cycle at G_0/G_1 (Newton, 2000).

Changes in T-cell localization or cell survival due to glucocorticoids may be sufficient to allow HSV to reactivate from latency. Restraint stress during latency has been shown to decrease lymphocyte numbers in the spleen and lymph nodes as well as disrupt the ability of $CD8^+$ T cells to lyse HSV-infected target cells (Bonneau *et al.*, 1991). Restraint stress during latency has also been associated with a reduction of IL-2, IL-4, IL-6 and IFN-γ production by splenic lymphocytes in response to HSV-1 antigens (Bonneau, 1996). In addition, glucocorticoids might directly influence viral gene expression in latently infected neurons. Activated glucocorticoid receptor binding to CREB-binding protein (CBP) can enhance an endogenous histone acetylation activity of CBP (Adcock, 2000). This could allow for previously inaccessible DNA sequences of the latent viral episome to become accessible to transcription factors. Thus stress might lead to HSV-1 reactivation from latency in humans by transiently compromising the function of $CD8^+$ T cells within the sensory ganglia, and by enhancing viral gene expression within latently infected neurons.

It emerges from the previous discussion that a bivalent interaction between a neuron and the HSV-1 genome probably does not constitute a full representation of HSV-1 latency. The establishment and maintenance of latency in at least some neurons appears to involve a tripartite interaction among a neuron, the viral genome and the host immune system. In particular, the emerging data suggest that maintaining the virus in a latent state in certain neurons is accomplished through the activity of IFN-γ and at least one other effector mechanism of HSV-1-specific $CD8^+$ T cells. Understanding the role of T lymphocytes in maintaining HSV-1 latency and the viral antigens that stimulate their activity might lead to new vaccine-based approaches to preventing recurrent herpetic disease.

REFERENCES

Adcock, I. M. (2000). Molecular mechanisms of glucocorticosteroid actions. *Pulm Pharmacol Ther* **13**, 115–126.

Ashwell, J. D., Lu, F. W. & Vacchio, M. S. (2000). Glucocorticoids in T cell development and function. *Annu Rev Immunol* **18**, 309–345.

Bonneau, R. H. (1996). Stress-induced effects on integral immune components involved in herpes simplex virus (HSV)-specific memory cytotoxic T lymphocyte activation. *Brain Behav Immun* **10**, 139–163.

Bonneau, R. H., Sheridan, J. F., Feng, N. G. & Glaser, R. (1991). Stress-induced effects on cell-mediated innate and adaptive memory components of the murine immune response to herpes simplex virus infection. *Brain Behav Immun* **5**, 274–295.

Burkham, J., Coen, D. M., Hwang, C. B. & Weller, S. K. (2001). Interactions of herpes simplex virus type 1 with ND10 and recruitment of PML to replication compartments. *J Virol* **75**, 2353–2367.

Cantin, E. M., Hinton, D. R., Chen, J. & Openshaw, H. (1995). Gamma interferon expression during acute and latent nervous system infection by herpes simplex virus type 1. *J Virol* **69**, 4898–4905.

Cantin, E., Tanamachi, B. & Openshaw, H. (1999). Role for gamma interferon in control of herpes simplex virus type 1 reactivation. *J Virol* **73**, 3418–3423.

Chee, A. V., Lopez, P., Pandolfi, P. P. & Roizman, B. (2003). Promyelocytic leukemia protein mediates interferon-based anti-herpes simplex virus 1 effects. *J Virol* **77**, 7101–7105.

Chelbi-Alix, M. K., Pelicano, L., Quignon, F., Koken, M. H., Venturini, L., Stadler, M., Pavlovic, J., Degos, L. & de The, H. (1995). Induction of the PML protein by interferons in normal and APL cells. *Leukemia* **9**, 2027–2033.

Chen, S. H., Kramer, M. F., Schaffer, P. A. & Coen, D. M. (1997). A viral function represses accumulation of transcripts from productive-cycle genes in mouse ganglia latently infected with herpes simplex virus. *J Virol* **71**, 5878–5884.

Chen, S. H., Garber, D. A., Schaffer, P. A., Knipe, D. M. & Coen, D. M. (2000). Persistent elevated expression of cytokine transcripts in ganglia latently infected with herpes simplex virus in the absence of ganglionic replication or reactivation. *Virology* **278**, 207–216.

Chen, S. H., Lee, L. Y., Garber, D. A., Schaffer, P. A., Knipe, D. M. & Coen, D. M. (2002). Neither LAT nor open reading frame P mutations increase expression of spliced or intron-containing ICP0 transcripts in mouse ganglia latently infected with herpes simplex virus. *J Virol* **76**, 4764–4772.

Croen, K. D., Ostrove, J. M., Dragovic, L. J., Smialek, J. E. & Straus, S. E. (1987). Latent herpes simplex virus in human trigeminal ganglia. Detection of an immediate early gene "anti-sense" transcript by in situ hybridization. *N Engl J Med* **317**, 1427–1432.

Davido, D. J., Leib, D. A. & Schaffer, P. A. (2002). The cyclin-dependent kinase inhibitor Roscovitine inhibits the transactivating activity and alters the posttranslational modification of herpes simplex virus type 1 ICP0. *J Virol* **76**, 1077–1088.

Davido, D. J., Von Zagorski, W. F., Maul, G. G. & Schaffer, P. A. (2003). The differential requirement for cyclin-dependent kinase activities distinguishes two functions of herpes simplex virus type 1 ICP0. *J Virol* **77**, 12603–12616.

Everett, R. D. (2000). ICP0, a regulator of herpes simplex virus during lytic and latent infection. *Bioessays* **22**, 761–770.

Everett, R. D., Freemont, P., Saitoh, H., Dasso, M., Orr, A., Kathoria, M. & Parkinson, J. (1998a). The disruption of ND10 during herpes simplex virus infection correlates with the Vmw110- and proteasome-dependent loss of several PML isoforms. *J Virol* **72**, 6581–6591.

Everett, R. D., Orr, A. & Preston, C. M. (1998b). A viral activator of gene expression functions via the ubiquitin-proteasome pathway. *EMBO J* **17**, 7161–7169.

Feldman, L. T., Ellison, A. R., Voytek, C. C., Yang, L., Krause, P. & Margolis, T. P. (2002). Spontaneous molecular reactivation of herpes simplex virus type 1 latency in mice. *Proc Natl Acad Sci U S A* **99**, 978–983.

Goldsmith, K., Chen, W., Johnson, D. C. & Hendricks, R. L. (1998). Infected cell protein (ICP)47 enhances herpes simplex virus neurovirulence by blocking the CD8$^+$ T cell response. *J Exp Med* **187**, 341–348.

Grotzinger, T., Sternsdorf, T., Jensen, K. & Will, H. (1996). Interferon-modulated expression of genes encoding the nuclear-dot-associated proteins Sp100 and promyelocytic leukemia protein (PML). *Eur J Biochem* **238**, 554–560.

Halford, W. P. & Schaffer, P. A. (2001). ICP0 is required for efficient reactivation of herpes simplex virus type 1 from neuronal latency. *J Virol* **75**, 3240–3249.

Halford, W. P., Gebhardt, B. M. & Carr, D. J. J. (1996). Persistent cytokine expression in trigeminal ganglion latently infected with herpes simplex virus type 1. *J Immunol* **157**, 3542–3549.

Halford, W. P., Gebhardt, B. M. & Carr, D. J. (1997). Acyclovir blocks cytokine gene expression in trigeminal ganglia latently infected with herpes simplex virus type 1. *Virology* **238**, 53–63.

Halford, W. P., Kemp, C. D., Isler, J. A., Davido, D. J. & Schaffer, P. A. (2001). ICP0, ICP4, or VP16 expressed from adenovirus vectors induces reactivation of latent herpes simplex virus type 1 in primary cultures of latently infected trigeminal ganglion cells. *J Virol* **75**, 6143–6153.

Harle, P., Sainz, B., Jr, Carr, D. J. & Halford, W. P. (2002). The immediate-early protein, ICP0, is essential for the resistance of herpes simplex virus to interferon-alpha/beta. *Virology* **293**, 295–304.

Hendricks, R. L. (1999). Immunopathogenesis of viral ocular infections. *Chem Immunol* **73**, 120–136.

Hill, A., Jugovic, P., York, I., Russ, G., Bennink, J., Yewdell, J., Ploegh, H. & Johnson, D. (1995). Herpes simplex virus turns off TAP to evade host immunity. *Nature* **375**, 411–415.

Khanna, K. M., Bonneau, R. H., Kinchington, P. R. & Hendricks, R. L. (2003). Herpes simplex virus-specific memory CD8$^+$ T cells are selectively activated and retained in latently infected sensory ganglia. *Immunity* **18**, 593–603.

Kramer, M. F. & Coen, D. M. (1995). Quantification of transcripts from the ICP4 and thymidine kinase genes in mouse ganglia latently infected with herpes simplex virus. *J Virol* **69**, 1389–1399.

Liesegang, T. J. (2001). Herpes simplex virus epidemiology and ocular importance. *Cornea* **20**, 1–13.

Lin, R., Noyce, R. S., Collins, S. E., Everett, R. D. & Mossman, K. L. (2004). The herpes simplex virus ICP0 RING finger domain inhibits IRF3- and IRF7-mediated activation of interferon-stimulated genes. *J Virol* **78**, 1675–1684.

Liu, T., Tang, Q. & Hendricks, R. L. (1996). Inflammatory infiltration of the trigeminal ganglion after herpes simplex virus type 1 corneal infection. *J Virol* **70**, 264–271.

Liu, T., Khanna, K. M., Chen, X., Fink, D. J. & Hendricks, R. L. (2000). CD8(+) T cells can block herpes simplex virus type 1 (HSV-1) reactivation from latency in sensory neurons. *J Exp Med* **191**, 1459–1466.

Liu, T., Khanna, K. M., Carriere, B. N. & Hendricks, R. L. (2001). Gamma interferon can prevent herpes simplex virus type 1 reactivation from latency in sensory neurons. *J Virol* **75**, 11178–11184.

Mandal, M., Bandyopadhyay, D., Goepfert, T. M. & Kumar, R. (1998). Interferon-induces expression of cyclin-dependent kinase-inhibitors p21WAF1 and p27Kip1 that prevent activation of cyclin-dependent kinase by CDK-activating kinase (CAK). *Oncogene* **16**, 217–225.

Maul, G. G. (1998). Nuclear domain 10, the site of DNA virus transcription and replication. *Bioessays* **20**, 660–667.

McEwen, B. S., Biron, C. A., Brunson, K. W. & 8 other authors (1997). The role of adrenocorticoids as modulators of immune function in health and disease: neural, endocrine and immune interactions. *Brain Res Brain Res Rev* **23**, 79–133.

Minami, M., Kita, M., Yan, X. Q., Yamamoto, T., Iida, T., Sekikawa, K., Iwakura, Y. & Imanishi, J. (2002). Role of IFN-gamma and tumor necrosis factor-alpha in herpes simplex virus type 1 infection. *J Interferon Cytokine Res* **22**, 671–676.

Newton, R. (2000). Molecular mechanisms of glucocorticoid action: what is important? *Thorax* **55**, 603–613.

Noisakran, S., Halford, W. P., Veress, L. & Carr, D. J. (1998). Role of the hypothalamic pituitary adrenal axis and IL-6 in stress-induced reactivation of latent herpes simplex virus type 1. *J Immunol* **160**, 5441–5447.

Padgett, D. A., Sheridan, J. F., Dorne, J., Berntson, G. G., Candelora, J. & Glaser, R. (1998). Social stress and the reactivation of latent herpes simplex virus type 1. *Proc Natl Acad Sci U S A* **95**, 7231–7235.

Pereira, D. B., Antoni, M. H., Danielson, A. & 8 other authors (2003). Stress as a predictor of symptomatic genital herpes virus recurrence in women with human immunodeficiency virus. *J Psychosom Res* **54**, 237–244.

Pereira, R. A. & Simmons, A. (1999). Cell surface expression of H2 antigens on primary sensory neurons in response to acute but not latent herpes simplex virus infection in vivo. *J Virol* **73**, 6484–6489.

Pereira, R. A., Tscharke, D. C. & Simmons, A. (1994). Upregulation of class I major histocompatibility complex gene expression in primary sensory neurons, satellite cells, and Schwann cells of mice in response to acute but not latent herpes simplex virus infection in vivo. *J Exp Med* **180**, 841–850.

Sainz, B., Loutsch, J. M., Marquart, M. E. & Hill, J. M. (2001). Stress-associated immuno-modulation and herpes simplex virus infections. *Med Hypotheses* **56**, 348–356.

Sawtell, N. M. (2003). Quantitative analysis of herpes simplex virus reactivation in vivo demonstrates that reactivation in the nervous system is not inhibited at early times postinoculation. *J Virol* **77**, 4127–4138.

Schang, L. M., Phillips, J. & Schaffer, P. A. (1998). Requirement for cellular cyclin-dependent kinases in herpes simplex virus replication and transcription. *J Virol* **72**, 5626–5637.

Schang, L. M., Bantly, A. & Schaffer, P. A. (2002). Explant-induced reactivation of herpes simplex virus occurs in neurons expressing nuclear cdk2 and cdk4. *J Virol* **76**, 7724–7735.

Shimeld, C., Whiteland, J. L., Nicholls, S. M., Grinfeld, E., Easty, D. L., Gao, H. & Hill, T. J. (1995). Immune cell infiltration and persistence in the mouse trigeminal ganglion after infection of the cornea with herpes simplex virus type 1. *J Neuroimmunol* **61**, 7–16.

Shimeld, C., Whiteland, J. L., Williams, N. A., Easty, D. L. & Hill, T. J. (1997). Cytokine production in the nervous system of mice during acute and latent infection with herpes simplex virus type 1. *J Gen Virol* **78**, 3317–3325.

Simmons, A. & Tscharke, D. C. (1992). Anti-CD8 impairs clearance of herpes simplex virus from the nervous system: implications for the fate of virally infected neurons. *J Exp Med* **175**, 1337–1344.

Stevens, J. G., Wagner, E. K., Devi-Rao, G. B., Cook, M. L. & Feldman, L. T. (1987). RNA complementary to a herpesvirus alpha gene mRNA is prominent in latently infected neurons. *Science* **235**, 1056–1059.

Stroop, W. G. & Schaefer, D. C. (1987). Herpes simplex virus, type 1 invasion of the rabbit and mouse nervous systems revealed by in situ hybridization. *Acta Neuropathol (Berl)* **74**, 124–132.

Taylor, J. L., Little, S. D. & O'Brien, W. J. (1998). The comparative anti-herpes simplex virus effects of human interferons. *J Interferon Cytokine Res* **18**, 159–165.

Theil, D., Derfuss, T., Paripovic, I., Herberger, S., Meinl, E., Schueler, O., Strupp, M., Arbusow, V. & Brandt, T. (2003). Latent herpesvirus infection in human trigeminal ganglia causes chronic immune response. *Am J Pathol* **163**, 2179–2184.

Hepatitis C virus disruption of interferon signalling pathways and evasion of innate intracellular antiviral defences

Stanley M. Lemon and Kui Li

Department of Microbiology & Immunology, Institute for Human Infections & Immunity,
University of Texas Medical Branch, Galveston, TX 77555-0428, USA

INTRODUCTION

A decade and a half after its discovery (Choo *et al.*, 1989; Kuo *et al.*, 1989), persistent infection with hepatitis C virus (HCV) remains the leading cause of chronic hepatitis in the United States and many other developed nations (Alter *et al.*, 1999; Thomas *et al.*, 2004). As many as 4 million Americans are infected with this positive-strand RNA virus, placing them at risk for progressive hepatic fibrosis as well as potentially fatal cirrhosis and liver cancer (Seeff, 1997; Alter *et al.*, 1999). Studies of the natural history of hepatitis C have long suggested that the disease progresses slowly, with only a minority of infected individuals likely to develop life-threatening liver disease (Seeff *et al.*, 1992). Nonetheless, as many as 12 000 deaths annually are attributed to HCV infection in the United States. In Japan, a doubling of the incidence of liver cancer during the latter decades of the 20th century has been related to the spread of HCV within the Japanese population following World War II (Yoshizawa, 2002). Similarly, HCV-related mortality is projected to increase substantially over the next decade in the United States as the duration of the disease in the affected cohorts becomes sufficiently long for cirrhosis to become clinically manifest (Wong *et al.*, 2000).

Despite promising evidence that immunization might be beneficial (Weiner *et al.*, 2001), there is no vaccine for prevention of HCV infection or disease. In addition, antiviral therapies for treatment of chronic hepatitis C are woefully inadequate, able to effect permanent cures in only about half of all patients infected with the most prevalent virus genotypes, and often associated with significant adverse side effects

SGM symposium 64: Molecular pathogenesis of virus infections.
Editors P. Digard, A. A. Nash & R. E. Randall. Cambridge University Press. ISBN 0 521 83248 9 ©SGM 2005

(McHutchison & Fried, 2003). There can be little doubt that HCV infection represents a continuing threat to public health.

The ability of HCV to cause disease is closely linked to its capacity to establish persistent infections that continue for many years in the majority of individuals. Although much has been learned about HCV since it was first identified (Major & Feinstone, 1997; Pawlotsky, 2004), and although a variety of mechanisms have been suggested to contribute to the long-term persistence of HCV infection (Racanelli & Rehermann, 2003), the reasons underlying the high proportion of acute HCV infections that become persistent remain ill defined. Nonetheless, immunological responses during the earliest stages of acute infection are likely to play an important role in determining whether the virus is eliminated or the infection goes on to become persistent. Although data from studies of persons with acute HCV infection remain extremely limited, vigorous and broadly directed CD8$^+$ and CD4$^+$ T lymphocyte responses appear to correlate with virus clearance in acute resolving infection (Lechner et al., 2000). Such responses are likely to require early host recognition of the infection, and proper initiation of the signalling cascades required for successful activation and priming of T lymphocytes.

Evolving evidence suggests that innate immune responses are closely linked to subsequent development of specific, adaptive immunity in viral infections (Iwasaki & Medzhitov, 2004; Hoebe et al., 2004), and it appears that HCV begins to evade the host's attempts at elimination of the virus at the very earliest stages of engagement of the immune system. The NS3/4A protease expressed by HCV disrupts the virus-activated intracellular signalling pathways that normally induce type 1 interferons (α/β) in response to virus infection, and that lead in turn to the expression of numerous antiviral cytokines and interferon-stimulated genes (ISGs) (Foy et al., 2003). More recent data suggest that this disruption of intracellular signalling extends to include the blockade of nuclear factor kappa B (NF-κB)-dependent infection responses as well (Li et al., 2005b; Foy et al., 2005). The disruption of these critical immune signalling pathways is likely to promote virus persistence by limiting innate defences, including the expression of ISGs with antiviral activity as well as the activation of natural killer (NK) cells. Moreover, impaired production of cytokines and chemokines may contribute to persistence by adversely influencing dendritic cell (DC) maturation and/or antigen cross-priming (Iwasaki & Medzhitov, 2004; Hoebe et al., 2004), events that are likely to be essential for the initiation of a robust and durable CD8$^+$ T lymphocyte response to HCV.

This review will focus on these early events in the interaction of HCV with the innate immune system, and in particular on recent work that has delineated an important role for the viral NS3/4A protease in disrupting the intracellular signalling pathways that normally lead to the expression of type 1 interferons and NF-κB-dependent proinflammatory cytokines in response to virus infection.

SIGNALLING PATHWAYS LEADING TO INDUCTION OF TYPE 1 INTERFERONS

The earliest stages of microbial recognition involve the binding of specific products of bacterial or virus replication (termed 'pathogen-associated molecular patterns' or PAMPs) by a diverse and most likely incompletely defined set of 'pathogen recognition receptors' (PRRs). Toll-like receptors (TLRs), a class of 10–13 PRRs, bind double-stranded RNA (dsRNA), bacterial lipopolysaccharides and other PAMPs to initiate signalling pathways that lead to activation of several transcriptional factors, including NF-κB and interferon regulatory factor 3 (IRF-3) (Beutler, 2004). Once activated, NF-κB and IRF-3 induce the expression of a large number of protective genes, including type 1 interferons and ISGs with direct antiviral activity.

dsRNA is typically produced during infections with RNA viruses, and has long been considered to be an important PAMP. Interferon-β synthesis is induced by dsRNA engagement of the extracellular, leucine-rich repeat domain of TLR3 (Alexopoulou *et al.*, 2001). TLR3 is expressed on the plasma membrane or in intracellular vesicles. Toll-IL1 receptor domain containing adaptor inducing interferon-β (TRIF or TICAM-1) (Yamamoto *et al.*, 2002; Oshiumi *et al.*, 2003) interacts through its Toll-IL1 receptor (TIR) domain with the TIR domain present in the cytoplasmic portion of TLR3 (see Fig. 1). TRIF acts as an adaptor, linking TLR3 to the kinases responsible for downstream activation of IRF-3 as well as NF-κB. TLR4, when engaged by bacterial lipopolysaccharides in association with the MD-2 protein, also utilizes TRIF to signal to IRF-3 and induce interferon-β. However, signalling through this pathway also requires a second adaptor protein, TRIF-related adaptor molecule (TRAM). Another TLR adaptor protein, myeloid differentiation factor 88 (MyD88), interacts with TLR4 to signal to NF-κB (McGettrick & O'Neill, 2004). While MyD88 serves as an adaptor for most if not all other TLRs, TRIF interacts only with TLR3 and TLR4. TRIF is the only known adaptor protein for TLR3, and is thus essential for dsRNA signalling through the TLR3 pathway. Although typically considered a bacterial PAMP receptor, TLR4 has also been implicated in recognition of respiratory syncytial virus infections (Haynes *et al.*, 2001).

The activation of IRF-3 plays a pivotal role in induction of interferon synthesis (Fig. 1). This transcription factor is normally latently expressed in the cytoplasm. It is activated by specific phosphorylation near its carboxyl-terminus, which results in its dimerization and translocation from cytoplasm to nucleus (Lin *et al.*, 1998; Yoneyama

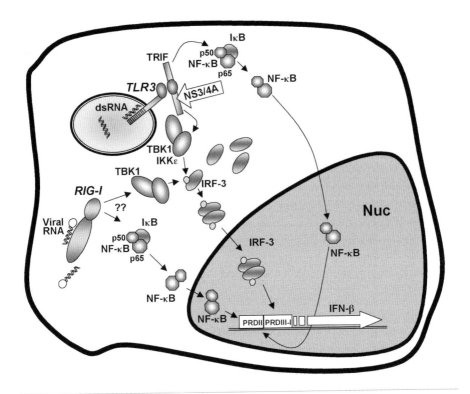

Fig. 1. Virus-induced TLR3- and RIG-I-dependent cellular signalling pathways leading to the activation of IRF-3 and NF-κB, with resulting stimulation of the interferon-β promoter and interferon-β transcription. PRD, positive regulatory (promoter) domain; see text for other abbreviations and a description of the pathways. The open arrow represents the site of TRIF cleavage by the NS3/4A protease, which appears to lead to the rapid degradation of the TLR3 adaptor protein.

et al., 1998). Both of the noncanonical IκB kinases, TANK-binding kinase 1 (TBK1) and IκB kinase epsilon (IKKε), have been implicated in IRF-3 phosphorylation following the engagement of TLR3 (Sharma *et al.*, 2003; Fitzgerald *et al.*, 2003). The induction of interferon-β synthesis is dependent upon coordinated assembly of activated IRF-3, NF-κB and ATF/c-jun on the interferon-β promoter (Maniatis *et al.*, 1998; Panne *et al.*, 2004). The expression and secretion of interferon-β leads in turn through autocrine/paracrine signalling mechanisms involving its recognition by type 1 interferon receptors and signalling through the Janus kinase/signal transducer and activator of transcription (Jak/STAT) pathway to activation of IRF-7 transcription. IRF-7 is phosphorylated by TBK1/IKKε, and then translocates to the nucleus, where it activates interferon-α synthesis and contributes to further activation of the interferon-β promoter. An interaction between IRF-7 and TRIF has been suggested to play a role in this process (Han *et al.*, 2004). Interferon-α/β subsequently induces the expression of many protective ISGs, cytokines and chemokines, while also up-regulating the expression of MHC molecules and costimulatory molecules important for adaptive immune responses. Activated IRF-3 is also capable of directly stimulating the expression of a smaller subset of ISGs (Grandvaux *et al.*, 2002).

Virus activation of IRF-3 can also occur through a second major pathway, recently recognized, that does not involve TLR3. This pathway, which is not dependent on any TLR, involves the apparent recognition of structured RNA by a cellular DExH/D box RNA helicase that functions as a PRR, retinoic-acid-inducible gene I (RIG-I) (Yoneyama *et al.*, 2004; Sumpter *et al.*, 2005). RIG-I signals the downstream activation of IRF-3 and NF-κB through amino-terminal caspase recruitment domain (CARD) homologues, and its helicase activity appears essential to its role in interferon signalling. The RIG-I pathway (Fig. 1) functions largely independently of TLR3 in cells derived from hepatocytes, the primary cell type targeted by HCV (Li *et al.*, 2005a). This is consistent with the original description of the RIG-I pathway by Yoneyama *et al.* (2004). However, other data suggest extensive cross-talk between these pathways, and even codependence of TLR3 and RIG-I in dsRNA-activated signalling in a subline of HeLa cells (Li *et al.*, 2005b). The RIG-I pathway is likely to act as an amplification loop for dsRNA activation of IRF-3, as RIG-I itself is strongly interferon inducible.

Finally, in some cell types, such as plasmacytoid dendritic cells (pDCs), yet a third pathway may lead to virus-induced expression of interferon. U-rich single-stranded viral RNAs (such as vesicular stomatitis virus and influenza virus RNAs) and CpG DNA (present in herpesvirus DNAs) can induce interferon-α synthesis by engagement of TLR7/8 and TLR9, respectively, in endosomes (Diebold *et al.*, 2004; Hochrein *et al.*, 2004; Crozat & Beutler, 2004). In this case, interferon-α synthesis occurs independently of interferon-β and involves direct interactions between TLR7, TLR8 or TLR9, the TLR adaptor protein MyD88, and IRF-7 (Kawai *et al.*, 2004; Honda *et al.*, 2004). IRF-3 does not appear to participate in this pathway, which leads predominantly to interferon-α synthesis with relatively little interferon-β produced. These redundant signalling pathways provide flexibility as well as specificity to the host cell in mediating the expression of ISGs, cytokines and chemokines. It is likely that subtle and perhaps other, not so subtle, differences in the sets of genes induced by each of these pathways contribute to more tailored, virus-specific host responses. Signalling through these pathways is also likely to be modulated by parallel activation of NF-κB, and to differ significantly from cell type to cell type.

CELL SYSTEMS FOR STUDYING INTERACTIONS OF HCV WITH SIGNALLING PATHWAYS

HCV is a single-stranded, positive-sense RNA virus with a genome length of approximately 9·6 kb (Pawlotsky, 2004; Major & Feinstone, 1997; Thomas *et al.*, 2004). It is

(a)

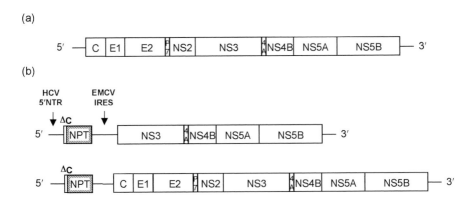

Fig. 2. (a) Organization of the 9·6 kb, positive-strand HCV genome. (b) Subgenomic (top) and genome-length (bottom) dicistronic HCV RNAs expressing neomycin phosphotransferase (NPT) and that are capable of autonomous replication in Huh7 hepatoma cells. See text for details.

classified within a separate genus of the *Flaviviridae*, the genus *Hepacivirus*. The HCV genome contains a large open reading frame (ORF) that spans most of the length of the genome, and commences downstream of a 342-nt-long 5′ nontranslated region (5′NTR) (Fig. 2a). The 5′NTR is highly structured and contains an internal ribosome entry site (IRES) that directs the cap-independent initiation of translation of the downstream ORF (Rijnbrand & Lemon, 2000). The ORF encodes a large polyprotein that undergoes co- and post-translational processing under the control of several cellular and viral proteases. This yields a series of structural proteins, derived from the amino-terminal third of the polyprotein, which include a core or nucleocapsid protein, two envelope glycoproteins, E1 and E2, and a small, 7 kDa protein that may be a viroporin (Griffin *et al.*, 2003) (Fig. 2a). At least six nonstructural replicative proteins are processed from the remainder of the polyprotein. These include NS2 (which with the adjacent NS3 sequence demonstrates *cis*-active, Zn^{++}-dependent protease activity at the NS2/NS3 cleavage site), NS3 (a chymotrypsin-like serine protease in its amino-terminal third, with NTPase/RNA helicase activity associated with the carboxyl-terminal two-thirds), NS4 (serine protease accessory factor), NS4B, NS5A and NS5B (the RNA-dependent RNA polymerase). There appear to be multiple interactions between these nonstructural proteins. NS3 forms a noncovalent complex with NS4A, resulting in the mature NS3/4A protease (Kwong *et al.*, 1999). This protease directs *cis*-cleavage of the polyprotein at the NS3/4A junction, and subsequent *trans*-cleavages at the NS4A/4B, NS4B/5A and NS5A/5B junctions, and thus plays a critical role in generating the proteins that assemble into the viral RNA replicase complex. The NS3/4A protease is a prime target for antiviral drug development, and early clinical trials with a prototype NS3/4A inhibitor, BILN2061, demonstrated dramatic antiviral properties (Lamarre *et al.*, 2003).

One of the great difficulties in working with HCV is the fact that the virus does not replicate with any degree of efficiency in cultured cells. However, several years ago, Lohmann *et al*. (1999) described subgenomic HCV replicons that were competent for replication in cultured cells, thereby providing the field with a major breakthrough. These replicons were derived from Con1, a genotype 1b strain of HCV, and were dicistronic in nature. They were constructed by removing most of the structural protein coding region of HCV, and inserting in its place sequence encoding the selectable antibiotic marker neomycin (Neo) phosphotransferase (NPT), followed by a heterologous IRES from the picornavirus encephalomyocarditis virus (EMCV) (Fig. 2b). Thus these replicons express NPT from the upstream cistron under control of the natural HCV IRES, allowing the selection of stable cell clones containing a substantial abundance of viral RNA and proteins, while the downstream cistron, encoding the essential nonstructural proteins, was under the translational control of the EMCV IRES. Since these HCV replicon RNAs lack most of the sequence encoding the structural proteins (core and E1, E2 and p7), they are not capable of producing infectious particles, despite robust RNA replication in Huh7 cells.

Replicon RNAs recovered from G418-resistant cells following transfection of subgenomic Con1 RNAs contained a variety of mutations within the NS3, NS5A or NS5B sequences that greatly enhanced the replication of the Con1 RNA in Huh7 cells (Blight *et al*., 2000; Lohmann *et al*., 2001). Such cell-culture-adaptive mutations appear to be required for efficient replication of Con1 replicons, and increase the efficiency of selection of G418-resistant cell clones under Neo pressure by several orders of magnitude. Interestingly, many of these cell-culture-adaptive mutations involve a segment of the NS5A coding region adjacent to the so-called 'interferon sensitivity determining region' (ISDR), a genome segment that has been implicated in interferon resistance in human infections (Blight *et al*., 2000; Enomoto *et al*., 1996). Similar subgenomic replicons, with different adaptive mutations, have now been constructed from other genotype 1b virus strains (Guo *et al*., 2001; Ikeda *et al*., 2002), as well as genotype 1a (Blight *et al*., 2003; Yi & Lemon, 2004) and genotype 2a (Kato *et al*., 2003) viruses.

In contrast to the Con1 replicons, subgenomic replicons derived from an infectious molecular clone of the genotype 1b, HCV-N virus, were found to be capable of efficient replication in Huh7 cells in the absence of specific cell-culture-adaptive mutations (Ikeda *et al*., 2002). Like Con1 replicons, G418-resistant cells selected following transfection with these RNAs contained abundant NS5A antigen and HCV RNA detectable by Northern analysis, but in at least one case no mutations in the NS3–NS5B segment of the polyprotein. However, the deletion of a unique 4-amino-acid insertion that is present within the NS5A protein in wild-type HCV-N drastically decreased the number of G418-resistant colonies obtained following transfection of Huh7 cells. This effect could be reversed by inclusion of a previously described Con1 cell-culture-adaptive mutation (S2005→I), confirming that this natural 4-amino-acid insertion has a controlling role in determining the replication capacity of wild-type HCV-N RNA in Huh7 cells (Ikeda *et al.*, 2002). Dicistronic selectable RNAs encoding the entire HCV-N polyprotein (under control of the EMCV IRES) were equally capable of replication and gave rise to G418-resistant cell clones following transfection of Huh7 cells (Fig. 2b) (Ikeda *et al.*, 2002). These cell lines have proven to be very useful for a number of studies, as they contain replicating viral RNAs expressing the entire complement of HCV proteins (Scholle *et al.*, 2004; Konan *et al.*, 2003; Disson *et al.*, 2004; Foy *et al.*, 2003).

To determine whether expression of the HCV protein in the context of ongoing viral RNA replication significantly alters the host cell transcriptome and cell cycle regulatory processes, two independent Huh7 cell clones were studied, each harbouring selectable, full-length, replicating HCV RNAs that expressed the entire HCV-N polyprotein (Huh7 2-3 and 3 cells), and paired with clonally related cells in which the viral RNA had been eliminated by a prior 2-week course of treatment with interferon-α (2-3c and 3c cells, respectively) (Scholle *et al.*, 2004). Surprisingly, high-density oligonucleotide microarray analyses revealed only subtle, coordinated differences between the mRNA profiles of cells containing a high abundance of the replicating viral RNA and their interferon-cured progeny. The degree of variation in the global transcript profiles was much greater between different Neo-resistant Huh7 cell clones containing the replicating viral RNA (e.g. Huh7 2-3 vs 3 cells) than between clonally related cells before and after being 'cured' of the replicating HCV RNA by interferon treatment (Huh7 2-3 or 2-3c cells) (Scholle *et al.*, 2004). Flow cytometric analysis also demonstrated no significant differences in cell cycle distribution among populations of asynchronously growing cells of both types, with cell lines containing replicating viral RNA able to re-enter the cell cycle after transient G_1 arrest in a fashion similar to those that were cured of the RNA. Except for the expression of viral proteins, therefore, the HCV RNA-positive and RNA-negative cell lines were virtually indistinguishable.

There was, however, a dramatic difference between Huh7 cells that did or did not contain replicating HCV RNA when they were challenged with infection by another virus. Sendai virus is a strong inducer of type 1 interferons and ISGs, because of its ability to activate IRF-3. High-density HG-U133A Affymetrix GeneChips were used for these studies, which compared Huh7 2-3 cells with their cured 2-3c counterparts before and after infection with Sendai virus. Infection of the cured cells with Sendai virus resulted in a >twofold change in the signal intensity obtained with 399 individual probe sets, most of which represent a single unique gene (out of approximately 10 000 probe sets for which a target transcript was detectably expressed in Huh7 cells) (K. Li and S. M. Lemon, unpublished). In contrast, identical Sendai virus infection of the 2-3 cells which contained the replicating genome-length HCV RNA resulted in equivalent increases in the signal intensity of only 120 probe sets. Moreover, in the cured cells, there were 78 probe sets with changes in signal intensity of >eightfold, and 42 with a change of >16-fold (at least half of which were recognized ISGs). In contrast, there were only four probe sets in the cells with ongoing HCV RNA replication that registered a change in signal intensity of >eightfold, and none were >16-fold (K. Li and S. M. Lemon, unpublished). These data confirm earlier suspicions that, while the cured 2-3c cells retain the capacity for interferon-mediated responses to virus infection, these protective cellular responses do not occur or are substantially muted in cells containing replicating HCV RNA (Scholle *et al.*, 2004).

THE NS3/4A PROTEASE BLOCKS VIRUS-MEDIATED ACTIVATION OF IRF-3

Recent studies have shown that Sendai virus infection induces phosphorylation and nuclear translocation of IRF-3 in normal Huh7 cells, but not in the Huh7 2-3 cells containing replicating genome-length RNA (Foy *et al.*, 2003). The capacity to activate IRF-3 was restored in clonally related 2-3c cells that had been cured of HCV RNA by prior treatment with interferon-α (Fig. 3). The ability of Sendai virus to activate the interferon-β promoter was also sharply reduced in 2-3 cells, but again restored in the 2-3c cells (Fig. 4). This was also the case with other IRF-3-dependent promoters, such as the ISG56 promoter and the promoter for the chemokine RANTES (Foy *et al.*, 2003). More recent studies suggest that the blockade extends to include NF-κB-dependent promoters as well (Fig. 4). These results are thus very consistent with the oligonucleotide microarray studies described earlier.

To determine which of the 10 proteins expressed by HCV were responsible for the blockade of IRF-3 activation in Huh7 2-3 cells, the viral proteins were ectopically expressed either singly or in combination with each other in normal Huh7 cells, followed by challenge with Sendai virus (Foy *et al.*, 2003). These experiments demonstrated that the IRF-3 blockade was uniquely reproduced by overexpression of NS3/4A, but not by any other protein or combination of proteins. Interestingly, expression of either NS3 or NS4A by itself failed to significantly impair the ability of Sendai virus to activate IRF-3, suggesting that the protease activity of the NS3/4A complex might be responsible for the disruption of IRF-3 signalling (Foy *et al.*, 2003). As indicated earlier, NS3/4A protein is a noncovalent enzyme complex with serine protease activity that normally directs post-translational cleavage of the HCV polyprotein and also possesses RNA helicase activity. The ability of NS3/4A to block Sendai virus activation of IRF-3 was shown to be dependent upon NS3/4A protease activity by

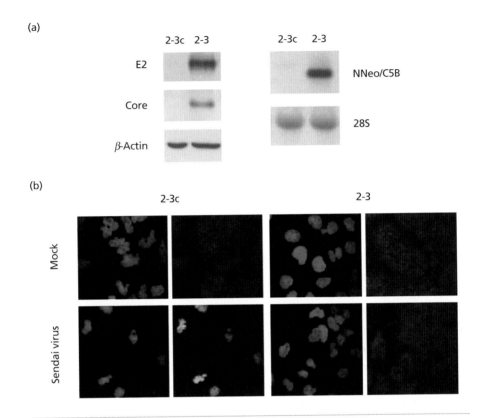

Fig. 3. (a) Immunoblots for the core and E2 envelope protein (left panels) and Northern blot showing genome-length NNeo/C-5B RNA (right panels) in Huh7 2-3 cells (Ikeda *et al.*, 2002). Note the absence of these proteins and RNA in the interferon-cured 2-3c cells (Scholle *et al.*, 2004). (b) Indirect immunofluorescence detection of IRF-3 in 2-3 cells containing replicating HCV RNA (right panel sets) and the cured 2-3c cells (left panel sets), either following mock infection (top panel sets) or infection with Sendai virus (bottom panel sets). IRF-3 antigen appears in green with FITC staining (right panels), while nuclei are visualized by DAPI staining (left panels). Note the translocation of IRF-3 to the nuclei in 2-3c cells following infection with Sendai virus, and the absence of this evidence of IRF-3 activation in the 2-3 cells containing replicating HCV RNA.

demonstrating its reversal by point mutations in NS3 that knock out the protease but not the helicase activity. Furthermore, IRF-3 activation could be restored by treatment of the cells with a specific, peptidomimetic ketoamide inhibitor of the protease, SCH6 (Foy *et al.*, 2003).

These results point to the strong likelihood that NS3/4A mediates proteolysis of one or more cellular proteins residing within the virus-activated signalling pathways upstream of IRF-3. As summarized earlier, activation of IRF-3 in response to viral infection appears to occur through two principal signalling pathways in human hepatocytes (Li *et al.*, 2005a). One pathway is activated by the binding of extracellular dsRNA (or

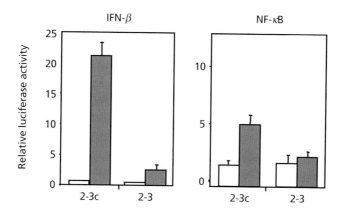

Fig. 4. Luciferase promoter reporter assays showing activities of the interferon-β promoter (left panel) or NF-κB-dependent PRDII promoter (right panel) following mock infection (white bars) or infection with Sendai virus (grey bars). Note the significant inhibition of activation of both promoters in the 2-3 cells containing replicating HCV RNA.

the dsRNA surrogate poly-I:C) to TLR3, while the other is activated by intracellular binding of structured RNA to the cellular DExD/H box RNA helicase, RIG-I. These pathways function largely independently of each other in cells derived from human hepatocytes, but appear to converge at the level of the TBK1 and IKKε kinases, which are thought to play important roles in the phosphorylation of IRF-3. Highlighting the importance of these pathways to the initiation of robust host defences against invading viral pathogens, the NS3/4A protease has acquired the capacity to suppress signalling through both of these redundant pathways. To accomplish this, NS3/4A appears to mediate the proteolytic cleavage of at least two distinct cellular proteins, each of which is critical for signal transduction through one of these pathways.

NS3/4A DISRUPTION OF TLR3 SIGNALLING

Normal Huh7 cells produce little if any interferon-β in response to stimulation with extracellular poly-I:C, a synthetic dsRNA surrogate that acts as a specific ligand for TLR3. This is due to a lack of expression of TLR3, as poly-I:C responsiveness can be restored by ectopic expression of TLR3 (Li *et al.*, 2005a). This may explain in part the permissiveness of these cells for HCV RNA replication, and why they have been favoured for studies of HCV replicons (Lohmann *et al.*, 1999). Thus, to investigate the interaction of HCV with the TLR3 pathway, it has been necessary to study alternative types of cells. The TLR3 pathway is normally functional in HeLa cells, but was shown to be blocked in SL1 cells (Li *et al.*, 2005b), a HeLa cell line supporting replication of a subgenomic replicon derived from the genotype 1b HCV-N virus (Zhu *et al.*, 2003). The inhibition of signalling through the TLR3 pathway could be reproduced in interferon-cured SL1c cells by ectopic expression of NS3/4A, but not by expression of an active-site NS3/4A mutant devoid of protease activity (Li *et al.*, 2005b). TLR3 signalling could also be restored by treatment with the specific NS3/4A inhibitor, SCH6.

Data suggest that the NS3/4A blockade of poly-I:C activation of IRF-3 through the TLR3 pathway is due to specific proteolysis of the TLR3 adaptor protein, TRIF (Li *et al.*, 2005b). As summarized earlier, TRIF is a proline-rich adaptor protein that associates with TLR3 through its TIR domain, thereby linking TLR3 to downstream kinases responsible for the activation of IRF-3, IRF-7 and also NF-κB (Yamamoto *et al.*, 2002; Han *et al.*, 2004; Oshiumi *et al.*, 2003). TRIF is cleaved *in vitro* by the viral protease at a site just upstream of the TIR domain that shares considerable homology with the NS4B/5A cleavage site in the viral polyprotein (Li *et al.*, 2005b). While specific cleavage of TRIF has not been observed in cultured cells, ectopic expression of NS3/4A greatly accelerates the degradation of TRIF in transfected cells. Significantly, immuno-blotting showed that the abundance of endogenous TRIF was reduced in SL1 cells supporting HCV RNA replication, compared with clonally related, interferon-cured SL1c cells (Li *et al.*, 2005b). *In vitro* assays also indicate that the velocity of NS3/4A proteolysis of TRIF peptide substrates approximates that of peptides representing the viral NS4B/5A substrate (Ferreon *et al.*, 2005). Thus it appears that the TLR3 pathway can be inhibited, at least in some cell types, by NS3/4A-protease-mediated cleavage of TRIF.

DISRUPTION OF RIG-I SIGNALLING BY NS3/4A
While the TLR3 pathway is defective in Huh7 hepatoma cells, the alternative RIG-I pathway leading to IRF-3 activation remains active in these cells (Sumpter *et al.*, 2005; Li *et al.*, 2005a). Thus initial observations concerning the blockade of Sendai virus-induced activation of IRF-3 in Huh7 2-3 cells (Foy *et al.*, 2003) suggest that RIG-I signalling is disrupted by the protease. This has recently been confirmed in experiments showing that RIG-I-mediated activation of IRF-3 is potently blocked by the NS3/4A protease (Foy *et al.*, 2005). RIG-I was found to be essential for triggering host responses to transfected HCV RNA in Huh7 cells, while the NS3/4A protease specifically inhibited activation of IRF-3 by a constitutively active form of RIG-I comprising its N-terminal CARD-like domains (Sumpter *et al.*, 2005; Foy *et al.*, 2005). The cellular protein that is putatively targeted for proteolysis by NS3/4A to block signalling through the RIG-I pathway has not been identified, however. Bacterially expressed NS3/4A protease has the ability to proteolytically cleave RIG-I *in vitro*, but immunoblotting has failed to show any reduction in the intracellular abundance of the intact RIG-I molecule in cells containing replicating HCV RNAs (H. Ishida & K. Li, unpublished data). Also, TRIF does not participate in RIG-I signalling in these cells (Li *et al.*, 2005a; Foy *et al.*, 2005). Thus it appears that a second, yet unidentified cellular signalling

protein is likely to be targeted for proteolysis by NS3/4A. This putative NS3/4A substrate appears to be efficiently targeted for proteolysis by the protease, as the RIG-I pathway is potently blocked by a very low abundance of NS3/4A expressed under control of the Tet-Off promoter in HepG2 hepatoma cells (K. Li, unpublished data).

The RIG-I pathway appears to be important for control of HCV RNA replication in Huh7 cells that are defective in TLR3 signalling. Activation of the RIG-I pathway leads to suppression of HCV RNA replication in Huh7 cells, while a highly permissive variant of Huh7 cells, Huh7.5 cells, was shown to be defective in RIG-I signalling (Sumpter *et al.*, 2005). These Huh7.5 cells, which were established by eliminating HCV RNA from a G418-resistant clone of replicon-bearing cells (Blight *et al.*, 2002), are thus defective for both TLR3 and RIG-I signalling, and do not show activation of IRF-3 on challenge with either highly structured RNAs or Sendai virus (Sumpter *et al.*, 2005). An analysis of cells harbouring genetically distinct HCV RNA replicon variants has also suggested that the ability of the HCV protease to regulate the RIG-I pathway correlates directly with replication efficiency and replicon fitness (Sumpter *et al.*, 2005).

EVOLUTION OF THE NS3/4A PROTEASE

Studies of the structure of NS3/4A protease have noted that the substrate-binding channel is remarkably shallow and featureless, and more solvent-exposed than is the case with other chymotrypsin-like serine proteases (Kim *et al.*, 1996). While this has made design of potent and selective small molecule inhibitors of the protease difficult, it is intriguing to speculate that the relatively open nature of the substrate-binding surface may be related to its ability to accommodate the molecular interactions necessary for cleavage of one or more of the cellular proteins participating in the signalling pathways described earlier. Although the TRIF cleavage site shows close homology to the HCV NS4B/5A cleavage site on the P-side of the scission, it lacks the P6 acidic residue common to the *trans*-cleavage sites within the viral polyprotein. This is replaced in TRIF with an eight-residue polyproline track that is likely to form a polyproline-II helix. NMR studies with a polyproline-II helix-forming peptide have suggested that this segment of TRIF interacts with a 3_{10} helix in the NS3 structure close to the protease active site (Ferreon *et al.*, 2005). Thus the molecular recognition of TRIF by NS3/4A appears to differ significantly from the manner in which NS3/4A recognizes viral substrates, as the 3_{10} helix does not appear to participate in these interactions (Ferreon *et al.*, 2005). This has potentially important implications for the design of protease inhibitors, and possibly the impact of mutations that confer resistance to protease inhibitors on the ability of NS3/4A to effectively cleave cellular targets. This is likely to be a fruitful area for further investigation.

DISRUPTION OF INNATE IMMUNITY AND IMPAIRMENTS IN ADAPTIVE IMMUNITY

The NS3/4A-mediated blockade of IRF-3 and NF-κB activation in infected hepatocytes (Foy *et al.*, 2003, 2005; Li *et al.*, 2005b) may be of fundamental importance to the outcome of acute HCV infection. First, it would prevent or delay normal defence mechanisms that the cell invokes in response to virus infection. As described earlier, these involve the induction of type 1 interferons and the subsequent expression of a wide array of IRF-3 and interferon-sensitive response element (ISRE)-dependent genes such as ISG56 and protein kinase R (PKR) that have direct, intracellular antiviral actions (Katze *et al.*, 2002). The importance of these responses is suggested by the ability of genes induced via the RIG-I-dependent pathway to limit HCV RNA replication in Huh7 cells (Foy *et al.*, 2003; Sumpter *et al.*, 2005). Equally, if not more, important, however, may be the impact of this disruption of innate immune signalling on the subsequent adaptive CD8$^+$ T cell responses that follow. There is growing recognition of the critical role of innate immune responses in shaping subsequent adaptive immunity (Hoebe *et al.*, 2004; Iwasaki & Medzhitov, 2004), and disruption of IRF-3 signalling by NS3/4A may well set the stage for a subsequent, inadequate T-lymphocyte response that is unable to eliminate the infection.

In addition to their direct intracellular antiviral effects, type 1 interferons exert critically important immunoregulatory effects, including stimulation of NK cytotoxicity and activation of γδ T cells, in addition to their roles in promoting maturation of DCs and the priming of T$_H$1 and cytotoxic T lymphocytes (Beignon *et al.*, 2003). Type 1 interferons promote antigen cross-priming, providing an essential 'licence' to DCs to present exogenous antigens produced in non-antigen presenting cells (APCs) to CD8$^+$ T cells (Le Bon *et al.*, 2003). Cross-priming is essential for the development of robust and durable CD8$^+$ T cell responses against antigens that are not expressed in APCs, such as DCs, and has been shown to occur with lymphocytic choriomeningitis virus by a mechanism that is dependent on type 1 interferons and independent of CD4$^+$ T cell help, or interaction of CD40 with CD40 ligand (Le Bon *et al.*, 2003).

While the specific events that lead to T-cell priming during acute HCV infection of the liver remain ill defined, the suppression of type 1 interferon synthesis by NS3/4A disruption of IRF-3 activation could thus have pleiotropic effects on the downstream development of adaptive T-cell immunity, perhaps setting the stage for eventual T-cell failure. Were HCV to actually replicate within the critical DC subsets involved in the presentation of HCV antigens, NS3/4A disruption of TLR3 and RIG-I signalling could have a profound effect by inhibiting the up-regulation of essential costimulatory CD80 and CD86 molecules, and thereby suppressing T-cell activation (Iwasaki & Medzhitov, 2004). Significant abnormalities in DC function have been described in hepatitis C, and, although far from confirmed, it has been suggested that the virus may be capable of replicating in certain types of DCs (Kanto *et al.*, 2004; Kaimori *et al.*, 2004).

As both TLR3 and RIG-I signalling also lead to NF-κB activation, the NS3/4A blockade of these pathways would impair the virus-induced expression of NF-κB-dependent proinflammatory cytokines, such as IL-6 and IL-12, and chemokines, such as RANTES, that promote the migration of immune cells to the site of infection (Hoebe *et al.*, 2004; Iwasaki & Medzhitov, 2004). The impaired or delayed expression of type 1 interferons, cytokines and chemokines by infected hepatocytes could contribute to the remarkable delay that has been observed in appearance of T-cell responses in acutely infected chimpanzee liver. These responses do not occur until 4–8 weeks after the onset of viraemia, despite the presence of viral antigens within the liver (Thimme *et al.*, 2002). The delay in these various responses may be fundamentally important to the establishment of persistent viral infection (Major *et al.*, 2004; Thimme *et al.*, 2002).

ACKNOWLEDGEMENTS

Supported in part by grants from the National Institutes of Health (U19-AI40035 and R21-DA018054) and the Advanced Technology Program of the Texas Higher Education Coordinating Board (004952-0067-2003). K. L. is the John Mitchell Hemophilia of Georgia Liver Scholar of the American Liver Foundation.

REFERENCES

Alexopoulou, L., Holt, A. C., Medzhitov, R. & Flavell, R. A. (2001). Recognition of double-stranded RNA and activation of NF-kappaB by Toll-like receptor 3. *Nature* **413**, 732–738.

Alter, M. J., Kruszon-Moran, D., Nainan, O. V., McQuillan, G. M., Gao, F., Moyer, L. A., Kaslow, R. A. & Margolis, H. S. (1999). The prevalence of hepatitis C virus infection in the United States, 1988 through 1994. *N Engl J Med* **341**, 556–562.

Beignon, A. S., Skoberne, M. & Bhardwaj, N. (2003). Type I interferons promote cross-priming: more functions for old cytokines. *Nat Immunol* **4**, 939–941.

Beutler, B. (2004). Inferences, questions and possibilities in Toll-like receptor signalling. *Nature* **430**, 257–263.

Blight, K. J., Kolykhalov, A. A. & Rice, C. M. (2000). Efficient initiation of HCV RNA replication in cell culture. *Science* **290**, 1972–1974.

Blight, K. J., McKeating, J. A. & Rice, C. M. (2002). Highly permissive cell lines for subgenomic and genomic hepatitis C virus RNA replication. *J Virol* **76**, 13001–13014.

Blight, K. J., McKeating, J. A., Marcotrigiano, J. & Rice, C. M. (2003). Efficient replication of hepatitis C virus genotype 1a RNAs in cell culture. *J Virol* **77**, 3181–3190.

Choo, Q.-L., Kuo, G., Weiner, A. J., Overby, L. R., Bradley, D. W. & Houghton, M. (1989). Isolation of a cDNA clone derived from a blood-borne non-A, non-B viral hepatitis genome. *Science* **244**, 359–362.

Crozat, K. & Beutler, B. (2004). TLR7: a new sensor of viral infection. *Proc Natl Acad Sci U S A* **101**, 6835–6836.

Diebold, S. S., Kaisho, T., Hemmi, H., Akira, S. & Reis e Sousa, C. (2004). Innate antiviral responses by means of TLR7-mediated recognition of single-stranded RNA. *Science* **303**, 1529–1531.

Disson, O., Haouzi, D., Desagher, S. & 7 other authors (2004). Impaired clearance of virus-infected hepatocytes in transgenic mice expressing the hepatitis C virus polyprotein. *Gastroenterology* **126**, 859–872.

Enomoto, N., Sakuma, I., Asahina, Y. & 7 other authors (1996). Mutations in the nonstructural protein 5A gene and response to interferon in patients with chronic hepatitis C virus 1b infection. *N Engl J Med* **334**, 77–81.

Ferreon, J. C., Ferreon, A. C. M., Li, K. & Lemon, S. M. (2005). Molecular determinants involved in TRIF proteolysis by the hepatitis C virus NS3/4A protease. Submitted.

Fitzgerald, K. A., McWhirter, S. M., Faia, K. L., Rowe, D. C., Latz, E., Golenbock, D. T., Coyle, A. J., Liao, S. M. & Maniatis, T. (2003). IKKepsilon and TBK1 are essential components of the IRF3 signaling pathway. *Nat Immunol* **4**, 491–496.

Foy, E., Li, K., Wang, C., Sumter, R., Ikeda, M., Lemon, S. M. & Gale, M. (2003). Regulation of interferon regulatory factor-3 by the hepatitis C virus serine protease. *Science* **300**, 1145–1148.

Foy, E., Li, K., Sumpter, R. & 8 other authors (2005). Control of antiviral defenses through hepatitis C virus disruption of RIG-I signaling. *Proc Natl Acad Sci U S A* (in press).

Grandvaux, N., Servant, M. J., tenOever, B., Sen, G. C., Balachandran, S., Barber, G. N., Lin, R. & Hiscott, J. (2002). Transcriptional profiling of interferon regulatory factor 3 target genes: direct involvement in the regulation of interferon-stimulated genes. *J Virol* **76**, 5532–5539.

Griffin, S. D., Beales, L. P., Clarke, D. S., Worsfold, O., Evans, S. D., Jaeger, J., Harris, M. P. & Rowlands, D. J. (2003). The p7 protein of hepatitis C virus forms an ion channel that is blocked by the antiviral drug, Amantadine. *FEBS Lett* **535**, 34–38.

Guo, J. T., Bichko, V. V. & Seeger, C. (2001). Effect of alpha interferon on the hepatitis C virus replicon. *J Virol* **75**, 8516–8523.

Han, K. J., Su, X., Xu, L. G., Bin, L. H., Zhang, J. & Shu, H. B. (2004). Mechanisms of the TRIF-induced interferon-stimulated response element and NF-kappaB activation and apoptosis pathways. *J Biol Chem* **279**, 15652–15661.

Haynes, L. M., Moore, D. D., Kurt-Jones, E. A., Finberg, R. W., Anderson, L. J. & Tripp, R. A. (2001). Involvement of toll-like receptor 4 in innate immunity to respiratory syncytial virus. *J Virol* **75**, 10730–10737.

Hochrein, H., Schlatter, B., O'Keeffe, M., Wagner, C., Schmitz, F., Schiemann, M., Bauer, S., Suter, M. & Wagner, H. (2004). Herpes simplex virus type-1 induces IFN-alpha production via Toll-like receptor 9-dependent and -independent pathways. *Proc Natl Acad Sci U S A* **101**, 11416–11421.

Hoebe, K., Janssen, E. & Beutler, B. (2004). The interface between innate and adaptive immunity. *Nat Immunol* **5**, 971–974.

Honda, K., Yanai, H., Mizutani, T. & 7 other authors (2004). Role of a transductional-transcriptional processor complex involving MyD88 and IRF-7 in Toll-like receptor signaling. *Proc Natl Acad Sci U S A* **101**, 15416–15421.

Ikeda, M., Yi, M., Li, K. & Lemon, S. M. (2002). Selectable subgenomic and genome-length dicistronic RNAs derived from an infectious molecular clone of the HCV-N strain of hepatitis C virus replicate efficiently in cultured Huh7 cells. *J Virol* **76**, 2997–3006.

Iwasaki, A. & Medzhitov, R. (2004). Toll-like receptor control of the adaptive immune responses. *Nat Immunol* **5**, 987–995.

Kaimori, A., Kanto, T., Kwang, L. C. & 10 other authors (2004). Pseudotype hepatitis C virus enters immature myeloid dendritic cells through the interaction with lectin. *Virology* **324**, 74–83.

Kanto, T., Inoue, M., Miyatake, H. & 9 other authors (2004). Reduced numbers and impaired ability of myeloid and plasmacytoid dendritic cells to polarize T helper cells in chronic hepatitis C virus infection. *J Infect Dis* **190**, 1919–1926.

Kato, T., Date, T., Miyamoto, M., Furusaka, A., Tokushige, K., Mizokami, M. & Wakita, T. (2003). Efficient replication of the genotype 2a hepatitis C virus subgenomic replicon. *Gastroenterology* **125**, 1808–1817.

Katze, M. G., He, Y. & Gale, M. (2002). Viruses and interferon: a fight for supremacy. *Nat Rev Immunol* **2**, 675–687.

Kawai, T., Sato, S., Ishii, K. J. & 9 other authors (2004). Interferon-alpha induction through Toll-like receptors involves a direct interaction of IRF7 with MyD88 and TRAF6. *Nat Immunol* **5**, 1061–1068.

Kim, J. L., Morgenstern, K. A., Lin, C. & 12 other authors (1996). Crystal structure of the hepatitis C virus NS3 protease domain complexed with a synthetic NS4A cofactor peptide. *Cell* **87**, 343–355.

Konan, K. V., Giddings, T. H., Ikeda, M., Li, K., Lemon, S. M. & Kirkegaard, K. (2003). Nonstructural protein precursor NS4A/B from hepatitis C virus alters function and ultrastructure of host secretory apparatus. *J Virol* **77**, 7843–7855.

Kuo, G., Choo, Q.-L., Alter, H. J. & 17 other authors (1989). An assay for circulating antibodies to a major etiologic virus of human non-A, non-B hepatitis. *Science* **244**, 362–364.

Kwong, A. D., Kim, J. L., Rao, G., Lipovsek, D. & Raybuck, S. A. (1999). Hepatitis C virus NS3/4A protease. *Antiviral Res* **41**, 67–84.

Lamarre, D., Anderson, P. C., Bailey, M. & 25 other authors (2003). An NS3 protease inhibitor with antiviral effects in humans infected with hepatitis C virus. *Nature* **426**, 186–189.

Le Bon, A., Etchart, N., Rossmann, C., Ashton, M., Hou, S., Gewert, D., Borrow, P. & Tough, D. F. (2003). Cross-priming of CD8+ T cells stimulated by virus-induced type I interferon. *Nat Immunol* **4**, 1009–1015.

Lechner, F., Wong, D. K., Dunbar, P. R. & 7 other authors (2000). Analysis of successful immune responses in persons infected with hepatitis C virus. *J Exp Med* **191**, 499– 1512.

Li, K., Chen, Z., Kato, N., Gale, M. & Lemon, S. M. (2005a). Distinct poly-I : C and virus-activated signaling pathways leading to interferon-β production in hepatocytes. Submitted.

Li, K., Foy, E., Ferreon, J. C., Nakamura, M., Ferreon, A. C. M., Ikeda, M., Ray, S. C., Gale, M. J. J. & Lemon, S. M. (2005b). Immune evasion by hepatitis C virus NS3/4A protease-mediated cleavage of the TLR adaptor protein TRIF. *Proc Natl Acad Sci U S A* (in press).

Lin, R., Heylbroeck, C., Pitha, P. M. & Hiscott, J. (1998). Virus-dependent phosphory-lation of the IRF-3 transcription factor regulates nuclear translocation, transactivation potential, and proteasome-mediated degradation. *Mol Cell Biol* **18**, 2986–2996.

Lohmann, V., Korner, F., Koch, J., Herian, U., Theilmann, L. & Bartenschlager, R. (1999). Replication of subgenomic hepatitis C virus RNAs in a hepatoma cell line. *Science* **285**, 110–113.

Lohmann, V., Korner, F., Dobierzewska, A. & Bartenschlager, R. (2001). Mutations in hepatitis C virus RNAs conferring cell culture adaptation. *J Virol* **75**, 1437–1449.

Major, M. E. & Feinstone, S. M. (1997). The molecular virology of hepatitis C. *Hepatology* **25**, 1527–1538.

Major, M. E., Dahari, H., Mihalik, K., Puig, M., Rice, C. M., Neumann, A. U. & Feinstone, S. M. (2004). Hepatitis C virus kinetics and host responses associated with disease and outcome of infection in chimpanzees. *Hepatology* **39**, 1709–1720.

Maniatis, T., Falvo, J. V., Kim, T. H., Kim, T. K., Lin, C. H., Parekh, B. S. & Wathelet, M. G. (1998). Structure and function of the interferon-beta enhanceosome. *Cold Spring Harb Symp Quant Biol* **63**, 609–620.

McGettrick, A. F. & O'Neill, L. A. (2004). The expanding family of MyD88-like adaptors in Toll-like receptor signal transduction. *Mol Immunol* **41**, 577–582.

McHutchison, J. G. & Fried, M. W. (2003). Current therapy for hepatitis C: pegylated interferon and ribavirin. *Clin Liver Dis* **7**, 149–161.

Oshiumi, H., Matsumoto, M., Funami, K., Akazawa, T. & Seya, T. (2003). TICAM-1, an adaptor molecule that participates in Toll-like receptor 3-mediated interferon-beta induction. *Nat Immunol* **4**, 161–167.

Panne, D., Maniatis, T. & Harrison, S. C. (2004). Crystal structure of ATF-2/c-Jun and IRF-3 bound to the interferon-beta enhancer. *EMBO J* **23**, 4384–4393.

Pawlotsky, J. M. (2004). Pathophysiology of hepatitis C virus infection and related liver disease. *Trends Microbiol* **12**, 96–102.

Racanelli, V. & Rehermann, B. (2003). Hepatitis C virus infection: when silence is deception. *Trends Immunol* **24**, 456–464.

Rijnbrand, R. C. & Lemon, S. M. (2000). Internal ribosome entry site-mediated translation in hepatitis C virus replication. *Curr Top Microbiol Immunol* **242**, 85–116.

Scholle, F., Li, K., Bodola, F., Ikeda, M., Luxon, B. A. & Lemon, S. M. (2004). Virus–host cell interactions during hepatitis C virus RNA replication: impact of polyprotein expression on the cellular transcriptome and cell cycle association with viral RNA synthesis. *J Virol* **78**, 1513–1524.

Seeff, L. B. (1997). Natural history of hepatitis C. *Hepatology* **26**, 21S–28S.

Seeff, L. B., Buskell-Bales, Z., Wright, E. C. & 10 other authors (1992). Long-term mortality after transfusion-associated non-A, non-B hepatitis. *N Engl J Med* **327**, 1906–1911.

Sharma, S., tenOever, B. R., Grandvaux, N., Zhou, G. P., Lin, R. & Hiscott, J. (2003). Triggering the interferon antiviral response through an IKK-related pathway. *Science* **300**, 1148–1151.

Sumpter, R., Loo, M. Y., Foy, E., Li, K., Yoneyama, M., Fujita, T., Lemon, S. M. & Gale, M. J. (2005). Regulating intracellular anti-viral defense and permissiveness to hepatitis C virus RNA replication through a cellular RNA helicase, RIG-I. *J Virol* (in press).

Thimme, R., Bukh, J., Spangenberg, H. C., Wieland, S., Pemberton, J., Steiger, C., Govindarajan, S., Purcell, R. H. & Chisari, F. V. (2002). Viral and immunological determinants of hepatitis C virus clearance, persistence, and disease. *Proc Natl Acad Sci U S A* **99**, 15661–15668.

Thomas, D. L., Ray, S. C. & Lemon, S. M. (2004). Hepatitis C. In *Principles and Practice of Infectious Disease*, pp. 1950–1981. Edited by G. L. Mandell & R. Dolin. New York: Churchill Livingstone.

Weiner, A. J., Paliard, X., Selby, M. J. & 17 other authors (2001). Intrahepatic genetic inoculation of hepatitis C virus RNA confers cross-protective immunity. *J Virol* **75**, 7142–7148.

Wong, J. B., McQuillan, G. M., McHutchison, J. G. & Poynard, T. (2000). Estimating future hepatitis C morbidity, mortality, and costs in the United States. *Am J Public Health* **90**, 1562–1569.

Yamamoto, M., Sato, S., Mori, K., Hoshino, K., Takeuchi, O., Takeda, K. & Akira, S. (2002). Cutting edge: a novel Toll/IL-1 receptor domain-containing adapter that preferentially activates the IFN-beta promoter in the Toll-like receptor signaling. *J Immunol* **169**, 6668–6672.

Yi, M. & Lemon, S. M. (2004). Adaptive mutations producing efficient replication of genotype 1a hepatitis C virus RNA in normal Huh7 cells. *J Virol* **78**, 7904–7915.

Yoneyama, M., Suhara, W., Fukuhara, Y., Fukuda, M., Nishida, E. & Fujita, T. (1998). Direct triggering of the type I interferon system by virus infection: activation of a transcription factor complex containing IRF-3 and CBP/p300. *EMBO J* **17**, 1087–1095.

Yoneyama, M., Kikuchi, M., Natsukawa, T., Shinobu, N., Imaizumi, T., Miyagishi, M., Taira, K., Akira, S. & Fujita, T. (2004). The RNA helicase RIG-I has an essential function in double-stranded RNA-induced innate antiviral responses. *Nat Immunol* **5**, 730–737.

Yoshizawa, H. (2002). Hepatocellular carcinoma associated with hepatitis C virus infection in Japan: projection to other countries in the foreseeable future. *Oncology* **62** (Suppl. 1), 8–17.

Zhu, Q., Guo, J. T. & Seeger, C. (2003). Replication of hepatitis C virus subgenomes in nonhepatic epithelial and mouse hepatoma cells. *J Virol* **77**, 9204–9210.

Human papillomaviruses and their effects on cell cycle control and apoptosis

L. Gray, C. Jolly and C. S. Herrington

Bute Medical School, University of St Andrews, Bute Medical Buildings, Westburn Lane, St Andrews, Fife KY16 9TS, UK

HUMAN PAPILLOMAVIRUS (HPV) INFECTION AND NEOPLASIA

HPVs are double-stranded DNA viruses that have been identified as an important, and probably essential, aetiological factor in squamous carcinogenesis (Southern & Herrington, 2000). The papillomavirus genome is circular, approximately 7·9 kb in length, and contains seven early and two late open reading frames (ORFs). Expression of the early genes occurs at the onset of infection, and the products of these genes mediate specific functions controlling virus replication and, in the case of the oncogenic viruses, cellular transformation. The E1 protein is involved in virus replication and genome maintenance. It is a nuclear phosphoprotein that binds specifically to the virus origin of replication, unwinding the DNA template and acting as a helicase (Wilson et al., 2002). The E2 protein is a transcriptional regulator and is also involved, with E1, in viral DNA replication (Longworth & Laimins, 2004). The E4 protein, which is formed as a fusion protein incorporating part of the E1 protein (E1^E4), is a late protein that is not involved in transformation by the virus or assembly of capsids (Doorbar, 1998). Its exact function in vivo is uncertain but it appears to be involved in the late stages of virus production. The function of the E5 protein is unclear but its interaction with the vacuolar ATPase may be pathogenetically important (DiMaio & Mattoon, 2001). The E6 and E7 genes encode the main transforming proteins (Southern & Herrington, 2000). These are capable of immortalization and neoplastic transformation under appropriate conditions. The L1 and L2 ORFs encode the major and minor capsid proteins, respectively, that make up the icosahedral capsid coat of the virus (Zhou & Frazer, 1996). A non-coding region referred to as the upstream regulatory region (URR), or long control region (LCR), that contains glucocorticoid

SGM symposium 64: Molecular pathogenesis of virus infections.
Editors P. Digard, A. A. Nash & R. E. Randall. Cambridge University Press. ISBN 0 521 83248 9 ©SGM 2005

response elements and the origin of DNA replication is present between the early and late gene regions (Moodley *et al.*, 2003).

HPVs infect only epithelial cells and, in particular, target squamous epithelium, which lines the skin and mucosal surfaces of the oral, and anogenital, regions (Stoler, 2000). Over 100 HPV types have been identified, based on DNA sequence, and construction of a phylogenetic tree has led to the definition of several groups of HPVs that are associated with different types of abnormality of this epithelium. Cutaneous HPV types are associated with skin warts. Mucosal HPV types are associated predominantly with anogenital lesions that vary from benign warts (associated with, for example, HPVs 6 and 11) to invasive cancer (associated with, for example, HPVs 16 and 18). These clinical associations most likely reflect biological differences between different groups of HPVs and therefore the study of HPV infection, and the properties of HPV proteins, is likely to provide significant insights into the neoplastic process.

The HPV life cycle is closely linked to the differentiation programme of squamous epithelium such that vegetative replication of HPVs does not occur in the absence of such differentiation (Bedell *et al.*, 1991). Specifically, virion production does not occur in basal keratinocytes but is restricted to the superficial cells of a differentiating squamous epithelium. Initial infection most likely occurs in basal epithelial cells, which are actively dividing, and which are exposed through small abrasions. The cellular receptor for HPV is not known, although there is some evidence that α_6-integrin may play a role (Evander *et al.*, 1997). Following initial infection, the HPV genome replicates with cellular DNA, then undergoes large-scale amplification in suprabasal, non-dividing, keratinocytes in the upper layers of the epithelium. Capsid protein is then synthesized and intact virions are released (Stanley, 2003). This process is associated with spatially regulated expression of viral genes and leads to a characteristic cyto-pathic effect, recognized morphologically as koilocytosis, nuclear enlargement, dyskeratosis, multinucleation and, in some cases, low-grade squamous intraepithelial lesions (SIL) – cervical intraepithelial neoplasia (CIN) grade 1.

This process may not complete normally and viral gene expression does not then take place in a spatially regulated manner (see later). This situation is associated most frequently with integration of the viral genome into host chromosomes and vegetative virus production does not occur under these circumstances. Such integration does not occur at consistent chromosomal sites but the virus breakpoint generally occurs in the E1/E2 region of the viral genome (Wentzensen *et al.*, 2004). This breakpoint is important because loss of E1/E2 function leads not only to failure of viral DNA replication but also to loss of transcriptional repression of E6/E7 by E2 (Choo *et al.*, 1987).

Constitutive expression of E6 and E7 oncoproteins is important for malignant transformation. The E7 protein of high-risk HPVs (e.g. HPV-16) can transform some established rodent cell lines (e.g. NIH3T3) and, in cooperation with an activated *ras* oncogene, can transform primary rodent cells (Munger *et al.*, 1992). Although the E7 protein can immortalize primary human epithelial cells (Halbert *et al.*, 1991), it cannot transform primary cells in the absence of other oncogenes. The E7 proteins encoded by the low-risk HPVs (e.g. HPV-6) have only weak immortalizing activity in similar assays (Halbert *et al.*, 1992). Although E6 is not considered to be a transforming protein, it can immortalize certain transformation-sensitive cell lines when overexpressed *in vitro* (Band *et al.*, 1993) and enhances the efficiency of the E7 immortalizing function (Halbert *et al.*, 1991).

The HPVs disrupt the cell cycle in order to direct squamous cells that have exited the cell cycle to synthesize viral DNA. To be entirely successful, the accumulation of abnormal cellular DNA in dividing cells needs to be prevented during this process. This is achieved more successfully by low-risk HPVs such as HPV-6, but there is increasing evidence that high-risk HPVs such as HPV-16 disrupt cell cycle control in such a way that aneuploid cells develop and are able to survive and replicate. The mechanisms by which this occurs involve subversion of host regulatory proteins such as cyclins, cyclin-dependent kinases (CDKs) and cyclin-dependent kinase inhibitors (CDKIs) involved in cell cycle control (Southern & Herrington, 2000); and interference with normal apoptotic pathways (Finzer *et al.*, 2002).

THE CELL CYCLE

The purpose of the cell cycle is to generate genetically identical daughter cells. This is achieved by replication of the genome through DNA synthesis (S phase), followed by segregation of chromosomes and equal divison of the mother cell at mitosis (M phase). Between these two phases, there are two gaps, termed G_1 (between M and S phase) and G_2 (between S and M phase), during which complex processes occur to provide the machinery required for DNA synthesis and mitosis, respectively. These processes have to be controlled very carefully in order for DNA synthesis to proceed accurately and for mitosis to deliver the correct chromosomes into each daughter cell. There are a number of evolutionarily conserved pathways involved in these processes and the major proteins involved fall into three main classes: the cyclins; the CDKs; and the CDKIs (Thomas *et al.*, 1998; Southern & Herrington, 1998b; Grana & Reddy, 1995). There are many different CDKs and cyclins, each controlling different stages of the cell cycle. When a cyclin activates a CDK, the cell progresses through the stage that the particular CDK regulates. Upon completion of the specific event and deactivation of the cyclin/CDK complex, this part of the cell cycle ceases and other downstream events occur to take the cell through the next stages. As well as the positive regulatory activity

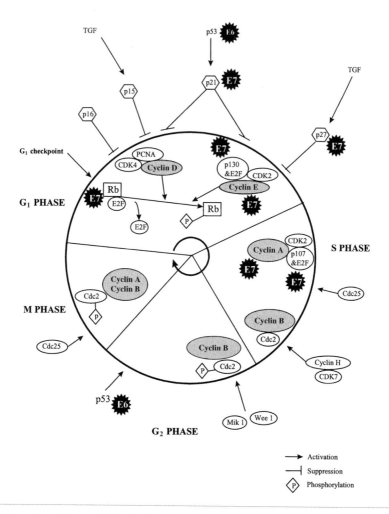

Fig. 1. Interactions between HPV and cell cycle control proteins. Further details are given in Southern & Herrington (2000). Reproduced from Southern & Herrington (2000) with permission from Blackwell Publishing.

of cyclins/CDKs, there are a number of direct negative regulatory mechanisms that inhibit CDK activity, including CDKIs.

HPVs target a number of host cell regulatory proteins in order to subvert cell cycle machinery and effect virus replication (see Fig. 1). Cell cycle progression is altered and cells that have exited the cell cycle (as normally occurs during squamous differentiation) are stimulated to synthesize a variety of cell cycle-associated proteins. Some viral proteins, particularly those from high-risk HPVs, interfere with surveillance mechanisms or checkpoints that prevent abnormal replication of cellular DNA, or detect abnormal chromosome segregation, allowing potential accumulation of genetic abnormalities.

Progression through G_1/S phase

Progression through G_1 and S phase is controlled by cyclins D, E and A. In epithelial cells, the regulatory protein cyclin D1 associates with CDK4 and CDK6, which initiate various phosphorylation events that allow the cell to progress through G_1 phase. pRB is the major target of CDK4/6 phosphorylation because hypophosphorylated pRB binds to and inactivates E2F, thereby preventing cell cycle progression. Phosphorylation of pRB disrupts this association, allowing E2F to activate synthesis of S-phase-associated genes. One of the major functions of the E7 oncoprotein, particularly of high-risk HPVs, is to bind to, and functionally inactivate, the hypophosphorylated pRB tumour suppressor gene product, releasing transcriptionally active E2F and hence allowing cell cycle progression (Helt & Galloway, 2003). There is therefore no requirement for cyclin D1 expression, or CDK4/6 activity, for cell cycle progression in E7-expressing cells. It is of note that cyclin D1 expression is rarely immunohistochemically detectable in naturally occurring low-grade cervical squamous intraepithelial lesions infected with high-risk HPV. However, cyclin D1 is expressed in low-grade lesions infected with low-risk HPVs (Southern & Herrington, 1998a), presumably reflecting a requirement for cyclin D1 in cells infected with these viruses. This most likely reflects the lower affinity of E7 proteins of low-risk HPVs for pRB. E7 binds to pRB through an LxCxE motif that also allows binding to other 'pocket' proteins, such as p107 and p130. The role of these interactions in the mechanism of action of E7 is less well understood (Helt & Galloway, 2003).

The CDKIs are also important at the G_1/S-phase transition. These molecules fall into two broad groups, namely the CIP/KIP family of inhibitors (p21, p27, p57) and the INK (inhibitors of CDK4) family (p15, p16, p18). p21 has been termed a 'universal inhibitor' as it interacts with a number of cyclin/CDK complexes involving principally cyclins D, E and A (Xiong *et al.*, 1993). The INK family have more specific effects, principally on the cyclin D/CDK4 complex. p21 gene transcription is p53-dependent and is one mechanism by which p53 induces cell cycle arrest (see later). However, the E7 protein of high-risk HPVs binds to p21 and can overcome p21-induced cell cycle arrest (Funk *et al.*, 1997; Jones *et al.*, 1997). p16 expression is inhibited by pRB (Li *et al.*, 1994). Therefore, inhibition of pRB by high-risk HPV E7 leads to up-regulation of p16 expression, which is, however, unable to induce cell cycle arrest because there is no requirement for cyclin D/CDK4/6 activity in E7-expressing cells. This is supported by the observation that cells infected with low-risk HPVs, whose E7 protein binds to pRB with only low affinity, do not up-regulate p16 expression. High-level expression of p16 has been used as a marker of high-risk HPV E7 activity in clinical samples (Klaes *et al.*, 2001).

As cells progress into late G_1 phase, cyclin E synthesis is initiated and this continues into S phase. High-risk HPV E7 up-regulates cyclin E expression (Zerfass *et al.*, 1995) and inhibits the effects of both p27 and p21 (Jones *et al.*, 1997; Zerfass-Thome *et al.*, 1996), thereby leading to enhanced S-phase activity. It is not clear whether this represents a true S phase, or simply up-regulation of S-phase-associated genes, but the net result is the creation of an environment conducive to viral DNA synthesis by the uncoupling of keratinocyte proliferation and differentiation. As the cell progresses into S phase, cyclin E is degraded and cyclin A, in association with CDK2, forms the major S-phase complex. Again, high-risk HPV E7 up-regulates cyclin A expression (Zerfass *et al.*, 1995) and hence promotes maintenance of expression of S-phase-associated genes. The concept that the papillomavirus E7 protein induces S-phase-associated genes to effect viral DNA replication is attractive. However, there is evidence from studies of differentiating squamous epithelium that the situation *in vivo* is not so straightforward. Thus coordinated induction of p21 and cyclin E by HPV-18 E7 leads to inhibition of DNA synthesis, possibly by formation of an inactive protein complex (Jian *et al.*, 1999). This phenomenon may be related to the point in the cell cycle at which HPV E7 expression occurs and keratinocytes appear to have two alternative fates, namely endo-reduplication of cellular DNA, which occurs in a low percentage of cells, and cell cycle arrest with accumulation of p27kip1, cyclin E and p21cip1 proteins, which occurs in the majority of cells and prevents further S-phase re-entry (Chien *et al.*, 2002).

One of the major cellular proteins targeted by E6 oncoproteins is the tumour suppressor p53 (Werness *et al.*, 1990). The tumour suppressor activity of this protein is well documented; loss of wild-type p53 function due to mutation within the p53 gene is an important event in the development of many human tumours (Hollstein *et al.*, 1991). p53 is central to the G_1/S checkpoint that controls entry of cells into S phase (Vousden & Lu, 2002). In normal cells, the response to DNA damage increases p53 levels and transcriptionally activates p21. p21 binds to and inactivates a number of cyclin/CDK complexes, including cyclin D1/CDK4, cyclin E/CDK2 and cyclin A/CDK2, preventing phosphorylation of pRB and release of E2F, thereby inducing G_1 arrest and preventing cells entering S phase. This allows time for DNA repair or disposal of cells via apoptosis. Cells lacking functional p53 may therefore accumulate genetic alterations leading to malignant transformation. Loss of functional p53 as a consequence of E6 binding results in failure of G_1/S arrest (Thomas & Laimins, 1998).

Progression through G_2/M phase

The main cyclin/CDK complex involved in mitosis is cyclin B/CDK1 (cdc2): this is also known as mitosis promoting factor (MPF). The cdc2 component of MPF is constitutively present and therefore it is the accumulation and destruction of cyclin B that is phase-specific. Several checkpoints are involved in the control of the G_2/M boundary (Thompson *et al.*, 1997; Thomas & Laimins, 1998). In particular, the G_2/M boundary checkpoint causes G_2 arrest in response to DNA damage. The checkpoint coupling S phase completion to initiation of mitosis is also defective in cells expressing HPV-16 E6, as is the mitotic spindle checkpoint.

A common characteristic of neoplastic cells is the acquisition of numerical and structural chromosome abnormalities. Accumulation of these abnormalities is termed chromosome instability, which is a specific form of the more general phenomenon of genetic instability. Both high-risk E6 and E7 can induce chromosome abnormalities but their effects are different. Stable expression of the E6 and E7 genes of HPV-16, but not those of HPV-6, can induce abnormal centrosome numbers (and hence chromosome instability) in primary keratinocytes, individually and in combination (Duensing *et al.*, 2000). However, transient transfection experiments showed that abnormal centrosome numbers could be induced by E7, but not E6, expression within 48 h (Duensing *et al.*, 2000). Such centrosome duplication occurred without morphological nuclear abnormality and contrasted with the effect of E6 expression, which led to centrosome accumulation associated with nuclear atypia and multinucleation (Duensing *et al.*, 2001). These findings suggest that both E6 and E7 can induce centrosome abnormalities but that E7 may act in such a way as to drive numerical chromosome instability (Duensing & Munger, 2002). Structural chromosome abnormalities are also associated with expression of HPV-16 E6 and E7 (Duensing & Munger, 2004). Chromosome instability can occur without viral integration. However, in the context of the whole viral genome, high-level chromosome instability occurred only after integration, consistent with observations in naturally occurring lesions, and supporting the hypothesis that viral integration is an important factor in neoplastic progression (Pett *et al.*, 2004).

APOPTOSIS

Any genetic damage that occurs in papillomavirus-infected cells should be detected by normal cell cycle control mechanisms, leading to cell death by apoptosis. However, there are not only defects in normal cell cycle checkpoints (see earlier) in cells infected with papillomaviruses, particularly of high-risk type, but apoptotic pathways are also abnormal. This makes intuitive sense given that HPVs would want to prevent cells from undergoing cell death in response to infection. There is evidence, however, that cells expressing HPV oncoproteins are sensitized to apoptosis in response to some apoptosis-inducing agents. Therefore, understanding the mechanisms by which HPV infection interferes with normal apoptotic pathways may help the design of therapeutic agents that can induce death of HPV-infected cells.

The normal process of apoptosis (Fig. 2)

Apoptosis can occur via two main routes: the death receptor pathway; or the mitochondrial pathway. Both pathways utilize the cysteine aspartyl-specific proteases (caspases), which dismantle normal cellular structures causing characteristic changes within the cell (Igney & Krammer, 2002). These enzymes are stored as inactive procaspases, which can be individually activated when required to undertake a specific part of the process.

Death receptor ('extrinsic') pathway. The receptors and ligands of the death receptor, or 'extrinsic', pathway are members of the tumour necrosis factor (TNF) receptor superfamily (Igney & Krammer, 2002). A specific ligand binds to the appropriate cell surface receptor such as CD95, TRAIL-R1 (TNF-related apoptosis-inducing ligand-R1) and TRAIL-R2. Ligand binding causes oligomerization of the intracellular 'death domain', which attracts the cytosolic adapter protein FADD (Fas-associated death domain protein) (Finzer et al., 2002). This in turn recruits the appropriate pro-caspases to the complex forming the DISC (death-inducing signalling complex). The pro-caspases recruited at this stage are caspase-8 and caspase-10, which can then be activated at the DISC under the control of the inhibitory proteins, to form the activated 'initiator' caspases. The production of initiator caspases can be inhibited by FLIPs (FADD-like interleukin-1β-converting enzyme-like protease), which function by competing with the pro-caspase.

The amount of activated caspase-8 produced can vary between cell types in the CD95 pathway. Type I cells produce sufficient amounts to activate downstream apoptotic mechanisms independently. Type II cells require the signal to be amplified, as DISC formation is temporarily delayed and occurs at a lower rate than in type I cells. The caspase-8 produced at the DISC causes the cleavage of cytosolic Bid, a member of the Bcl-2 family. The truncated form of the protein is translocated to the mitochondrion.

Mitochondrial ('intrinsic') pathway. The mitochondrial pathway is regulated by members of the Bcl-2 family, which can be pro-apoptotic (e.g. Bax) or anti-apoptotic (e.g. Bcl-2). The main pro-apoptotic proteins used within this pathway include Bax, Bid, Bad and Bim. The release of these signals is controlled by upstream regulators such as p53.

Final common pathway. The final stage of apoptosis is undertaken by caspases and is initiated by either the death receptor or the mitochondrial pathways. When apoptosis is initiated, the signal proteins insert into the mitochondrial membranes with consequent permeabilization of the outer (OMM) and inner (IMM) mitochondrial membranes (ZamZami & Kroemer, 2001). The effect on the IMM is more transient because of its vital function in the maintenance of the electrochemical gradient, important in cellular energy production. The exact nature of the interaction between the Bcl-2 family, and other mitochondrial proteins, and the membranes is still unclear. The initial permeabilization is of the IMM, which may lead to the activation of a channel that controls the permeability of both membranes. Alternatively, initial IMM permeabilization may lead to the disruption of osmotic balance, causing swelling of the matrix and ultimately lysis of the OMM. Increased permeability of the mitochondrial membranes allows the release of apoptogenic factors such as cytochrome c, Smac (second mitochondria-derived activator of caspase)/DIABLO (direct IAP binding protein with low pI) and AIF (apoptosis-inducing factor) from within the intra-membranous space where they are normally safely stored. Within the cytoplasm, cytochrome c, pro-caspase-9 and ATP join with APAF1 (apoptotic protease activating factor-1) to form the apoptosome. It is within this structure that the initiator caspase-9 is activated.

This common mitochondrial pathway can be inhibited in several ways. One route is to use the two anti-apoptotic proteins, Bcl-2 and Bcl-x_L, that are located within the mitochondrial membrane (ZamZami & Kroemer, 2001) and can directly inhibit any increase in mitochrondrial membrane permeability. An alternative route used in the non-death receptor pathway utilizes extracellular survival signals that bind cytosolic PI3K (phosphatidylinositol 3-kinase), which in turn activates protein kinase B (PKB, also known as AKT) to deactivate the Bad protein via phosphorylation (Igney & Krammer, 2002).

The activated initiator caspases-8/9/10 from both pathways are then able to move to the next step in the apoptotic cascade, by cleaving and hence activating the executioner caspases. The main proteins used in this role are caspases-3/6/7, which, once activated, amplify the cascade by cleaving each other. Apoptosis can be inhibited at this stage by the use of IAPs (inhibitor of apoptosis proteins): these bind to the caspases and prevent further function, possibly by acting as a ubiquitin ligase (Igney & Krammer, 2002). In turn, these IAPs can be inhibited by the mitochondrially produced Smac/DIABLO, which bind IAPs in order to allow the caspases to function.

The activated executioner caspases eventually cleave their cellular substrates ('death substrates'), leading to the characteristic cellular changes seen in apoptosis; for example, chromatin condensation and nuclear shrinkage caused by the cleavage of nuclear lamins by disinhibited endonuclease enzymes. The outcome of intracellular breakdown is the exposure of phagocytic signals to attract phagocytes to dispose of cellular debris.

Disruption of apoptosis by HPV proteins (Fig. 2)

Several of the HPV-encoded proteins interfere specifically with apoptotic pathways. A central function of the E6 protein is to inactivate pro-apoptotic proteins such as p53, Bak or Bax proteolytically via the ubiquitin-proteasome pathway (Finzer *et al.*, 2002) but it also sensitizes cells to pro-apoptotic stimuli. These effects are, however, dependent not only on the model system investigated but also on the nature of the stimulus. For example, human mammary epithelial cells are sensitized to apoptosis by E6 when exposed to tamoxifen or DNA-damaging agents (Seewaldt *et al.*, 2001). By contrast, lens fibre cells are protected from apoptosis (Pan & Griep, 1995).

The situation is complicated still further by the fact that the full-length E6 transcript, which encodes a ~16 kDa protein, can undergo post-transcriptional splicing to encode a truncated ~8 kDa version of the protein, termed E6*I, which corresponds to the N-terminal part of the full-sized E6 protein (Filippova *et al.*, 2004). The relative amount of each of these two proteins has important consequences for the ability of both infected and normal keratinocytes to undergo apoptosis (Finzer *et al.*, 2002). This is thought to be mainly due to the ability of E6*I to interact with E6 and the ubiquitin-protein ligase E6-AP (Pim & Banks, 1999). E6*I in general acts to bind to and inhibit the biological activities of the full-length protein and hence modulates the effects of E6 on p53. However, E6*I also has anti-proliferative effects and can induce apoptosis (Pim & Banks, 1999).

E6 can protect cells from TNF-α-mediated cell death in a p53-independent manner to prevent apoptosis (Filippova *et al.*, 2002). HPV-16 E6 does not affect the levels of TNF-receptor. Instead it binds the intracellular death domain to prevent it interacting with the FADD protein via the C-terminal region of the receptor protein and hence prevents transmission of the signal into the cell, subsequently preventing the activation of pro-caspase-8. The full-length E6 protein but not the truncated form is able to bind directly

Fig. 2. Interactions between HPV proteins and apoptotic pathways. The two main apoptotic routes, namely the death receptor (extrinsic) and mitochondrial (intrinsic) pathways, are depicted. These pathways converge in the final common pathway, which ultimately leads to the cleavage of death substrates and cell destruction via phagocytosis. The HPV E6 protein functions to protect cells from apoptosis via several mechanisms, which include binding to the intracellular death domain to prevent interaction of this molecule with FADD or binding directly to FADD or the death ligands TNF/CD95, thus preventing death receptor induced apoptosis. E6 also inactivates pro-apoptotic proteins (Bax/Bim) which function within the mitochondrial pathway. The HPV E7 protein decreases activation of the initiator pro-caspase-8, which forms part of the death receptor pathway. HPV E5 also inhibits the function of the death receptor pathway via a reduction in the expression of the Fas death receptor and impaired formation of the DISC. The HPV E1^E4 protein appears to bind to mitochondria, initiating the detachment of mitochondria from microtubules and inducing apoptosis. The HPV E2 protein also appears to induce the apoptotic machinery, this time via activation of the initiator caspase-8, which is specific to the extrinsic pathway.

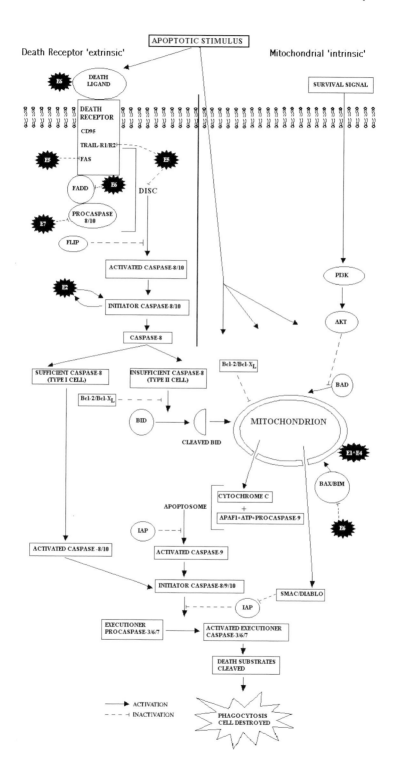

to the death effector domain of FADD. This prevents death receptor induced apoptosis via TNF or CD95, which share the same binding surface on the FADD protein. This binding also leads to the accelerated degradation of FADD in proportion to the amount of E6 expressed in the host cell. The ability of E6 to bind TNF and CD95 is probably an important factor in the ability of the virus to limit the activation of the immune response and hence the persistence of the virus within the host cell (Filippova *et al.*, 2004).

The outcome of E6 gene expression in a particular cell type is therefore likely to represent a complex interaction between the effects of E6 and its splice variants both on each other and on cellular functions.

E7 oncoproteins have a wide range of effects on the host cell (Finzer *et al.*, 2002; Munger *et al.*, 2001). E7 decreases activation of the initiator pro-caspase-8, hence protecting against death receptor mediated apoptosis (Duensing & Munger, 2004). Experimentally, E7 immortalized human fibroblasts are almost resistant to TNF-α/ cycloheximide-mediated cell death (Thompson *et al.*, 2001), but E7-immortalized keratinocytes are not immune to TRAIL-induced apoptosis (Basile *et al.*, 2001).

HPV E1^E4 is the most abundantly expressed viral protein in HPV-infected epithelia. As discussed earlier, although several activities have been attributed to this protein, its precise role in the viral life cycle remains unclear. Recently, it has been reported that the HPV-16 E1^E4 protein binds to mitochondria, after binding to and collapsing the cytokeratin network (Raj *et al.*, 2004). Once bound, the E1^E4 protein induces the detachment of mitochondria from microtubules and as a result the organelles form a single large cluster adjacent to the nucleus. This is followed by a marked reduction in the mitochondrial membrane potential and induction of apoptosis (Raj *et al.*, 2004). The apoptotic induction exhibited by the 'late' E1^E4 protein through mitochondrial binding may be employed to render the tough, rigid, keratin-rich squamous cells of the upper epithelial layers more malleable, thereby easing the exit of newly synthesized HPV particles for subsequent infection rounds during the vegetative cycle.

The HPV E2 protein is a transcription factor that represses the expression of viral oncogenes and activates viral DNA replication during the vegetative viral cycle. It has been shown that, above a threshold level of expression, HPV-18 E2 induces apoptosis in a manner that is independent of other viral functions (Demeret *et al.*, 2003). The N-terminus of E2 was found to be responsible for this apoptotic activity but does not seem to have any involvement in the transcriptional properties of the protein. E2 appears to trigger cell death through activation of the initiator caspase-8, which is specific to the extrinsic apoptotic pathway. During apoptosis, E2 is itself cleaved by caspases and therefore acts as both inducer and target of caspase activity. The ability of E2 to initiate cell death may counteract the proliferative functions of the HPV oncogenes, thus highlighting the importance of E2 inactivation, for example by viral integration, for carcinogenic progression. Such is the apoptotic effect of E2 that it has been suggested that the protein itself could be used to treat cervical cancer/other HPV-associated diseases. Herpes simplex virus VP22–HPV E2 fusion proteins were shown to induce apoptosis in transiently transfected HPV-transformed cervical carcinoma cell lines (Roeder *et al.*, 2004).

The weak oncogenic properties exhibited by the E5 protein are exemplified by the ability of E5 to enhance HPV-16-induced immortalization of human keratinocytes (Conrad-Stöppler *et al.*, 1996), and to allow anchorage-independent growth of human keratinocyte colonies in soft agarose (Chen *et al.*, 1996). HPV-16 E5 has been shown to inhibit ligand-mediated apoptosis. FasL or TRAIL-induced cell death was strongly suppressed in HaCaT cells stably infected with the E5 protein in both mono-layer (Kabsch & Alonso, 2002) and raft (Kabsch *et al.*, 2004) culture. HPV-16 E5 reduced Fas expression by a factor of approximately two, but did not down-regulate TRAIL receptor expression, but instead impaired formation of the DISC triggered by TRAIL. It would seem likely that the anti-apoptotic mechanisms employed by E5 are cell type-specific because mouse fibroblasts expressing HPV-16 E5 were protected from TRAIL-induced apoptosis but not from FasL-induced apoptosis (Kabsch *et al.*, 2004).

The finding that E5 is often deleted in cervical carcinoma cells (Rösl & Schwarz, 1997) perhaps suggests that the physiological relevance of the anti-apoptotic mechanisms may be restricted to the vegetative viral life cycle where temporal blockage of apoptosis would protect replicative intermediates and immature virions from caspase-induced DNA cleavage. In contrast to this, large amounts of E5 mRNA have been found in abnormal cervical smears (Biswas *et al.*, 1997; Kell *et al.*, 1994) and low-grade cervical intraepithelial neoplasia (CIN1) (Stoler *et al.*, 1992), perhaps suggesting a role for E5 in the first steps of cellular transformation in which the suppression of apoptosis is relevant for the progression to cancer.

CONCLUSION

The HPV genome encodes only a small number of proteins but these proteins exert numerous effects on cellular pathways that control cell cycle progression and apoptosis. The aim of these interactions is to subvert host cell functions and allow replication of the viral genome with subsequent production of virions. However, the high-risk HPV types induce changes that are associated with the accumulation of cells with chromosome abnormalities, and that inhibit the elimination of these cells by apoptosis.

Under these circumstances, progression to neoplasia may occur. The study of these events has provided considerable insight into the mechanisms of neoplastic progression and will guide the development of therapeutic intervention in HPV-associated disease.

REFERENCES

Band, V. S., Dalal, S., Delmolino, L. & Androphy, E. J. (1993). Enhanced degradation of p53 protein in HPV 6 and BPV1 E6 immortalised human mammary epithelial cells. *EMBO J* **12**, 1847–1853.

Basile, J. R., Zacny, V. & Munger, K. (2001). The cytokines tumor necrosis factor-alpha and TNF-related apoptosis-inducing ligand differentially modulate proliferation and apoptotic pathways in human keratinocytes expressing the human papillomavirus-16 E7 oncoprotein. *J Biol Chem* **276**, 22522–22528.

Bedell, M. A., Hudson, J. B., Golub, T. D., Turyk, M. E., Hosken, M., Wilbanks, G. D. & Laimins, L. A. (1991). Amplification of human papillomavirus in vitro is dependent on epithelial differentiation. *J Virol* **65**, 2254–2260.

Biswas, C., Kell, B., Mant, C., Jewers, J. R., Cason, J., Muir, P., Raju, K. S. & Nest, J. M. (1997). Detection of human papillomavirus type 16 early-gene transcription by reverse transcription-PCR is associated with abnormal cervical cytology. *J Clin Microbiol* **35**, 1560–1564.

Chen, S.-L., Huang, C.-H., Tsai, T.-C., Lu, K.-Y. & Tsao, Y.-P. (1996). The regulation mechanism of c-jun and junB by human papillomavirus type 16 E5 oncoprotein. *Arch Virol* **141**, 791–800.

Chien, W. M., Noya, F., Benedict-Hamilton, H. M., Broker, T. R. & Chow, L. T. (2002). Alternative fates of keratinocytes transduced by human papillomavirus type 18 E7 during squamous differentiation. *J Virol* **76**, 2964–2972.

Choo, K. B., Pan, C. C. & Han, S. H. (1987). Integration of human papillomavirus type 16 into cellular DNA of cervical carcinoma: preferential deletion of the E2 gene and invariable retention of the long control region and the E6/E7 open reading frames. *Virology* **161**, 259–261.

Conrad-Stöppler, M., Straight, S. W., Tsao, G., Schlegel, R. & McCance, D. J. (1996). The E5 gene of HPV-16 enhances keratinocyte immortalization by full-length DNA. *Virology* **223**, 251–254.

Demeret, C., Garcia-Carranca, A. & Thierry, F. (2003). Transcription-independent triggering of the extrinsic pathway of apoptosis by human papillomavirus 18 E2 protein. *Oncogene* **22**, 168–175.

DiMaio, D. & Mattoon, D. (2001). Mechanisms of cell transformation by papillomavirus E5 proteins. *Oncogene* **20**, 7866–7873.

Doorbar, J. (1998). Late stages of the papillomavirus life cycle. *Papillomavirus Rep* **9**, 119–126.

Duensing, S. & Munger, K. (2002). Human papillomaviruses and centrosome duplication errors: modeling the origins of chromosome instability. *Oncogene* **21**, 6241–6248.

Duensing, S. & Munger, K. (2004). Mechanisms of genomic instability in human cancer: insights from studies with human papillomavirus proteins. *Int J Cancer* **109**, 157–162.

Duensing, S., Lee, L. Y., Duensing, A., Basile, J., Piboonniyom, S., Gonzalez, S., Crum, C. P. & Munger, K. (2000). The human papillomavirus type 16 E6 and E7 oncoproteins cooperate to induce mitotic defects and genomic instability by uncoupling centrosome duplication from the cell division cycle. *Proc Natl Acad Sci U S A* **97**, 10002–10007.

Duensing, S., Duensing, A., Crum, C. P. & Munger, K. (2001). Human papillomavirus type 16 E7 oncoprotein-induced abnormal centrosome synthesis is an early event in the evolving malignant phenotype. *Cancer Res* **61**, 2356–2360.

Evander, M., Frazer, I. H., Payne, E., Qi, Y. M., Hengst, K. & McMillan, N. A. (1997). Identification of the alpha 6 integrin as a candidate receptor for papillomaviruses. *J Virol* **71**, 2449–2456.

Filippova, M., Song, H., Connolly, J. L., Dermody, T. S. & Duerksen-Hughes, P. J. (2002). The human papillomavirus 16 E6 protein binds to tumour necrosis factor (TNF) R1 and protects cells from TNF-induced apoptosis. *J Biol Chem* **277**, 21730–21739.

Filippova, M., Parkhurst, L. & Duerksen-Hughes, P. J. (2004). The human papillomavirus 16 E6 protein binds to Fas-associated death domain and protects cells from Fas-triggered apoptosis. *J Biol Chem* **279**, 25729–25744.

Finzer, P., Aguillar-Lemarroy, A. & Rosl, F. (2002). The role of human papillomavirus oncoproteins E6 and E7 in apoptosis. *Cancer Lett* **188**, 15–24.

Funk, J. O., Waga, S., Harry, J. B., Espling, E., Stillman, B. & Galloway, D. A. (1997). Inhibition of CDK activity and PCNA-dependent DNA replication by p21 is blocked by interaction with the HPV-16 E7 oncoprotein. *Genes Dev* **11**, 2090–2100.

Grana, X. & Reddy, P. (1995). Cell cycle control in mammalian cells: role of cyclins, cyclin dependent kinases (CDKs), growth suppressor genes and cyclin-dependent kinase inhibitors (CKIs). *Oncogene* **11**, 211–219.

Halbert, C. L., Demers, G. W. & Galloway, D. A. (1991). The E7 gene of human papillomavirus type 16 is sufficient for immortalisation of human cells. *J Virol* **65**, 473–478.

Halbert, C. L., Demers, G. W. & Galloway, D. A. (1992). The E6 and E7 genes of human papillomavirus type 6 have weak immortalising activity in human epithelial cells. *J Virol* **66**, 2125–2134.

Helt, A. M. & Galloway, D. A. (2003). Mechanisms by which DNA tumor virus oncoproteins target the Rb family of pocket proteins. *Carcinogenesis* **24**, 159–169.

Hollstein, M., Sidransky, D., Vogelstein, B. & Harris, C. C. (1991). p53 mutations in human cancers. *Science* **253**, 49–53.

Igney, F. H. & Krammer, P. H. (2002). Death and anti-death: tumour resistance to apoptosis. *Nat Rev Cancer* **2**, 277–286.

Jian, Y., Van Tine, B. A., Chien, W. M., Shaw, G. M., Broker, T. R. & Chow, L. T. (1999). Concordant induction of cyclin E and p21cip1 in differentiated keratinocytes by the human papillomavirus E7 protein inhibits cellular and viral DNA synthesis. *Cell Growth Differ* **10**, 101–111.

Jones, D. L., Alani, R. M. & Munger, K. (1997). The human papillomavirus E7 oncoprotein can uncouple cellular differentiation and proliferation in human keratinocytes by abrogating p21^Cip1-mediated inhibition of cdk2. *Genes Dev* **11**, 2101–2111.

Kabsch, K. & Alonso, A. (2002). The human papillomavirus type 16 E5 protein impairs TRAIL- and FasL-mediated apoptosis in HaCaT cells by different mechanisms. *J Virol* **76**, 12162–12172.

Kabsch, K., Mossadegh, N., Kohl, A., Komposch, G., Schenkel, J., Alonso, A. & Tomakidi, P. (2004). The HPV-16 E5 protein inhibits TRAIL- and FasL-mediated apoptosis in human keratinocyte raft cultures. *Intervirology* **47**, 48–56.

Kell, B., Jewers, R. J., Cason, J., Pakarian, F., Kaye, J. N. & Best, J. M. (1994). Detection of E5 oncoprotein in human papillomavirus type 16-positive cervical scrapes using antibodies raised to synthetic peptides. *J Gen Virol* **75**, 2451–2456.

Klaes, R., Friedrich, T., Spitkovsky, D., Ridder, R., Rudy, W., Petry, U., Dallenbach-Hellweg, G., Schmidt, D. & von Knebel Doeberitz, M. (2001). Overexpression of p16(INK4A) as a specific marker for dysplastic and neoplastic epithelial cells of the cervix uteri. *Int J Cancer* **92**, 276–284.

Li, Y., Nichols, M. A., Shay, J. W. & Xiong, Y. (1994). Transcriptional repression of the D-type cyclin-dependent kinase inhibitor p16 by the retinoblastoma susceptibility gene product pRb. *Cancer Res* **54**, 6078–6082.

Longworth, M. S. & Laimins, L. A. (2004). Pathogenesis of human papillomaviruses in differentiating epithelia. *Microbiol Mol Biol Rev* **68**, 362–372.

Moodley, M., Moodley, J., Chetty, R. & Herrington, C. S. (2003). The role of steroid contraceptive hormones in the pathogenesis of invasive cervical cancer. *Int J Gynecol Cancer* **13**, 103–110.

Munger, K., Scheffner, M., Huibregste, J. & Howley, P. M. (1992). Interaction of HPV E6 and E7 oncoproteins with tumour suppressor gene products. *Cancer Surv* **12**, 197–217.

Munger, K., Basile, J. R., Duensing, S., Eichten, A., Gonzalez, A. L., Grace, M. & Zacny, V. L. (2001). Biological activities and molecular targets of the human papillomavirus E7 oncoprotein. *Oncogene* **20**, 7888–7898.

Pan, H. & Griep, A. E. (1995). Temporal distinct patterns of p53-dependent and p53-independent apoptosis during mouse lens development. *Genes Dev* **9**, 2157–2169.

Pett, M. R., Alazawi, W. O., Roberts, I., Dowen, S., Smith, D. I., Stanley, M. A. & Coleman, N. (2004). Acquisition of high-level chromosomal instability is associated with integration of human papillomavirus type 16 in cervical keratinocytes. *Cancer Res* **64**, 1359–1368.

Pim, D. & Banks, L. (1999). HPV 18 E6*I protein modulates the E6-directed degradation of p53 by binding to full-length HPV 18 E6. *Oncogene* **18**, 7403–7408.

Raj, K., Berguerand, S., Southern, S., Doorbar, J. & Beard, P. (2004). E1^E4 protein of human papillomavirus type 16 associates with mitochondria. *J Virol* **78**, 7199–7207.

Roeder, G. E., Parish, J. L., Stern, P. L. & Gaston, K. (2004). Herpes simplex virus VP22-human papillomavirus E2 fusion proteins produced in mammalian or bacterial cells enter mammalian cells and induce apoptosis. *Biotechnol Appl Biochem* **40**, 157–165.

Rösl, F. & Schwarz, E. (1997). Regulation of E6 and E7 oncogene transcription. In *Papillomaviruses in Human Cancer: the Role of E6 and E7 Oncoproteins*, pp. 25–70. Edited by M. Tommasino. Molecular Biology Intelligence Unit. Austin, TX: R. G. Landes.

Seewaldt, V., Mrozek, K., Dietze, E. C., Parker, M. & Caldwell, L. E. (2001). Human papillomavirus type 16 E6 inactivation of p53 in normal human mammary epithelial cells promotes tamoxifen-mediated apoptosis. *Cancer Res* **61**, 616–624.

Southern, S. A. & Herrington, C. S. (1998a). Differential cell cycle regulation by low- and high-risk human papillomaviruses in low-grade squamous intraepithelial lesions of the cervix. *Cancer Res* **58**, 2941–2945.

Southern, S. A. & Herrington, C. S. (1998b). Molecular events in uterine cervical cancer. *Sex Transm Infect* **74**, 101–110.

Southern, S. A. & Herrington, C. S. (2000). Disruption of cell cycle control by human papillomaviruses with special reference to cervical carcinoma. *Int J Gynecol Cancer* **10**, 263–274.

Stanley, M. (2003). Genital human papillomavirus infections – current and prospective therapies. *J Natl Cancer Inst Monogr* **31**, 117–122.

Stoler, M. H. (2000). Human papillomaviruses and cervical neoplasia: a model for carcinogenesis. *Int J Gynecol Pathol* **19**, 16–28.

Stoler, M. H., Rhodes, C. R., Whitbeck, A., Wolinsky, S. M., Chow, L. T. & Broker, T. R. (1992). Human papillomavirus type 16 and 18 gene expression in cervical neoplasias. *Hum Pathol* **23**, 117–128.

Thomas, J. T. & Laimins, L. A. (1998). Human papillomavirus oncoproteins E6 and E7 independently abrogate the mitotic spindle checkpoint. *J Virol* **72**, 1131–1137.

Thomas, J. T., Laimins, L. A. & Ruesch, M. N. (1998). Perturbation of cell cycle control by E6 and E7 oncoproteins of human papillomaviruses. *Papillomavirus Rep* **9**, 59–64.

Thompson, D. A., Belinsky, G., Chang, T. H. T., Jones, D. L., Schlegel, R. & Munger, K. (1997). The human papillomavirus-16 E6 oncoprotein decreases the vigilance of mitotic checkpoints. *Oncogene* **15**, 3025–3035.

Thompson, D. A., Zacny, V., Belinsky, G. S., Classon, M., Jones, D. L., Schlegel, R. & Munger, K. (2001). The HPV E7 oncoprotein inhibits tumor necrosis factor alpha-mediated apoptosis in normal human fibroblasts. *Oncogene* **20**, 3629–3640.

Vousden, K. H. & Lu, X. (2002). Live or let die: the cell's response to p53. *Nat Rev Cancer* **2**, 594–604.

Wentzensen, N., Vinokurova, S. & von Knebel Doeberitz, M. (2004). Systematic review of genomic integration sites of human papillomavirus genomes in epithelial dysplasia and invasive cancer of the female lower genital tract. *Cancer Res* **64**, 3878–3884.

Werness, B. A., Levine, A. J. & Howley, P. M. (1990). Association of human papillomavirus type 16 and 18 E6 proteins with p53. *Science* **248**, 76–79.

Wilson, B. G., West, M., Woytek, K. & Rangasamy, D. (2002). Papillomavirus E1 proteins: form, function and features. *Virus Genes* **24**, 275–290.

Xiong, Y., Hannon, G. J., Zhang, H., Casso, D., Kobayashi, R. & Beach, D. (1993). p21 is a universal inhibitor of cyclin kinases. *Nature* **366**, 701–704.

ZamZami, N. & Kroemer, G. (2001). The mitochondrion in apoptosis: how Pandora's Box opens. *Nat Rev Mol Biol* **2**, 67–71.

Zerfass, K., Schulze, A., Spitkovsky, D., Friedman, V., Henglein, B. & Jansen-Durr, P. (1995). Sequential activation of cyclin E and cyclin A gene expression by human papillomavirus type 16 E7 through sequences necessary for transformation. *J Virol* **69**, 6389–6399.

Zerfass-Thome, K., Zwerschke, W., Mannhardt, B., Tindle, R., Botz, J. W. & Jansen-Durr, P. (1996). Inactivation of the cdk inhibitor p27KIP1 by the human papillomavirus type 16 E7 oncoprotein. *Oncogene* **13**, 2323–2330.

Zhou, J. & Frazer, I. H. (1996). Papovaviridiae: capsid structure and capsid protein function. In *Papillomavirus Reviews: Current Research on Papillomaviruses*, pp. 93–100. Edited by C. Lacey. Leeds: Leeds University Press.

Intracellular antiviral defence mechanisms: the power of interferon-regulated restriction factors

Otto Haller, Friedemann Weber and Georg Kochs

Abteilung Virologie, Institut für Medizinische Mikrobiologie und Hygiene, Universität Freiburg, D-79008 Freiburg, Germany

INTRODUCTION

Viruses are intracellular pathogens that multiply at the expense of their host cells. For productive infection, they require a supportive cellular environment that allows completion of the virus life cycle with formation of new infectious particles. In the past, major efforts were devoted to defining and characterizing the specific host cell factors necessary for virus growth and release. In many cases we now have a detailed knowledge about the viral and cellular factors involved and how they cooperate in virus replication. Much less is known, however, about cellular factors restricting virus growth in target cells. Defining cellular restriction factors and how viruses deal with them is important for a better understanding of viral pathogenesis and may have practical implications in designing new preventive and therapeutic strategies.

Initially, mouse models of genetically determined resistance against specific viral pathogens proved most fruitful for studying virus growth restricting factors. For example, some inbred mouse strains are less susceptible to infection by certain viruses than others. In most cases, the degree of antiviral resistance is controlled by several genes, but occasionally a single gene is responsible and provides the restrictive element (Guenet & Bonhomme, 2003). Of particular interest is the Friend virus susceptibility gene locus *Fv-1*. The *Fv-1* gene product blocks some murine leukaemia viruses soon after cell entry, presumably by targeting a major component of the incoming virus, the capsid protein (Best *et al.*, 1996). A broad antiretroviral activity, termed lentivirus susceptibility factor 1 (Lv1), was subsequently detected in primate cells and was also found to act against incoming viral capsids (Cowan *et al.*, 2002). Finally, human cells have an

SGM symposium 64: Molecular pathogenesis of virus infections.
Editors P. Digard, A. A. Nash & R. E. Randall. Cambridge University Press. ISBN 0 521 83248 9 ©SGM 2005

antiretroviral activity called restriction factor 1 (Ref-1) (Towers *et al.*, 2000). The identity of these retroviral restriction factors has recently been determined. Surprisingly, human Ref-1 and primate Lv1 encode the same antiviral protein, namely human TRIM5α and monkey TRIM5α, respectively (Hatziioannou *et al.*, 2004; Keckesova *et al.*, 2004; Yap *et al.*, 2004). TRIM family members are defined by three different protein motifs: a RING motif, one or two B boxes and a coiled-coil domain. TRIM5α impairs the growth of highly divergent retroviruses in human and other primate cells, including human immunodeficiency virus type 1 (HIV-1) (Greene, 2004). TRIM proteins appear to represent a new class of antiviral restriction factors that contribute to intracellular immunity. These findings open up a new era of research into innate immunity and cell autonomous antiviral factors and are of utmost importance for a better understanding of HIV-1 susceptibility and AIDS progression in human populations.

Another telling case is inborn resistance against influenza and influenza-like viruses found in wild mice and some inbred mouse strains (Haller, 1981; Haller *et al.*, 1987; Lindenmann, 1964). Resistance is inherited as an autosomal dominant trait and is specific for members of the orthomyxovirus family. Subsequent work revealed that this resistance is brought about by a single gene, *Mx1* (for orthomyxovirus resistance gene 1), localized on mouse chromosome 16 (Reeves *et al.*, 1988), and that the Mx1 protein has intrinsic antiviral activity (Arnheiter *et al.*, 1990; Staeheli *et al.*, 1986). Unexpectedly, the *Mx1* gene turned out to belong to the so-called interferon (IFN)-stimulated genes (ISGs) and is strictly regulated by type I (α and β) IFNs (Haller *et al.*, 1980). Hence gene expression is normally silent but rapidly induced upon infection with a variety of viruses through the action of virus-induced type I IFNs, making it an excellent marker for type I IFN activity (Antonelli *et al.*, 1999; Roers *et al.*, 1994).

In influenza virus-susceptible mice, the *Mx1* gene is defective. Most inbred strains of mice carry nonfunctional Mx1 alleles (Staeheli *et al.*, 1988). Why intact *Mx1* genes are absent in most inbred mouse strains is still an open question but would seem to indicate that most inbred strains share the distal part of chromosome 16 with a common ancestor mouse. In humans, an Mx homologue with antiviral activity, called MxA (Aebi *et al.*, 1989; Staeheli & Haller, 1985), is encoded by the human *MX1* gene in a region of chromosome 21 that is syntenic with mouse chromosome 16 (Horisberger *et al.*, 1988).

Like in the case of *Mx1*, a single autosomal dominant gene locus, designated *Flv/Wnv*, is responsible for natural resistance of mice against infection with West Nile virus (WNV) and other flaviviruses. The gene was recently identified as *Oas1b*, a member of a large IFN-regulated gene family encoding 2′–5′-oligoadenylate synthetases (2–5 OAS) known to play an important role in antiviral defence (Mashimo *et al.*, 2002; Perelygin *et al.*, 2002). The intact *Oas1b* gene is again found in wild mice and some rare inbred

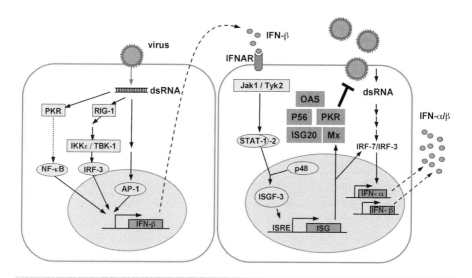

Fig. 1. Type I IFN gene expression and mode of action. Left panel: dsRNA, a characteristic by-product of virus replication, leads to activation of the transcription factors NF-κB, IRF-3 and AP-1. The cooperative action of these factors is required for full activation of the IFN-β promoter. IRF-3 is phosphorylated by the kinases IKKε and TBK-1, which in turn are activated by the RNA helicase RIG-I. Right panel: Newly synthesized IFN-β binds to the type I IFN receptor (IFNAR) and activates the expression of numerous ISGs via the Jak-STAT pathway. IRF-7 amplifies the IFN response by inducing the expression of several IFN-α subtypes. Mx, OAS, PKR, ISG20 and p56 are proteins with antiviral activity.

strains but not in most laboratory strains which carry a nonsense mutation in the distal part of chromosome 5. In contrast to *Mx1*, comparisons of the mouse and human genomes did not reveal a direct equivalent of the mouse *Oas1b* gene in humans (Brinton & Perelygin, 2003).

Additional examples of genetic resistance are known in mice in which single genes play a major role (Casanova *et al.*, 2002). In many instances, cellular antiviral proteins appear to be up-regulated or activated in response to type I IFNs. Here we summarize recent advances in our understanding of some of these IFN-regulated antiviral defence mechanisms and discuss how viruses manage to subvert these restriction elements.

HOW VIRUSES INDUCE TYPE I IFN PRODUCTION

Type I IFNs are produced by cells in direct response to virus infection and comprise a large number of IFN-α subspecies and a single IFN-β, as well as some additional family members (Roberts *et al.*, 2003; van Pesch *et al.*, 2004). The recently discovered IFN-λ1, IFN-λ2 and IFN-λ3 (also termed IL-28A, IL-28B and IL-29) are strikingly similar to the type I IFNs but use distinct receptors to mediate their antiviral activity (Kotenko *et al.*, 2003; Sheppard *et al.*, 2003). Induction of type I IFN gene expression is transcriptionally regulated and is best understood for IFN-β (Fig. 1). The IFN-β promoter has binding sites for several transcription factors which cooperate for maximal promoter activation. There is general agreement that interferon regulatory factor 3 (IRF-3) plays a central role (Hiscott *et al.*, 1999). IRF-3 needs to be phosphorylated to become active. The enzymes responsible for IRF-3 phosphorylation have recently been demonstrated to be the IKK-like kinases IKK*ε* and TBK-1 (Fitzgerald *et al.*, 2003; Sharma *et al.*, 2003). These kinases appear to be activated by the RNA helicase RIG-I (Yoneyama *et al.*, 2004) and, presumably, some Toll-like receptors (TLRs) (Beutler, 2004). RIG-I senses dsRNA molecules and possibly ribonucleoprotein (RNP) complexes in the cytoplasm of infected cells. Phosphorylated IRF-3 homodimerizes and moves into the nucleus, where it recruits the transcriptional coactivator CREB-binding protein (CBP) to initiate IFN-*β* mRNA synthesis (Hiscott *et al.*, 1999). In addition, NF-*κ*B and ATF-2/cJUN (AP-1) are activated as a more general stress response. Together these transcription factors strongly up-regulate IFN-*β* expression (Fig. 1).

A second IRF family member, IRF-7, is expressed in most cells at very low amounts. It needs to be induced by IFN to reach sufficient levels and is then activated by virus infection in much the same way as IRF-3 (tenOever *et al.*, 2004). IRF-7 is part of a positive feedback loop leading to amplification of IFN gene expression. Activated IRF-7 cooperates with IRF-3 and stimulates expression of the numerous IFN-*α* genes leading to a broad IFN-*α* response (Levy *et al.*, 2003). In specialized IFN-*α*-producing cells, e.g. plasmacytoid dendritic cells, IRF-7 is constitutively present at high levels and is directly activated in response to signals from certain TLRs which stimulate immediate IFN-*α* synthesis (Kawai *et al.*, 2004).

INTERFERON-INDUCED ANTIVIRAL PROTEINS IN INTRACELLULAR DEFENCE

IFN-*α*/*β* activate a common type I IFN receptor which sends a signal to the nucleus through the so-called Jak-STAT pathway (Fig. 1). The STAT proteins are latent cytoplasmic transcription factors which become phosphorylated by the Janus kinases Jak1 and Tyk2 (Samuel, 2001; Stark *et al.*, 1998). Phosphorylated STAT-1 and STAT-2 recruit a third factor, IRF-9 (or p48), to form a complex known as IFN-stimulated gene factor 3 (ISGF-3), which translocates to the nucleus and binds to the IFN-stimulated response element (ISRE) in the promoter region of so-called ISGs.

Type I IFNs activate the expression of more than 300 ISGs which have antiviral, antiproliferative and immunomodulatory functions (de Veer *et al.*, 2001; Der *et al.*, 1998). Up to now, only a few antiviral proteins have been thoroughly studied. Among these are the Mx proteins (Haller & Kochs, 2002; Horisberger *et al.*, 1983), the protein kinase R (PKR) (Williams, 1999) and the 2–5 OAS/RNaseL system (Silverman, 1994).

Mx proteins have been discovered as mediators of genetic resistance against orthomyxoviruses in mice (see earlier) and belong to the superfamily of dynamin-like large GTPases (Haller & Kochs, 2002; Horisberger *et al.*, 1983). In comparison to the mouse protein, the human MxA GTPase has a broader antiviral activity and inhibits the multiplication of several RNA viruses, including representative members of the *Orthomyxo-*, *Paramyxo-*, *Rhabdo-* and *Bunyaviridae* families (Haller *et al.*, 1998; Haller & Kochs, 2002). Their importance for host survival has been amply demonstrated (Arnheiter *et al.*, 1996; Hefti *et al.*, 1999; Pavlovic *et al.*, 1995) but their exact mode of action remains to be fully explored. The human MxA protein accumulates in the cytoplasm of IFN-treated cells and blocks replication of the infecting virus soon after cell entry. It has been shown to target the viral capsids by recognizing the major capsid component, the viral ribonucleoprotein of some orthomyxo- and bunyaviruses (Haller & Kochs, 2002; Kochs & Haller, 1999; Reichelt *et al.*, 2004). Although still incompletely understood, this mode of action has a striking resemblance to the proposed antiretroviral effect of TRIM5 (see earlier). Nuclear mouse and rat Mx1 proteins inhibit RNA viruses with a nuclear replication step by inactivating the transcriptional activity of their RNA-dependent RNA polymerases (Krug *et al.*, 1985; Pavlovic *et al.*, 1992).

GBP-1 belongs to the dynamin superfamily of large GTPases like Mx. However, it is predominantly induced by IFN-γ and its antiviral activity against vesicular stomatitis virus (VSV) is comparatively weak (Anderson *et al.*, 1999). GBP-1 was the first large GTPase to be crystallized and its 3D structure provided a valuable model for other family members (Prakash *et al.*, 2000).

The relevance of the OAS/RNaseL and PKR systems in the IFN response to viral infection is also well documented in both tissue culture and animal experiments. Mice lacking one of these components show increased virus susceptibilities (Zhou *et al.*, 1997, 1999). Nevertheless, cells from so-called triple knockout mice lacking PKR, RNaseL and Mx are still capable of mounting a limited IFN-induced antiviral state, indicating that additional antiviral pathways exist (Zhou *et al.*, 1999).

Additional proteins with potentially important antiviral activities are ISG20 (Espert *et al.*, 2003), P56 (Guo *et al.*, 2000; Hui *et al.*, 2003) and promyelocytic leukaemia protein (PML) (Regad & Chelbi-Alix, 2001). ISG20 is an IFN-induced 3'–5' exonuclease that specifically degrades ssRNA *in vitro*. Expression of ISG20 leads to a reduction in VSV gene expression and blocks virus replication in cell culture (Espert *et al.*, 2003). P56 binds the eukaryotic initiation factor 3e (eIF3e) subunit of the eukaryotic translation initiation factor eIF3 and is likely to suppress viral as well as cellular RNA translation (Hui *et al.*, 2003; Wang *et al.*, 2003).

The PML protein is a TRIM family member (also called TRIM19) which shuttles between the cytoplasm and the nucleus, where it forms a specialized subnuclear compartment known as nuclear domain-10 (ND10) or PML nuclear body. ND10 contains several other characteristic proteins, including Sp100, ISG20, Daxx and p53 (Maul *et al.*, 2000; Regad & Chelbi-Alix, 2001). Localization of PML and Sp100 to ND10 depends on modification with the small ubiquitin-related modifier SUMO-1 (Muller *et al.*, 1998; Sternsdorf *et al.*, 1997). The expression of some of these ND10 components is enhanced by IFN, suggesting a role in intracellular antiviral defence. Accordingly, overexpression of PML has been found to suppress replication of several viruses, including VSV, influenza A virus (FLUAV) (Chelbi-Alix *et al.*, 1998), lymphocytic choriomeningitis virus (Djavani *et al.*, 2001) and human foamy retrovirus (HFV) (Regad *et al.*, 2001). In the case of HFV, PML directly binds to the viral transactivator Tas, thereby preventing recruitment to its proviral DNA-binding site. HIV-1 infection induces PML to move into the cytoplasm and to influence early steps of HIV replication (Turelli *et al.*, 2001). Interestingly, cells from wild-type and PML knockout mice proved to be equally permissive for herpes simplex virus 1 (HSV-1) and FLUAV, suggesting that PML itself has no antiviral activity against these viruses (Chee *et al.*, 2003; Engelhardt *et al.*, 2004). Accordingly, overexpression of PML does not inhibit HSV-1 infection (Chelbi-Alix & de The, 1999; Hagglund & Roizman, 2004). Recent findings suggest, however, that the IFN responsiveness of cells may be influenced by PML and PML nuclear bodies (Chee *et al.*, 2003).

VIRUS ESCAPE FROM INTRACELLULAR RESTRICTION

It is becoming increasingly clear that there is coevolution between virus factors promoting virus growth and host factors governing intracellular defence mechanisms, presumably in a species-specific manner (Bossert & Conzelmann, 2002; Chatziandreou *et al.*, 2004; Parisien *et al.*, 2002; Young *et al.*, 2001). A better understanding of the intricate interplay between viruses and cell-autonomous immune defences should shed new light on viral pathogenesis and trans-species transmission. An interesting case is the so-called arthropod-borne viruses (arboviruses). They replicate in their blood-feeding arthropod hosts and are then transmitted by bite to vertebrate hosts. To continue the transmission cycle, virus replication in the vertebrate host must produce a long-lasting viraemia with a high virus load to allow infection of other blood-feeding arthopods. To succeed, arboviruses seem to have evolved specific viral proteins to counteract the IFN system present in the vertebrate host. As invertebrates do not have equivalent IFN genes, these viral IFN antagonists are most likely an adaptation to the mammalian host, allowing virus transmission. Several bunyaviruses with such an IFN-antagonistic activity have recently been studied in our laboratory. The nonstructural NSs protein of Rift Valley fever virus (RVFV; a phlebovirus of the *Bunyaviridae* family) is known to be an important pathogenicity factor in vertebrates

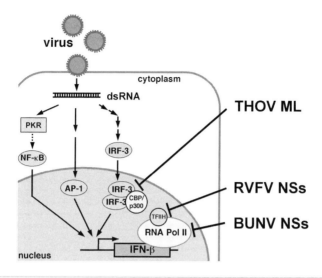

Fig. 2. Strategies used by Thogoto virus (THOV), Rift Valley fever virus (RVFV) and Bunyamwera virus (BUNV) to suppress IFN induction. The ML protein of THOV targets IRF-3 and interferes with the recruitment of the transcriptional coactivator CBP by IRF-3. The NSs protein of RVFV is an inhibitor of host cell transcription. It binds to the p44 subunit of the general transcription factor TFIIH. This interaction abolishes the proper assembly of TFIIH and leads in turn to an inactive RNA polymerase II. The NSs protein of BUNV disturbs the phosphorylation of RNA polymerase II. The action of both NSs proteins results in inhibition of IFN-β gene transcription as well as in down-regulation of other cellular genes.

(Vialat *et al.*, 2000). RVFV variants that lacked NSs proved to be excellent IFN inducers and were highly attenuated in IFN-competent mice but not in IFN-deficient animals (Bouloy *et al.*, 2001). Consequently, NSs was found to suppress IFN production both in tissue culture and in the infected animal (Billecocq *et al.*, 2004; Bouloy *et al.*, 2001). NSs was capable of suppressing IFN gene expression by blocking overall host cell transcription (Billecocq *et al.*, 2004). Further work by the groups of Michèle Bouloy and Jean-Marc Egly revealed that NSs targets the p44 subunit of the essential transcription factor TFIIH, thereby preventing its proper assembly. As a consequence, the cellular RNA polymerase II is nonfunctional (Le May *et al.*, 2004). Surprisingly, the NSs protein of Bunyamwera virus (BUNV; an orthobunyavirus of the *Bunyaviridae* family), which is totally unrelated to the NSs of RVFV, also inhibits IFN induction (Weber *et al.*, 2002) by targeting the RNA polymerase II, albeit by a different mechanism. BUNV NSs was recently demonstrated to inhibit phosphorylation of the large subunit of the RNA polymerase II (Thomas *et al.*, 2004). Thus both bunyaviral NSs proteins exert a general stop of mRNA transcription in their host cells (Fig. 2).

Another interesting example of adaptation to the IFN system of mammals is found in Thogoto virus (THOV), an influenza-like virus transmitted by ticks. THOV was first

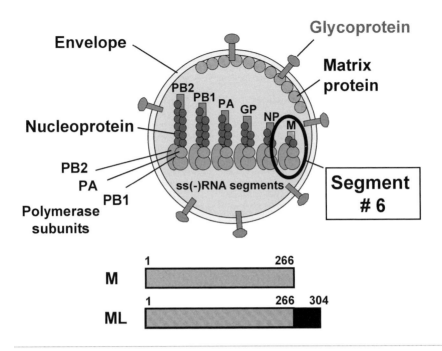

Fig. 3. IFN antagonist ML of THOV. The sixth genomic segment of THOV encodes two transcripts: a spliced mRNA that encodes the matrix (M) protein and an unspliced mRNA that encodes a C-terminally extended M protein, named ML (for matrix protein long). ML is an IFN antagonist and viral pathogenicity factor (see text).

isolated in the Thogoto forest near Nairobi, Kenya (Haig *et al.*, 1965), and has since been found in ticks and their vertebrate hosts in other parts of the world, including southern Europe (Davies *et al.*, 1986). THOV has six genomic segments of single-stranded negative-sense RNA (Fig. 3). Its replication occurs in the cell nucleus and is sensitive to inhibition by IFN-induced mouse Mx1 and human MxA GTPases (Haller *et al.*, 1995; Kochs & Haller, 1999). The Mx block inhibits a very early step in the viral multiplication cycle that affects steps between virus entry and primary transcription of the incoming genome. Since the virus can not transcribe and replicate its genome in the presence of the IFN-induced Mx antiviral state, generation of Mx escape mutants is virtually impossible. Therefore, the prime strategy of THOV is to suppress IFN production in the vertebrate host, thereby avoiding Mx expression in potential target cells. It was recently found that the virus has an accessory protein with IFN-antagonistic activity. The sixth genomic segment of THOV encodes two transcripts: a spliced mRNA that encodes the matrix (M) protein and an unspliced mRNA that encodes a C-terminally extended M protein (Kochs *et al.*, 2000), named ML (for matrix protein long; Fig. 3). Recombinant mutant viruses were generated that lacked ML. These ML-deficient viruses were strong IFN inducers but otherwise showed no obvious growth deficits in IFN-incompetent cells or animals (Hagmaier *et al.*, 2003). Experimental infections of Mx1-positive mice with wild-type or ML-deficient THOV showed that ML was capable of blocking IFN production, allowing the wild-type virus to grow in the infected host. In contrast, the mutant virus devoid of ML was highly attenuated although it showed normal growth properties in mice lacking Mx1 (Pichlmair *et al.*, 2004). Interestingly, ML inhibits the transcriptional activity of IRF-3 which is required for IFN gene expression. Recent data show that the ML protein of THOV does not inhibit phosphorylation or nuclear transport of IRF-3 but seems to interfere with IRF-3 dimerization and recruitment of the transcriptional coactivator CBP by activated IRF-3 (Jennings *et al.*, 2005) (Fig. 2).

The IFN-specific transcription factor IRF-3 is known to be a central player in IFN gene expression and, not surprisingly, is affected by viral proteins from many unrelated viruses. Current examples are the VP35 protein of Ebola virus (Basler *et al.*, 2003), the NS3/4A protease of hepatitis C virus (Foy *et al.*, 2003), the NS1/NS2 complex of respiratory syncytial virus (Bossert *et al.*, 2003; Spann *et al.*, 2004; Valarcher *et al.*, 2003), the V proteins of paramyxoviruses (He *et al.*, 2002; Poole *et al.*, 2002) and the human cytomegalovirus structural protein pp65 (Abate *et al.*, 2004). These specialized viral proteins prevent the initial phosphorylation of IRF-3 and, as a consequence, IRF-3 dimerization and nuclear transport.

A unique strategy is employed by human herpesvirus 8, the causative agent of Kaposi's sarcoma. It displays viral IRF homologues, termed vIRFs, which either mimic or exert a dominant-negative effect over their cellular counterparts (Lubyova *et al.*, 2004; Lubyova & Pitha, 2000).

Not surprisingly, viruses have more tricks up their sleeves. A prominent target is the danger signal dsRNA. Thus the NS1 protein of FLUAV binds dsRNA and prevents induction of IFNs and activation of dsRNA-dependent effector mechanisms (Bergmann *et al.*, 2000; Ludwig *et al.*, 2002; Talon *et al.*, 2000; Wang *et al.*, 2000). Some viruses try to suppress the specific IFN signalling pathway to prevent ISG expression by targeting components of the Jak-STAT pathway. Others have means to directly suppress the very antiviral effector proteins that are induced by IFNs and directed against them. Quite often a combination of several strategies is used (for recent reviews, see Basler & Garcia-Sastre, 2002; Garcia-Sastre, 2001; Goodbourn *et al.*, 2000; Krug *et al.*, 2003; Weber *et al.*, 2004). It is evident that long-term survival of viruses depends on a balance between virus-promoting and virus-inhibiting factors. Our present knowledge of the IFN system and viral countermeasures is still limited. However, the hope is that a better understanding of the intricate interplay between viruses and the innate immune defences will help to design new strategies for antiviral treatments and therapies. In addition, a better understanding of the viral anti-IFN mechanisms will be useful for developing new candidate vaccines.

ACKNOWLEDGEMENTS

We thank Heinz Arnheiter for valuable comments. Our own work described in the text was supported by grants from the Deutsche Forschungsgemeinschaft.

REFERENCES

Abate, D. A., Watanabe, S. & Mocarski, E. S. (2004). Major human cytomegalovirus structural protein pp65 (ppUL83) prevents interferon response factor 3 activation in the interferon response. *J Virol* **78**, 10995–11006.

Aebi, M., Fah, J., Hurt, N., Samuel, C. E., Thomis, D., Bazzigher, L., Pavlovic, J., Haller, O. & Staeheli, P. (1989). cDNA structures and regulation of two interferon-induced human Mx proteins. *Mol Cell Biol* **9**, 5062–5072.

Anderson, S. L., Carton, J. M., Lou, J., Xing, L. & Rubin, B. Y. (1999). Interferon-induced guanylate binding protein-1 (GBP-1) mediates an antiviral effect against vesicular stomatitis virus and encephalomyocarditis virus. *Virology* **256**, 8–14.

Antonelli, G., Simeoni, E., Turriziani, O., Tesoro, R., Redaelli, A., Roffi, L., Antonelli, L., Pistello, M. & Dianzani, F. (1999). Correlation of interferon-induced expression of MxA mRNA in peripheral blood mononuclear cells with the response of patients with chronic active hepatitis C to IFN-alpha therapy. *J Interferon Cytokine Res* **19**, 243–251.

Arnheiter, H., Skuntz, S., Noteborn, M., Chang, S. & Meier, E. (1990). Transgenic mice with intracellular immunity to influenza virus. *Cell* **62**, 51–61.

Arnheiter, H., Frese, M., Kambadur, R., Meier, E. & Haller, O. (1996). Mx transgenic mice – animal models of health. *Curr Top Microbiol Immunol* **206**, 119–147.

Basler, C. F. & Garcia-Sastre, A. (2002). Viruses and the type I interferon antiviral system: induction and evasion. *Int Rev Immunol* **21**, 305–337.

Basler, C. F., Mikulasova, A., Martinez-Sobrido, L., Paragas, J., Muhlberger, E., Bray, M., Klenk, H. D., Palese, P. & Garcia-Sastre, A. (2003). The Ebola virus VP35 protein inhibits activation of interferon regulatory factor 3. *J Virol* **77**, 7945–7956.

Bergmann, M., Garcia-Sastre, A., Carnero, E., Pehamberger, H., Wolff, K., Palese, P. & Muster, T. (2000). Influenza virus NS1 protein counteracts PKR-mediated inhibition of replication. *J Virol* **74**, 6203–6206.

Best, S., Le Tissier, P., Towers, G. & Stoye, J. P. (1996). Positional cloning of the mouse retrovirus restriction gene Fv1. *Nature* **382**, 826–829.

Beutler, B. (2004). Inferences, questions and possibilities in Toll-like receptor signalling. *Nature* **430**, 257–263.

Billecocq, A., Spiegel, M., Vialat, P., Kohl, A., Weber, F., Bouloy, M. & Haller, O. (2004). NSs protein of Rift Valley fever virus blocks interferon production by inhibiting host gene transcription. *J Virol* **78**, 9798–9806.

Bossert, B. & Conzelmann, K. K. (2002). Respiratory syncytial virus (RSV) nonstructural (NS) proteins as host range determinants: a chimeric bovine RSV with NS genes from human RSV is attenuated in interferon-competent bovine cells. *J Virol* **76**, 4287–4293.

Bossert, B., Marozin, S. & Conzelmann, K. K. (2003). Nonstructural proteins NS1 and NS2 of bovine respiratory syncytial virus block activation of interferon regulatory factor 3. *J Virol* **77**, 8661–8668.

Bouloy, M., Janzen, C., Vialat, P., Khun, H., Pavlovic, J., Huerre, M. & Haller, O. (2001). Genetic evidence for an interferon-antagonistic function of rift valley fever virus nonstructural protein NSs. *J Virol* **75**, 1371–1377.

Brinton, M. A. & Perelygin, A. A. (2003). Genetic resistance to flaviviruses. *Adv Virus Res* **60**, 43–85.

Casanova, J. L., Schurr, E., Abel, L. & Skamene, E. (2002). Forward genetics of infectious diseases: immunological impact. *Trends Immunol* **23**, 469–472.

Chatziandreou, N., Stock, N., Young, D., Andrejeva, J., Hagmaier, K., McGeoch, D. J. & Randall, R. E. (2004). Relationships and host range of human, canine, simian and porcine isolates of simian virus 5 (parainfluenza virus 5). *J Gen Virol* **85**, 3007–3016.

Chee, A. V., Lopez, P., Pandolfi, P. P. & Roizman, B. (2003). Promyelocytic leukemia protein mediates interferon-based anti-herpes simplex virus 1 effects. *J Virol* **77**, 7101–7105.

Chelbi-Alix, M. K. & de The, H. (1999). Herpes virus induced proteasome-dependent degradation of the nuclear bodies-associated PML and Sp100 proteins. *Oncogene* **18**, 935–941.

Chelbi-Alix, M. K., Quignon, F., Pelicano, L., Koken, M. H. & de The, H. (1998). Resistance to virus infection conferred by the interferon-induced promyelocytic leukemia protein. *J Virol* **72**, 1043–1051.

Cowan, S., Hatziioannou, T., Cunningham, T., Muesing, M. A., Gottlinger, H. G. & Bieniasz, P. D. (2002). Cellular inhibitors with Fv1-like activity restrict human and simian immunodeficiency virus tropism. *Proc Natl Acad Sci U S A* **99**, 11914–11919.

Davies, C. R., Jones, L. D. & Nuttall, P. A. (1986). Experimental studies on the transmission cycle of Thogoto virus, a candidate orthomyxovirus, in *Rhipicephalus appendiculatus*. *Am J Trop Med Hyg* **35**, 1256–1262.

Der, S. D., Zhou, A., Williams, B. R. & Silverman, R. H. (1998). Identification of genes differentially regulated by interferon alpha, beta, or gamma using oligonucleotide arrays. *Proc Natl Acad Sci U S A* **95**, 15623–15628.

de Veer, M. J., Holko, M., Frevel, M., Walker, E., Der, S., Paranjape, J. M., Silverman, R. H. & Williams, B. R. (2001). Functional classification of interferon-stimulated genes identified using microarrays. *J Leukoc Biol* **69**, 912–920.

Djavani, M., Rodas, J., Lukashevich, I. S., Horejsh, D., Pandolfi, P. P., Borden, K. L. & Salvato, M. S. (2001). Role of the promyelocytic leukemia protein PML in the interferon sensitivity of lymphocytic choriomeningitis virus. *J Virol* **75**, 6204–6208.

Engelhardt, O. G., Sirma, H., Pandolfi, P. P. & Haller, O. (2004). Mx1 GTPase accumulates in distinct nuclear domains and inhibits influenza A virus in cells that lack promyelocytic leukaemia protein nuclear bodies. *J Gen Virol* **85**, 2315–2326.

Espert, L., Degols, G., Gongora, C., Blondel, D., Williams, B. R., Silverman, R. H. & Mechti, N. (2003). ISG20, a new interferon-induced RNase specific for single-stranded RNA, defines an alternative antiviral pathway against RNA genomic viruses. *J Biol Chem* **278**, 16151–16158.

Fitzgerald, K. A., McWhirter, S. M., Faia, K. L., Rowe, D. C., Latz, E., Golenbock, D. T., Coyle, A. J., Liao, S. M. & Maniatis, T. (2003). IKKepsilon and TBK1 are essential components of the IRF3 signaling pathway. *Nat Immunol* **4**, 491–496.

Foy, E., Li, K., Wang, C., Sumpter, R., Jr, Ikeda, M., Lemon, S. M. & Gale, M., Jr (2003). Regulation of interferon regulatory factor-3 by the hepatitis C virus serine protease. *Science* **300**, 1145–1148.

Garcia-Sastre, A. (2001). Inhibition of interferon-mediated antiviral responses by influenza A viruses and other negative-strand RNA viruses. *Virology* **279**, 375–384.

Goodbourn, S., Didcock, L. & Randall, R. E. (2000). Interferons: cell signalling, immune modulation, antiviral response and virus countermeasures. *J Gen Virol* **81**, 2341–2364.

Greene, W. C. (2004). Redistricting the retroviral restriction factors. *Nat Med* **10**, 778–780.

Guenet, J. L. & Bonhomme, F. (2003). Wild mice: an ever-increasing contribution to a popular mammalian model. *Trends Genet* **19**, 24–31.

Guo, J., Hui, D. J., Merrick, W. C. & Sen, G. C. (2000). A new pathway of translational regulation mediated by eukaryotic initiation factor 3. *EMBO J* **19**, 6891–6899.

Hagglund, R. & Roizman, B. (2004). Role of ICP0 in the strategy of conquest of the host cell by herpes simplex virus 1. *J Virol* **78**, 2169–2178.

Hagmaier, K., Jennings, S., Buse, J., Weber, F. & Kochs, G. (2003). Novel gene product of Thogoto virus segment 6 codes for an interferon antagonist. *J Virol* **77**, 2747–2752.

Haig, D. A., Woodall, J. P. & Danskin, D. (1965). Thogoto virus: a hitherto underscribed agent isolated from ticks in Kenya. *J Gen Microbiol* **38**, 389–394.

Haller, O. (1981). Inborn resistance of ice to orthomyxoviruses. *Curr Top Microbiol Immunol* **92**, 25–52.

Haller, O. & Kochs, G. (2002). Interferon-induced mx proteins: dynamin-like GTPases with antiviral activity. *Traffic* **3**, 710–717.

Haller, O., Arnheiter, H., Lindenmann, J. & Gresser, I. (1980). Host gene influences sensitivity to interferon action selectively for influenza virus. *Nature* **283**, 660–662.

Haller, O., Acklin, M. & Staeheli, P. (1987). Influenza virus resistance of wild mice: wild-type and mutant Mx alleles occur at comparable frequencies. *J Interferon Res* **7**, 647–656.

Haller, O., Frese, M., Rost, D., Nuttall, P. A. & Kochs, G. (1995). Tick-borne thogoto virus infection in mice is inhibited by the orthomyxovirus resistance gene product Mx1. *J Virol* **69**, 2596–2601.

Haller, O., Frese, M. & Kochs, G. (1998). Mx proteins: mediators of innate resistance to RNA viruses. *Rev Sci Tech* **17**, 220–230.

Hatziioannou, T., Perez-Caballero, D., Yang, A., Cowan, S. & Bieniasz, P. D. (2004). Retrovirus resistance factors Ref1 and Lv1 are species-specific variants of TRIM5alpha. *Proc Natl Acad Sci U S A* **101**, 10774–10779.

He, B., Paterson, R. G., Stock, N., Durbin, J. E., Durbin, R. K., Goodbourn, S., Randall, R. E. & Lamb, R. A. (2002). Recovery of paramyxovirus simian virus 5 with a V protein lacking the conserved cysteine-rich domain: the multifunctional V protein blocks both interferon-beta induction and interferon signaling. *Virology* **303**, 15–32.

Hefti, H. P., Frese, M., Landis, H., Di Paolo, C., Aguzzi, A., Haller, O. & Pavlovic, J. (1999). Human MxA protein protects mice lacking a functional alpha/beta interferon system against La crosse virus and other lethal viral infections. *J Virol* **73**, 6984–6991.

Hiscott, J., Pitha, P., Genin, P., Nguyen, H., Heylbroeck, C., Mamane, Y., Algarte, M. & Lin, R. (1999). Triggering the interferon response: the role of IRF-3 transcription factor. *J Interferon Cytokine Res* **19**, 1–13.

Horisberger, M. A., Staeheli, P. & Haller, O. (1983). Interferon induces a unique protein in mouse cells bearing a gene for resistance to influenza virus. *Proc Natl Acad Sci U S A* **80**, 1910–1914.

Horisberger, M. A., Wathelet, M., Szpirer, J., Szpirer, C., Islam, Q., Levan, G., Huez, G. & Content, J. (1988). cDNA cloning and assignment to chromosome 21 of IFI-78K gene, the human equivalent of murine Mx gene. *Somat Cell Mol Genet* **14**, 123–131.

Hui, D. J., Bhasker, C. R., Merrick, W. C. & Sen, G. C. (2003). Viral stress-inducible protein p56 inhibits translation by blocking the interaction of eIF3 with the ternary complex eIF2.GTP.Met-tRNAi. *J Biol Chem* **278**, 39477–39482.

Jennings, S., Martínez-Sobrido, L., García-Sastre, A., Weber, F. & Kochs, G. (2005). Thogoto virus ML protein suppresses IRF3 function. *Virology* **331**, 63–72.

Kawai, T., Sato, S., Ishii, K. J. & 9 other authors (2004). Interferon-alpha induction through Toll-like receptors involves a direct interaction of IRF7 with MyD88 and TRAF6. *Nat Immunol* **5**, 1061–1068.

Keckesova, Z., Ylinen, L. M. & Towers, G. J. (2004). The human and African green monkey TRIM5alpha genes encode Ref1 and Lv1 retroviral restriction factor activities. *Proc Natl Acad Sci U S A* **101**, 10780–10785.

Kochs, G. & Haller, O. (1999). Interferon-induced human MxA GTPase blocks nuclear import of Thogoto virus nucleocapsids. *Proc Natl Acad Sci U S A* **96**, 2082–2086.

Kochs, G., Weber, F., Gruber, S., Delvendahl, A., Leitz, C. & Haller, O. (2000). Thogoto virus matrix protein is encoded by a spliced mRNA. *J Virol* **74**, 10785–10789.

Kotenko, S. V., Gallagher, G., Baurin, V. V. & 7 other authors (2003). IFN-lambdas mediate antiviral protection through a distinct class II cytokine receptor complex. *Nat Immunol* **4**, 69–77.

Krug, R. M., Shaw, M., Broni, B., Shapiro, G. & Haller, O. (1985). Inhibition of influenza viral mRNA synthesis in cells expressing the interferon-induced Mx gene product. *J Virol* **56**, 201–206.

Krug, R. M., Yuan, W., Noah, D. L. & Latham, A. G. (2003). Intracellular warfare between human influenza viruses and human cells: the roles of the viral NS1 protein. *Virology* **309**, 181–189.

Le May, N., Dubaele, S., De Santis, L. P., Billecocq, A., Bouloy, M. & Egly, J. M. (2004). TFIIH transcription factor, a target for the Rift Valley hemorrhagic fever virus. *Cell* **116**, 541–550.

Levy, D. E., Marie, I. & Prakash, A. (2003). Ringing the interferon alarm: differential regulation of gene expression at the interface between innate and adaptive immunity. *Curr Opin Immunol* **15**, 52–58.

Lindenmann, J. (1964). Inheritance of resistance to influenza virus in mice. *Proc Soc Exp Biol Med* **116**, 506–509.

Lubyova, B. & Pitha, P. M. (2000). Characterization of a novel human herpesvirus 8-encoded protein, vIRF-3, that shows homology to viral and cellular interferon regulatory factors. *J Virol* **74**, 8194–8201.

Lubyova, B., Kellum, M. J., Frisancho, A. J. & Pitha, P. M. (2004). Kaposi's sarcoma-associated herpesvirus-encoded vIRF-3 stimulates the transcriptional activity of cellular IRF-3 and IRF-7. *J Biol Chem* **279**, 7643–7654.

Ludwig, S., Wang, X., Ehrhardt, C. & 7 other authors (2002). The influenza A virus NS1 protein inhibits activation of Jun N-terminal kinase and AP-1 transcription factors. *J Virol* **76**, 11166–11171.

Mashimo, T., Lucas, M., Simon-Chazottes, D., Frenkiel, M. P., Montagutelli, X., Ceccaldi, P. E., Deubel, V., Guenet, J. L. & Despres, P. (2002). A nonsense mutation in the gene encoding 2′-5′-oligoadenylate synthetase/L1 isoform is associated with West Nile virus susceptibility in laboratory mice. *Proc Natl Acad Sci U S A* **99**, 11311–11316.

Maul, G. G., Negorev, D., Bell, P. & Ishov, A. M. (2000). Review: properties and assembly mechanisms of ND10, PML bodies, or PODs. *J Struct Biol* **129**, 278–287.

Muller, S., Matunis, M. J. & Dejean, A. (1998). Conjugation with the ubiquitin-related modifier SUMO-1 regulates the partitioning of PML within the nucleus. *EMBO J* **17**, 61–70.

Parisien, J. P., Lau, J. F. & Horvath, C. M. (2002). STAT2 acts as a host range determinant for species-specific paramyxovirus interferon antagonism and simian virus 5 replication. *J Virol* **76**, 6435–6441.

Pavlovic, J., Haller, O. & Staeheli, P. (1992). Human and mouse Mx proteins inhibit different steps of the influenza virus multiplication cycle. *J Virol* **66**, 2564–2569.

Pavlovic, J., Arzet, H. A., Hefti, H. P., Frese, M., Rost, D., Ernst, B., Kolb, E., Staeheli, P. & Haller, O. (1995). Enhanced virus resistance of transgenic mice expressing the human MxA protein. *J Virol* **69**, 4506–4510.

Perelygin, A. A., Scherbik, S. V., Zhulin, I. B., Stockman, B. M., Li, Y. & Brinton, M. A. (2002). Positional cloning of the murine flavivirus resistance gene. *Proc Natl Acad Sci U S A* **99**, 9322–9327.

Pichlmair, A., Buse, J., Jennings, S., Haller, O., Kochs, G. & Staeheli, P. (2004). Thogoto virus lacking interferon-antagonistic protein ML is strongly attenuated in newborn Mx1-positive but not Mx1-negative mice. *J Virol* **78**, 11422–11424.

Poole, E., He, B., Lamb, R. A., Randall, R. E. & Goodbourn, S. (2002). The V proteins of simian virus 5 and other paramyxoviruses inhibit induction of interferon-beta. *Virology* **303**, 33–46.

Prakash, B., Praefcke, G. J., Renault, L., Wittinghofer, A. & Herrmann, C. (2000). Structure of human guanylate-binding protein 1 representing a unique class of GTP-binding proteins. *Nature* **403**, 567–571.

Reeves, R. H., O'Hara, B. F., Pavan, W. J., Gearhart, J. D. & Haller, O. (1988). Genetic mapping of the Mx influenza virus resistance gene within the region of mouse chromosome 16 that is homologous to human chromosome 21. *J Virol* **62**, 4372–4375.

Regad, T. & Chelbi-Alix, M. K. (2001). Role and fate of PML nuclear bodies in response to interferon and viral infections. *Oncogene* **20**, 7274–7286.

Regad, T., Saib, A., Lallemand-Breitenbach, V., Pandolfi, P. P., de The, H. & Chelbi-Alix, M. K. (2001). PML mediates the interferon-induced antiviral state against a complex retrovirus via its association with the viral transactivator. *EMBO J* **20**, 3495–3505.

Reichelt, M., Stertz, S., Krijnse-Locker, J., Haller, O. & Kochs, G. (2004). Missorting of LaCrosse virus nucleocapsid protein by the interferon-induced MxA GTPase involves smooth ER membranes. *Traffic* **5**, 772–784.

Roberts, R. M., Ezashi, T., Rosenfeld, C. S., Ealy, A. D. & Kubisch, H. M. (2003). Evolution of the interferon tau genes and their promoters, and maternal-trophoblast interactions in control of their expression. *Reprod Suppl* **61**, 239–251.

Roers, A., Hochkeppel, H. K., Horisberger, M. A., Hovanessian, A. & Haller, O. (1994). MxA gene expression after live virus vaccination: a sensitive marker for endogenous type I interferon. *J Infect Dis* **169**, 807–813.

Samuel, C. E. (2001). Antiviral actions of interferons. *Clin Microbiol Rev* **14**, 778–809.

Sharma, S., TenOever, B. R., Grandvaux, N., Zhou, G. P., Lin, R. & Hiscott, J. (2003). Triggering the interferon antiviral response through an IKK-related pathway. *Science* **300**, 1148–1151.

Sheppard, P., Kindsvogel, W., Xu, W. & 23 other authors (2003). IL-28, IL-29 and their class II cytokine receptor IL-28R. *Nat Immunol* **4**, 63–68.

Silverman, R. H. (1994). Fascination with 2-5A-dependent RNase: a unique enzyme that functions in interferon action. *J Interferon Res* **14**, 101–104.

Spann, K. M., Tran, K. C., Chi, B., Rabin, R. L. & Collins, P. L. (2004). Suppression of the induction of alpha, beta, and gamma interferons by the NS1 and NS2 proteins of human respiratory syncytial virus in human epithelial cells and macrophages. *J Virol* **78**, 4363–4369.

Staeheli, P. & Haller, O. (1985). Interferon-induced human protein with homology to protein Mx of influenza virus-resistant mice. *Mol Cell Biol* **5**, 2150–2153.

Staeheli, P., Haller, O., Boll, W., Lindenmann, J. & Weissmann, C. (1986). Mx protein: constitutive expression in 3T3 cells transformed with cloned Mx cDNA confers selective resistance to influenza virus. *Cell* **44**, 147–158.

Staeheli, P., Grob, R., Meier, E., Sutcliffe, J. G. & Haller, O. (1988). Influenza virus-susceptible mice carry Mx genes with a large deletion or a nonsense mutation. *Mol Cell Biol* **8**, 4518–4523.

Stark, G. R., Kerr, I. M., Williams, B. R., Silverman, R. H. & Schreiber, R. D. (1998). How cells respond to interferons. *Annu Rev Biochem* **67**, 227–264.

Sternsdorf, T., Jensen, K. & Will, H. (1997). Evidence for covalent modification of the nuclear dot-associated proteins PML and Sp100 by PIC 1/SUMO-1. *J Cell Biol* **139**, 1621–1634.

Talon, J., Horvath, C. M., Polley, R., Basler, C. F., Muster, T., Palese, P. & Garcia-Sastre, A. (2000). Activation of interferon regulatory factor 3 is inhibited by the influenza A virus NS1 protein. *J Virol* **74**, 7989–7996.

tenOever, B. R., Sharma, S., Zou, W. & 9 other authors (2004). Activation of TBK1 and IKK epsilon kinases by vesicular stomatitis virus infection and the role of viral ribonucleoprotein in the development of interferon antiviral immunity. *J Virol* **78**, 10636–10649.

Thomas, D., Blakqori, G., Wagner, V., Banholzer, M., Kessler, N., Elliott, R. M., Haller, O. & Weber, F. (2004). Inhibition of RNA polymerase II phosphorylation by a viral interferon antagonist. *J Biol Chem* **279**, 31471–31477.

Towers, G., Bock, M., Martin, S., Takeuchi, Y., Stoye, J. P. & Danos, O. (2000). A conserved mechanism of retrovirus restriction in mammals. *Proc Natl Acad Sci U S A* **97**, 12295–12299.

Turelli, P., Doucas, V., Craig, E., Mangeat, B., Klages, N., Evans, R., Kalpana, G. & Trono, D. (2001). Cytoplasmic recruitment of INI1 and PML on incoming HIV preintegration complexes: interference with early steps of viral replication. *Mol Cell* **7**, 1245–1254.

Valarcher, J. F., Furze, J., Wyld, S., Cook, R., Conzelmann, K. K. & Taylor, G. (2003). Role of alpha/beta interferons in the attenuation and immunogenicity of recombinant bovine respiratory syncytial viruses lacking NS proteins. *J Virol* **77**, 8426–8439.

van Pesch, V., Lanaya, H., Renauld, J. C. & Michiels, T. (2004). Characterization of the murine alpha interferon gene family. *J Virol* **78**, 8219–8228.

Vialat, P., Billecocq, A., Kohl, A. & Bouloy, M. (2000). The S segment of rift valley fever phlebovirus (Bunyaviridae) carries determinants for attenuation and virulence in mice. *J Virol* **74**, 1538–1543.

Wang, C., Pflugheber, J., Sumpter, R., Jr, Sodora, D. L., Hui, D., Sen, G. C. & Gale, M., Jr (2003). Alpha interferon induces distinct translational control programs to suppress hepatitis C virus RNA replication. *J Virol* **77**, 3898–3912.

Wang, X., Li, M., Zheng, H., Muster, T., Palese, P., Beg, A. A. & Garcia-Sastre, A. (2000). Influenza A virus NS1 protein prevents activation of NF-kappaB and induction of alpha/beta interferon. *J Virol* **74**, 11566–11573.

Weber, F., Bridgen, A., Fazakerley, J. K., Streitenfeld, H., Randall, R. E. & Elliott, R. M. (2002). Bunyamwera bunyavirus nonstructural protein NSs counteracts the induction of alpha/beta interferon. *J Virol* **76**, 7949–7955.

Weber, F., Kochs, G. & Haller, O. (2004). Inverse interference: how viruses fight the interferon system. *Viral Immunol* **17**, 498–515.

Williams, B. R. (1999). PKR; a sentinel kinase for cellular stress. *Oncogene* **18**, 6112–6120.

Yap, M. W., Nisole, S., Lynch, C. & Stoye, J. P. (2004). Trim5alpha protein restricts both HIV-1 and murine leukemia virus. *Proc Natl Acad Sci U S A* **101**, 10786–10791.

Yoneyama, M., Kikuchi, M., Natsukawa, T., Shinobu, N., Imaizumi, T., Miyagishi, M., Taira, K., Akira, S. & Fujita, T. (2004). The RNA helicase RIG-I has an essential function in double-stranded RNA-induced innate antiviral responses. *Nat Immunol* **5**, 730–737.

Young, D. F., Chatziandreou, N., He, B., Goodbourn, S., Lamb, R. A. & Randall, R. E. (2001). Single amino acid substitution in the V protein of simian virus 5 differentiates its ability to block interferon signaling in human and murine cells. *J Virol* **75**, 3363–3370.

Zhou, A., Paranjape, J., Brown, T. L. & 8 other authors (1997). Interferon action and apoptosis are defective in mice devoid of 2′,5′-oligoadenylate-dependent RNase L. *EMBO J* **16**, 6355–6363.

Zhou, A., Paranjape, J. M., Der, S. D., Williams, B. R. & Silverman, R. H. (1999). Interferon action in triply deficient mice reveals the existence of alternative antiviral pathways. *Virology* **258**, 435–440.

Secreted tumour necrosis factor inhibitors encoded by poxviruses

M. Begoña Ruiz-Argüello,[1,2] Alí Alejo[1,3] and
Antonio Alcami[1,3]

[1]Department of Medicine, University of Cambridge, Addenbrooke's Hospital, Cambridge, UK

[2]Centro de Investigación en Sanidad Animal (INIA), Valdeolmos, Madrid, Spain

[3]Department of Molecular and Cellular Biology, Centro Nacional de Biotecnología (CSIC), Campus Universidad Autónoma, Cantoblanco 28049 Madrid, Spain

Tumour necrosis factor (TNF) is a multifunctional cytokine produced as an early response to virus infection that triggers local expression of chemokines and cytokines to promote the attraction and activation of leukocytes at the site of infection. At later stages, TNF facilitates transition from innate to acquired immunity by enhancing antigen presentation and T-cell costimulation (Beutler, 1992; Wallach *et al.*, 1999a). In this chapter, we address the molecular mechanisms employed by viruses to counteract the effect of TNF and describe in detail a unique strategy used by poxviruses to neutralize TNF, namely the expression of secreted proteins that bind TNF.

TNF: STRUCTURE AND BIOLOGICAL ACTIVITY

The major cellular source of TNF is activated mononuclear phagocytes, although antigen-stimulated T cells, natural killer cells and mast cells can also secrete this protein. TNF is synthesized as a nonglycosylated type II membrane protein, with an intracellular N-terminus and a large extracellular C-terminus. Membrane TNF can be cleaved by a membrane-associated metalloproteinase to release a trimeric 51 kDa circulating TNF protein (Locksley *et al.*, 2001; Wallach *et al.*, 1999a).

The 17 350 Da TNF monomer forms an elongated, antiparallel beta-pleated sheet sandwich with a 'jelly roll' topology. Three monomers associate intimately about a threefold axis of symmetry to form a compact bell-shaped trimer. The general site of interaction with the receptor is at the 'base' of the trimer, allowing simultaneous binding of the cytokine to three receptor molecules (Eck & Sprang, 1989).

SGM symposium 64: Molecular pathogenesis of virus infections.
Editors P. Digard, A. A. Nash & R. E. Randall. Cambridge University Press. ISBN 0 521 83248 9 ©SGM 2005

The principal physiological function of TNF is to stimulate the recruitment of neutrophils and monocytes to sites of infection and to activate these cells to eradicate microbes (Beutler, 1992). TNF mediates these effects by several actions on vascular endothelial cells and leukocytes. The actions of TNF on endothelium and leukocytes are critical for local inflammatory responses to microbes. TNF also contributes to local inflammatory reactions that are injurious to the host (e.g. in autoimmune diseases). In severe infections, TNF is produced in large amounts and causes systemic clinical and pathological abnormalities. If the stimulus for TNF production is sufficiently strong, the quantity of the cytokine produced is so large that it enters the bloodstream and acts at distant sites as an endocrine hormone.

TNF was originally identified as a substance present in the serum of animals treated with bacterial endotoxin that caused the necrosis of tumours *in vivo*. It was first isolated and characterized in 1985 (Aggarwal *et al.*, 1985) and it is now considered the prototype of a superfamily of structurally related ligands that have a wide variety of activities with 19 members having been identified so far that signal through 32 receptors (Locksley *et al.*, 2001) (Table 1).

TNF superfamily (TNFSF) members are generally expressed by cells of the immune system, including B cells, T cells, natural killer cells, monocytes and dendritic cells. After binding to their receptors, members of the TNFSF mediate apoptosis (such as TNF, LTα, FasL, TRAIL, TL1A, TWEAK and LIGHT), survival (such as RANKL and BAFF), differentiation (such as TNF or RANKL) or proliferation (such as TNF, CD27L, CD30L, CD40L, OX40L, 4-1BBL, APRIL and BAFF) (Aggarwal, 2003; Kwon *et al.*, 2003; Pfeffer, 2003).

THE TNFR SUPERFAMILY (TNFRSF)

Mammalian TNF signals through two distinct cell surface receptors (TNFRs): TNFRI (also called p55 receptor), the primary receptor for soluble TNF, and TNFRII (also called p75 receptor), the main receptor for membrane TNF. TNFRI is constitutively expressed by most cell types, whereas expression of TNFRII is highly regulated and restricted to endothelial cells and cells of the immune system. Both TNFRI and TNFRII are also the cognate receptors for lymphotoxin (LTα) (Locksley *et al.*, 2001).

The cell surface receptors for the TNFSF of proteins are type I membrane glycoproteins that can also be grouped into the TNFRSF. Members of this superfamily are composed of an N-terminal extracellular ligand-binding domain linked by a single trans-membrane segment to a C-terminal cytoplasmic signalling domain (Loetscher *et al.*, 1990; Schall *et al.*, 1990).

Table 1. TNFSF and TNFRSF members

4-1BBL, 4-1BB ligand; APRIL, a proliferation-inducing ligand; BAFF, B cell activating factor belonging to the TNF family; BCMA, B-cell maturation antigen; CD, cluster of differentiations; CD27L, CD27 ligand; CD30L, CD30 ligand; CD40L, CD40 ligand; DcR, decoy receptor; DR, death domain receptor; EDA1, ectodermal dysplasia 1 anhidrotic; FasL, Fas ligand; GITRL, glucocorticoid-induced TNFR; HVEM, herpes virus entry mediator; LIGHT, lymphotoxin-inducible expression, competes with herpes simplex virus glycoprotein D for HVEM T-lymphocyte receptor; LTα, lymphotoxin alpha; LTβ, lymphotoxin beta; OX40L, OX40 ligand; RANKL, receptor activator of NF-κB ligand; TACI, transmembrane activator and calcium-modulating cyclophilin ligand (CAML) interactor; TL1A, TNF-like cytokine; TNF, tumour necrosis factor; TRAIL, TNF-like apoptosis inducing ligand; TWEAK, TNF-like and weak inducer of apoptosis. Numbers in parentheses refer to the standard nomenclature number of the TNFRSF members. A more complete downloadable table can be accessed at http://www.irp.niams.nih.gov/ImageStore/Test/WORD/AB/IRG/tnfchart.doc

Symbol	Common name	Binds to receptor(s)	Key functions
TNFSF1	LTα	TNFRII (1B), TNFRI (1A), HVEM (14)	Lymphoid organ formation
TNFSF2	TNF	TNFRII (1B), TNFRI (1A)	Inflammation
TNFSF3	LTβ	As a $\beta2\alpha1$ heterotrimer with LTα binds to LTβ receptor (3)	Lymphoid organ formation
TNFSF4	OX40L	OX40 (4)	CD4 T cell expansion, survival, and Th2 development
TNFSF5	CD40L	CD40 (5)	Costimulation and differentiation of B cells and antigen-presenting cell (APC)
TNFSF6	FasL	Fas (6), DcR3 (6B)	Mediator of CD4(+) T cell apoptosis due to restimulation and apoptosis in other cell types
TNFSF7	CD27L	CD27 (7)	T-cell costimulation
TNFSF8	CD30L	CD30 (8)	T- and B-cell costimulation
TNFSF9	4-1BBL	4-1BB (9)	T-cell costimulation
TNFSF10	TRAIL	DR4 (10A), DR5 (10B), DcR1 (10C), DcR2 (10D)	Dendritic cell apoptosis, natural killer-cell mediated tumour cell killing
TNFSF11	RANKL	RANK (11A)	Mediates osteoclast formation and bone remodelling. Stimulation of APC
TNFSF12	TWEAK	TWEAK-R (12A)	Potential role in inflammation and lymphocyte function
TNFSF13	APRIL	TACI (13B), BCMA (17)	Promotes T-independent type-2 responses through interactions with TACI
TNFSF13B	BAFF	TACI (3B), BAFF-R (13C), BCMA(17)	Promotes B-cell maturation, plasmablast survival
TNFSF14	LIGHT	HVEM (14), LT-βR (3), DcR3 (6B)	CD8 T cell and APC costimulation
TNFSF15	TL1A	DR3 (25)	Recently identified ligand for DR3 (TNFRSF25)
TNFSF18	GITRL	GITR (18)	T-cell costimulation (+)CD25(+) regulatory T cells
ED1	EDA1	EDAR	Tooth, hair and sweat gland formation

The ligand-binding domain comprises four, or in some cases three, cysteine-rich domains (CRDs) of about 40 residues containing six conserved cysteines which are all involved in internal disulfide bridges. A consensus pattern [Prosite signature PDOC00561:C - x(4,6) - [FYH] - x(5,10) - C- x(0,2) – C - x(2,3) – C -x(7,11) - C- x(4,6) - [DNEQSKP] - x(2) - C] has been defined to identify proteins belonging to this superfamily (Banner *et al.*, 1993).

The extracellular domain of TNFRI has been shown to be an elongated rod of four CRDs arranged linearly along the long axis of the molecule which are numbered consecutively from the N-terminus. The disulfide bonds are arranged as rungs of a ladder (with a 10 Å step) along the length of the long molecular axis (Naismith *et al.*, 1995). In the crystal structure of the TNFRI–LTα complex, LTα binds three TNFRI molecules, each at an intersubunit site on the trimeric cytokine. The second and third CRDs of the TNFRI extracellular domain form virtually all of the contacts with LTα. Presumably this complex brings the receptor intracellular domains together below the membrane in the correct steric manner for signal initiation (Banner *et al.*, 1993). It has been shown that CRD three in TNFRI and CRD four in TNFRII are directly involved in binding of TNF, suggesting that the modes of interaction between TNF and its dual receptors are different (Chen *et al.*, 1995). The active form of the receptor is a trimer and evidence suggests that these trimeric TNFR complexes are pre-assembled on the cell surface in the absence of ligand (Chan *et al.*, 2000).

The transmembrane domain of TNFRs is helical and highly conserved throughout evolution, suggesting an important function for this region of the molecule. This notion is supported by the observation that replacement of the transmembrane domain of some unrelated receptors abolishes signalling (Declercq *et al.*, 1995).

The intracellular signalling domains of different TNFRSF members vary in sequence, characteristics and size, and can be divided into two groups. About one-third of TNFRSF members contain a death domain (DD), a sequence of approximately 80 amino acids that can bind to other DD-containing proteins to recruit caspases resulting in apoptotic cell death (Singh *et al.*, 1998; Wallach *et al.*, 1999b). Receptors that contain DDs include TNFRI, Fas, DR4, DR5, DcR2, DcR3 and cytopathic avian receptor-1. The second group comprises receptors, such as TNFRII, that lack this DD and mainly signal through the recruitment of TNFR-associated factors (TRAFs), which activate transcription factors such as nuclear factor κB (NF-κB), an important regulator for the expression of cell survival and proinflammatory genes.

TNFRSF members may also exist in solution, as a consequence of either an alternative splicing event, as in 4-1BB (Michel *et al.*, 1998), or proteolytic cleavage from the cellular membrane, such as soluble TNFRI and TNFRII (Kohno *et al.*, 1990; Nophar *et al.*, 1990). Other members of the TNFRSF exist naturally as non-signalling, decoy receptors. Some examples are the DcR1, which is membrane-associated by an inositol link but lacks a cytoplasmic tail, and the alternative receptor for the RANKL, osteoprotegerin (Simonet *et al.*, 1997), which acts only in its soluble form. Soluble TNFRs can capture TNF before their engagement to cellular TNFRs, inhibiting signalling and therefore acting as regulating decoy receptors. These alternative forms of TNFRs are likely to be involved in the regulation of the activity of membrane-bound TNFRs.

MODULATION OF THE ACTIVITY OF TNFSF MEMBERS BY VIRUSES

Successful transmission by viruses in the face of strong innate and acquired host immunity requires the ability to evade, obstruct or subvert critical elements that mediate host antiviral responses. To that end, viruses encode multiple classes of immunomodulatory proteins that have evolved specifically to inhibit such diverse processes as apoptosis, the production of interferons (IFNs), chemokines and inflammatory cytokines, and the activity of cytotoxic T lymphocytes, natural killer cells, complement and antibodies (Alcami & Koszinowski, 2000; Tortorella *et al.*, 2000).

As TNFSF members have a crucial role in antiviral immune responses, viruses have developed a variety of mechanisms to target virtually every step of TNFSF signalling: from ligand expression to inhibition of ligand–receptor binding and the modulation of downstream signalling pathways (Benedict *et al.*, 2002).

The interaction of TNFRI, TNFRII and Fas with their respective ligands induces the binding of cytoplasmic factors that interact through DDs and death effector domains (DEDs) present in adaptor proteins, such as TNF-receptor-activated death domain (TRADD) and Fas-associated death domain (FADD), and in some caspases. These factors initiate a cascade of interactions that culminate in the activation of caspases and the induction of apoptosis. The poxvirus molluscum contagiosum virus and Kaposi's sarcoma-associated herpesvirus encode vFLIPs (FLICE or caspase-8-inhibitory proteins) that have DDs and prevent signal transduction through receptors for TNF and FasL (Bertin *et al.*, 1997; Thome & Tschopp, 2001). Adenoviruses produce a number of proteins encoded by the E3 region that inhibit TNF- and FasL-mediated apoptosis (Wold *et al.*, 1999). The adenovirus receptor internalization and degradation (RID) complex, formed by the E3/10.4K and E3/14.5K proteins, induces internalization and subsequent degradation in lysosomes of TNFR and Fas. As a result of this, the infected cell becomes resistant to TNFR- and Fas-mediated apoptosis (Burgert *et al.*, 2002; McNees & Gooding, 2002).

Human cytomegalovirus, a β-herpesvirus, contains an ORF, UL144, the product of which has amino acid sequence similarity to TNFRs. UL144 is detected in low passage clinical virus samples that have been freshly isolated from patients, but not in laboratory strains of human cytomegalovirus. Interestingly, the UL144 gene product is not secreted, and attempts to identify a specific ligand for this receptor among the TNFSF members have been unsuccessful (Benedict *et al.*, 1999; Lurain *et al.*, 1999). The function of this protein remains unknown, but it might influence the ability of cells that are infected with human cytomegalovirus to respond to signals induced by TNF-related ligands.

Some viruses may subvert cytokine-mediated signalling for their own benefit. This is illustrated by the latent membrane protein 1 (LMP1) of Epstein–Barr virus (EBV), a virus that establishes latency in B cells. LMP1 is a membrane protein that lacks sequence similarity to members of the TNFRSF but has a short cytoplasmic tail that binds TRAFs. During the immune response, binding of trimeric TNF to its receptor induces receptor multimerization and recruitment of TRAFs followed by the initiation of a signalling cascade leading to either B-cell proliferation or the induction of apoptosis. In EBV-infected cells, the recruitment of TRAFs by LMP1 induces biological responses such as cell proliferation that may enhance virus replication (Farrell, 1998). This virus takes advantage of a B-cell activation pathway to ensure survival of the infected cells that carry the latent viral genome.

The strategy used by many poxviruses to inhibit TNF is the expression of secreted TNFRs (vTNFRs) or binding proteins (vTNFBPs) that neutralize the activity of TNF (Alcami, 2003; Seet *et al.*, 2003). Related proteins identified are a secreted homologue of CD30, a member of the TNFRSF, encoded by poxviruses and a set of iridoviral proteins with sequence similarity to host TNFRs (Fig. 1).

POXVIRUS SECRETED DECOY TNFRS
Poxviruses are large complex enveloped viruses with a linear dsDNA ranging from 130 to 300 kbp and encoding approximately 200 potential proteins (Moss, 2001). Virus replication and the formation of virus particles take place in the cytoplasm of the infected cell. In order to evade the host defence mechanisms, poxviruses express several secreted proteins that function as soluble cytokine receptors or binding proteins and block cytokine activity and modulate virus virulence.

One of these strategies is the expression of soluble secreted TNFRs that prevent the binding of TNF to its cellular receptor, blocking its activity (Fig. 1). Many poxvirus species have been found to encode at least one such active TNFR. These receptors can be classified into two groups. Those that share significant sequence similarity with

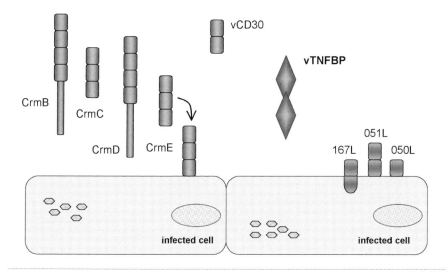

Fig. 1. Viral TNF receptor homologues and binding proteins.

cellular TNFRs, named vTNFRs, include the proteins S-T2, M-T2, cytokine response modifier B (CrmB), CrmC, CrmD and CrmE, which are differentially distributed among poxviral species. The second group includes the 2L protein of the yata-poxviruses and their swinepox virus (SPV) and deerpox virus (DPV) orthologues, and do not present amino acid sequence similarity to any known cellular proteins; they are referred to as vTNFBPs. Distribution of the secreted decoy TNFR genes in poxviruses is shown in Table 2.

Here we will describe the main characteristics of this important group of immuno-modulatory proteins and discuss their potential implications in virus pathogenesis.

Shope fibroma virus (SFV) S-T2

The first evidence that viruses might encode soluble TNFRs came from the report describing the cloning of a human TNFR gene (Smith *et al.*, 1990). In this paper, the authors described a high amino acid sequence similarity (almost 40 % identity) between the human TNFR and a putatively secreted protein encoded by ORF T2 from SFV (S-T2), a leporipoxvirus, and proposed a possible immunomodulatory role for such a viral protein. ORF T2 is present in two copies in the viral genome, due to its location within the terminal inverted repeats, and S-T2 is expressed as an early gene product (Upton *et al.*, 1987). It is a 461-amino-acid protein with a signal peptide, four CRDs with 20 cysteines and highest sequence similarity to the cellular TNFRII, as well as a long C-terminal extension with no described similarity to proteins in the database.

Table 2. Distribution of the secreted decoy TNFR genes in poxviruses

CPV, cowpox virus; crm, cytokine response modifier; DPV, deerpox virus; EV, ectromelia virus; SPV, swinepox virus; vTNFBP, TNF-binding protein; TPV, Tanapox virus; VaV, variola virus; VV, vaccinia virus; YLDV, Yaba-like disease virus.

Poxvirus	vTNFR				vCD30	vTNFBP
	CrmB	CrmC	CrmD	CrmE		
CPV Brighton Red	V005	V191	V221	–	V015	–
CPV GRI-90	I4R	A56R	K2R	K3R	D13L	–
EV Naval	–	–	E3	–	E9	–
VaV Bangladesh	G2R	–	–	–	–	–
VV Western Reserve	–	–	–	–	–	–
VV Lister	–	A53R	–	–	–	–
VV USSR	–	crmC homologue	–	crmE homologue	–	–
TPV	–	–	–	–	–	2L
YLDV	–	–	–	–	–	2L
SPV	–	–	–	–	–	003/148
DPV W83	–	–	–	–	–	008
DPV W84	–	–	–	–	005	008

The S-T2 protein was shown to be a secreted 58 kDa glycoprotein that binds TNF and it was proposed to function as a TNF scavenger that prevents TNF binding to cellular receptors (Smith *et al.*, 1991).

Myxoma virus (MV) M-T2

MV, a leporipoxvirus which specifically infects rabbits, encodes an ORF whose product, M-T2, is related to S-T2 and which is also present in two copies in the genome. M-T2 is a secreted 40·5 kDa TNFRII homologue expressed early during infection and forms dimers (Schreiber *et al.*, 1996; Upton *et al.*, 1991). The protein has a signal peptide, four CRDs containing 21 cysteines, and an extended C-terminal region with no amino acid sequence similarity to host proteins.

The secreted form of the M-T2 protein binds specifically to rabbit TNF with an affinity comparable to that of cellular receptors (K_D 170–195 pM). The affinity for mouse TNF is approximately 10-fold lower (K_D 1·5 nM) and no binding of human TNF was detected (Schreiber *et al.*, 1996). It was found that the complete protein is able to inhibit TNF-induced cytotoxicity and that the first three CRDs are required to bind TNF and inhibit its activity. Additionally, the intracellular M-T2 protein was shown to prevent apoptosis in infected T lymphocytes and this activity required only two CRDs. The molecular mechanism by which M-T2 prevents apoptosis has not been elucidated (Schreiber *et al.*, 1996, 1997).

Chimeric dimers of TNFRI have been shown to bind soluble TNF and LTα at 2000-fold higher affinity compared with monomers, and the dimers had greater capacity to block cytolytic activity of TNF and LTα *in vitro* and *in vivo* (Ashkenazi *et al.*, 1991; Lesslauer *et al.*, 1991; Peppel *et al.*, 1991). Greater blocking of TNF cytolysis *in vitro* by dimers of the MV M-T2 protein has also been shown (Schreiber *et al.*, 1996).

The secreted M-T2 protein is one of the viral immunomodulatory proteins that has been shown to contribute to virus pathogenesis *in vivo*. These experiments were performed in European rabbits, which are not the native host for MV, and die rapidly from myxomatosis, and thus the role of M-T2 during MV infection of its natural host will have to be elucidated (Upton *et al.*, 1991). Insertional inactivation of both copies of the M-T2 ORF from MV resulted in a virus that was attenuated, with five of eight infected European rabbits making a complete recovery, in contrast with the uniformly lethal infection seen with wild-type MV. Rabbits infected with the MV M-T2 mutant gave smaller primary lesions with less pathology than was seen for wild-type MV. Secondary lesions were only seen in 50 % of animals infected with the mutant and were smaller, and supervening Gram-negative bacterial infections were much reduced compared with wild-type MV-infected animals.

Crmb

The *crmB* gene is present in two copies in cowpox virus (CPV) strain Brighton Red (BR) and was described as an early gene that encodes a protein of 355 amino acids (Hu *et al.*, 1994). CrmB is the orthopoxvirus orthologue of the leporipoxvirus S-T2 and M-T2 proteins. The predicted amino acid sequence contains a signal peptide, three potential N-glycosylation sites, four CRDs (with 21 cysteines) and a long C-terminal region of 161 amino acids. The CRD-containing region of CPV BR CrmB is 42 % identical at the amino acid level to the extracellular domain of human TNFRII, while the C-terminal extension shows no recognizable amino acid sequence similarity with the cytoplasmic domains of members of the TNFRSF. However, it has approximately 50 % amino acid sequence similarity to the C-terminal region of the M-T2 and CrmD viral proteins.

The 48 kDa secreted protein was shown to bind both TNF and LTα (Hu *et al.*, 1994) although in another report binding to LTα was not detected (Alcami *et al.*, 1999). The CPV BR CrmB protein binds to human, mouse and rat TNF (with a K_D of ~700 pM for human TNF), and inhibits the binding of TNF to cell surface receptors, acting as a decoy receptor (Alcami *et al.*, 1999).

Sequence analysis showed that the *crmB* gene is also present in variola virus (VaV), camelpox virus (CaPV) and monkeypox virus (with 85 % amino acid similarity to CPV BR) but is predicted to be inactive in 12 strains of ectromelia virus (EV) and 18 strains of vaccinia virus (VV) tested so far (Alcami *et al.*, 1999; Gubser & Smith, 2002; Loparev *et al.*, 2001; Massung *et al.*, 1993; Reading *et al.*, 2002; Ribas *et al.*, 2003; Shchelkunov *et al.*, 2002).

A recombinant CPV in which both copies of the *crmB* gene are inactivated by insertion of a selectable marker showed no marked differences compared to wild-type CPV in a standard pock formation assay performed on chick embryo chorioallantoic membranes, an assay that measures the inflammatory response (Hu *et al.*, 1994). A role for CrmB *in vivo* has been shown by an increased virulence in a mouse intranasal model of a VV Western Reserve (WR) recombinant expressing CPV CrmB (Reading *et al.*, 2002).

In VaV, CrmB is the sole vTNFR predicted to be active (Massung *et al.*, 1993). The recent analysis of the gene expression profile of peripheral blood mononuclear cells from VaV-infected macaques showed that there were only small increases in levels of expression of a few of the TNF-regulated genes (Rubins *et al.*, 2004). Furthermore, TNF levels in serum were not increased in the face of an overwhelming and fatal VaV infection. The absence of a TNF response during infection at the level of gene expression and protein secretion is consistent with the activity of CrmB as a TNF decoy receptor.

CrmC

Smith *et al.* (1996) identified a gene, *crmC*, present in a single copy in CPV BR, which is transcribed during the late phase of virus replication. This gene encodes a protein of 186 amino acids with a hydrophobic N-terminal signal peptide, four CRDs with 24 cysteines, the fourth one of which is truncated, and two potential N-glycosylation sites. Among the mammalian TNFRs, the CrmC protein is most closely related to the extra-cellular region of human TNFRII (37·8 % amino acid identity). The CrmC protein is secreted from cells infected with CPV or recombinant VV expressing CPV CrmC as a 25 kDa protein, suggesting that post-translational modifications, particularly glyco-sylation, occur.

CrmC binds human and murine TNF with high affinity, but not human LTα or other ligands tested (LTα1/β2, LTα2/β1, TRAIL, CD40L, CD27L, CD30L, 4-1BBL, OX40L or FasL) (Smith *et al.*, 1996). The K_D for murine TNF (200 pM) is similar to that of the cellular receptor. CrmC inhibits murine TNF-mediated cytotoxicity but not human LTα-induced cytotoxicity.

In vivo experiments showed normal red pock formation in chorioallantoic membranes of chick embryos after 3 days of infection with an insertional mutant of CPV lacking CrmC expression (Smith *et al.*, 1996). A function for CrmC in virus pathogenesis was shown by the increase in virus virulence after intranasal inoculation of a VV WR recombinant expressing the CPV or VV Lister CrmC protein (Reading *et al.*, 2002).

The study of vTNFRs in 15 VV strains showed that only strains Lister, USSR and Evans encode an active CrmC protein (Alcami *et al.*, 1999).

CrmD

CrmD was the third poxviral vTNFR to be described (Loparev *et al.*, 1998). The protein CrmD contains a signal peptide, a 151-amino-acid region comprising four CRDs with 21 cysteines, and a C-terminal region that is highly diverged from the cellular TNFR C-terminal region sequence involved in signal transduction. The EV CrmD shows 38 and 47 % amino acid sequence identity with TNFRI and TNFRII, respectively, and the highest sequence similarity corresponds to the first three CRDs. The CrmD protein is a late secreted 46 kDa protein. CrmD forms disulfide-linked complexes that can bind ligand, perhaps increasing ligand-binding efficiency (Loparev *et al.*, 1998).

The proteins encoded by the CPV BR and EV Moscow *crmD* genes were expressed using a prokaryotic system and found to specifically bind TNF and LTα as well as protect cells from TNF- or LTα-induced necrosis (Loparev *et al.*, 1998).

EV Hampstead CrmD was expressed in eukaryotic systems as a secreted 46 kDa protein. The predicted molecular mass is 35 kDa, suggesting that the protein undergoes post-translational modifications, probably glycosylation of any or all of the four putative N-glycosylation sites. EV CrmD binds murine TNF with very high affinity (K_D ~20 pM). The affinity for human and rat TNF is lower (K_D ~200 pM) in accordance with EV being a mouse pathogen. No binding to other ligands of the TNFSF was detected, including LTα, LTα1/β2, LTα2/β1, CD40L, CD30L, 4-1BBL, BAFF, TWEAK or GITRL. The TNF-binding domain has been mapped to the four N-terminal CRDs, whereas the C-terminal domain is not involved in or required for TNF binding (Saraiva, 2002). EV CrmD is able to inhibit TNF binding to cell surface receptors acting as a decoy receptor and protects cells from TNF-induced necrosis. However, in contrast to Loparev *et al.* (1998), Saraiva (2002) did not find inhibition of LTα activity by EV CrmD.

The CrmD protein was initially found to be intact in CPV BR, which shows 44 and 22 % amino acid sequence identity to CPV BR CrmB and CrmC, respectively. Four strains of EV, Moscow, MP-3, MP-4 and Munich-SF, contained an intact *crmD* gene (97 % amino acid identity to the CPV *crmD* product) and lacked any other cognates of *crmB*, *crmC* or *crmE* (Loparev *et al.*, 1998; Saraiva & Alcami, 2001). An extensive study of immunomodulatory genes encoded by 12 EV isolates found that CrmD is the only vTNFR encoded by EV (Ribas *et al.*, 2003). *crmD* was truncated in three other strains of CPV and absent in CaPV, VaV, monkeypox virus and VV strains Copenhagen and WR, as well as in the strains of VV that encode TNF-binding activity (Lister, Evans and USSR) (Alcami *et al.*, 1999; Loparev *et al.*, 1998; Massung *et al.*, 1993). In fact, this gene is usually found only in orthopoxviruses that lack *crmB* and *crmC* (Alcami *et al.*, 1999; Reading *et al.*, 2002).

There is no experimental evidence published so far that demonstrates a role of CrmD in virus pathogenesis *in vivo*.

CrmE

The CPV *crmE* gene encodes a secreted protein of 167 amino acids and a predicted molecular mass of 18 kDa, with a hydrophobic signal peptide, one potential N-glycosylation site and four CRDs characteristic of the TNFRs, although the fourth one contains only two cysteines and is therefore truncated (Saraiva & Alcami, 2001). Treatment with tunicamycin has no effect on CrmE expression or size, while monensin impairs protein secretion, suggesting that CrmE is not N-glycosylated but that O-glycosylation is important for its correct folding and secretion. The CrmE protein encoded by CPV strains GRI90 and elephantpox is related to those of cellular TNFRs (with 27·5 and 31·1 % amino acid identity to the TNFRI and TNFRII, respectively) and CPV-encoded vTNFRs (with 42·5, 32·9 and 37·1 % amino acid identity to CrmB, CrmC and CrmD, respectively) (Saraiva & Alcami, 2001; Shchelkunov *et al.*, 1998).

CrmE binds human, murine and rat TNF *in vitro* but not LTα or any of the seven other members of the TNFSF tested (LTα1/β2, LTα2/β1, TRAIL, CD40L, 4-1BBL, GITRL, BAFF, RANKL or TWEAK) (Saraiva & Alcami, 2001). It is effective in protecting cells only against human TNF (Reading *et al.*, 2002; Saraiva & Alcami, 2001).

The *crmE* gene has been found in various orthopoxviruses but was only functional in CPV strain elephantpox and VV strain USSR (Reading *et al.*, 2002; Saraiva & Alcami, 2001). The *crmE* gene is inactive in CaPV and in 12 EV isolates because of mutations that introduce stop codons or frameshifts (Ribas *et al.*, 2003; Saraiva & Alcami, 2001).

TNF-binding activity at the surface of cells infected with VV Lister, USSR and Evans has been reported (Alcami *et al.*, 1999), and a more recent publication indicates that CrmE encodes this surface binding activity in VV USSR (Reading *et al.*, 2002). The mechanism by which CrmE from VV may be anchored at the cell surface has not been elucidated.

Deletion of the *crmE* gene from VV strain USSR caused a minor attenuation of the virus in a mouse intranasal model (Reading *et al.*, 2002). A clearer effect on virus pathogenesis was observed with increased virulence in a mouse intranasal model of a VV WR recombinant expressing CPV and VV USSR CrmE (Reading *et al.*, 2002). However, it is not clear why CrmE contributes to virus virulence in a mouse model since CrmE has been found not to inhibit the activity of murine TNF *in vitro*. The possibility that CrmE may bind other members of the TNFSF has been suggested, but this has not been detected so far (Reading *et al.*, 2002; Saraiva & Alcami, 2001).

Poxviral vTNFBPs

The yatapoxviruses have a restricted host range, infecting only primates including humans, and comprise Yaba-like disease virus (YLDV), Tanapox virus (TPV) and Yaba monkey tumour virus (YMTV). As the genome sequence of TPV is greater than 98 % identical to that of YLDV it should be considered a strain of YLDV.

TPV causes a mild, self-limiting disease in humans characterized by transient fever, one or more nodular skin lesions and swelling of regional lymph nodes. It had been found that supernatants from TPV-infected human endothelial cells contained a 38 kDa polypeptide that binds to TNF, as well as interleukin 2 (IL-2), IL-5 and IFN-γ. Moreover, these supernatants were able to inhibit TNF-induced expression of specific cell adhesion molecules (E-selectin, ICAM-1, VCAM-1) (Paulose *et al.*, 1998). However, sequencing of two members of the yatapoxviruses, YLDV and YMTV, did not reveal any obvious TNFR homologues (Brunetti *et al.*, 2003a; Lee *et al.*, 2001).

More recently, a vTNFBP from TPV was unequivocally identified and purified by affinity chromatography from TPV-infected cell supernatants (Brunetti *et al.*, 2003b) (Fig. 1). N-terminal sequencing of this 45 kDa protein identified it as the 2L protein product from TPV, YMTV and YLDV, which shows no sequence similarity to any known cellular TNFR. The product of the SPV003/148 gene from SPV and the product of ORF 008 from DPV are also related to yatapox 2L proteins, sharing approximately 31 % amino acid identity including six conserved cysteines that likely form three pairs of disulfide bonds (Afonso *et al.*, 2002, 2005).

The 2L protein is probably glycosylated and was shown to be expressed and secreted from YMTV (37 kDa)- and TPV (45 kDa)-infected cells. Using purified TPV 2L protein, it was shown that 2L binds to human TNF but not to IL-2, IL-5 or IFN-γ (Brunetti *et al.*, 2003b). Additionally, this protein does not bind to human LTα or other members of the human TNFSF, including FasL, APRIL, CD40L, TRAIL, EDA1, BAFF, TL1A, TWEAK, CD30L, 4-1BBL, LIGHT, GITRL, RANKL, OX40L and CD27L. The TPV 2L protein binds with high affinity human TNF (K_D 43 pM), which stems from a relatively fast association rate ($3 \cdot 38 \times 10^6 \, M^{-1} s^{-1}$) and an extremely slow dissociation ($1 \cdot 45 \times 10^{-4} \, M^{-1} s^{-1}$), giving an estimated half-life of the complex of about 6 h. Additionally, the TPV 2L protein was found to block binding of human TNF to TNFRI and TNFRII with an IC_{50} of 10–20 nM. The data indicated that the ratio of TPV 2L to TNF in the complex may be close to 1:1 stoichiometry and that TPV 2L is probably monomeric. Finally, it was shown that TPV 2L is able to inhibit human-TNF-induced cytotoxicity, but not that induced by murine or rabbit TNF.

vCD30

In addition to poxvirus homologues of TNFRII, a homologue of CD30 (TNFSF8) (vCD30) has been identified in CPV, EV and DPV (Afonso *et al.*, 2005; Panus *et al.*, 2002; Saraiva *et al.*, 2002) (Fig. 1). These viruses express a 12 kDa, non-N-glycosylated soluble protein with sequence similarity to mouse (60·7 % amino acid identity) and human (56·7 % amino acid identity) CD30. The vCD30 interacts with high affinity with CD30L, a TNFSF member expressed at the surface of immune cells, effectively antagonizing the ability of CD30L to signal via cell surface CD30. EV vCD30 efficiently down-regulates Th1-mediated inflammation in a murine model of antigen-induced granuloma (Saraiva *et al.*, 2002). A role for vCD30 in poxvirus pathogenesis has not yet been established; however, the ability of vCD30 to induce reverse signalling through CD30L is a new immunomodulatory mechanism that has not been described previously for a viral soluble cytokine receptor (Alcami, 2003). The conservation of this CD30 homologue in the poxviral genome suggests a previously undefined role for CD30/CD30L in mounting antiviral immune responses.

IRIDOVIRAL TNFRS

Putative TNFRs have been found in the genomes of iridoviruses, which are distantly related to the poxviruses. Iridoviruses are animal viruses that infect only invertebrates and poikilothermic vertebrates such as fish, amphibians and reptiles (Williams, 1996). They are complex, icosahedral enveloped viruses (120–300 nm diameter) with a circularly permuted dsDNA genome that replicate in the cytoplasm, and have a nuclear stage. Overall, iridoviruses have a major impact as pathogens of feral, cultured and ornamental fish.

The first iridovirus genome completely sequenced was that of lymphocystis disease virus 1 (LCDV-1) (Tidona & Darai, 1997) and it revealed the presence of a gene with similarity to poxviral and cellular TNFRs, named ORF167L (Fig. 1). The product of this gene contains only one copy of the CRD at its N-terminus plus an incomplete CRD containing only three of the six cysteine residues. Additionally, the protein contains a copy of the CUB domain at its C-terminus, a domain that has been described in functionally diverse, mostly developmentally regulated proteins. The complete protein encoded by ORF167L is 267 amino acids, has no signal peptide and contains a putative transmembrane region. Although the functionality of the protein has not been addressed yet, it was proposed to play a role in immune modulation or apoptotic events in the infected fish.

The sequence of ORF 167L was shown to be highly conserved among different viruses from the genus *Lymphocystisvirus*. Analysis of the complete genome sequence of Singapore grouper iridovirus, a related member of the *Ranavirus* genus, revealed the presence of three distinct ORFs whose products contain CRDs (Song *et al.*, 2004). While ORF 096R can be considered the orthologue of LCDV-1 ORF167L, the products of ORFs 050L and 051L contain only one or two CRDs, respectively, with no additional CUB domain. All of them are predicted to contain a signal peptide plus transmembrane domains. Moreover, the protein encoded by ORF050L has been detected in samples from purified virus particles. The functionality and role of these putative vTNFRs remains to be addressed.

CONCLUDING REMARKS

The expression of secreted proteins that bind TNF is a strategy acquired by many poxviruses to modulate the activity of this important mediator of the immune response (Alcami, 2003; Seet *et al.*, 2003). It is interesting that the origin of these TNF inhibitors may be different. The orthopoxviruses, including VV, VaV, CPV and EV, encode homologues of the extracellular binding domain of cellular TNFRs, and the genes for these are likely to be acquired from the host genome during virus–host co-evolution. By contrast, the yatapoxviruses encode a protein with amino acid sequence unrelated to TNFRs and likely to have different protein structure that still binds to the same host ligand. Maybe a strong evolutionary pressure to counteract the antiviral effects of TNF has forced poxviruses to acquire this strategy and the different virus species have addressed this in different ways. The reason for the variety of proteins may depend on the specific interaction of different poxviruses with their hosts, their replication strategy and their opportunities to incorporate specific host genes into their genomes.

The finding of four different TNFR homologues (CrmB, CrmC, CrmD and CrmE) that bind the same ligand(s) encoded in the genome of orthopoxviruses is remarkable and suggests that these proteins, although related, may have different functions that we do not fully understand. Some of these proteins are expressed at early or late times of infection and may bind LTα or not. The distribution of the genes encoding these proteins in orthopoxvirus genomes suggests that an ancestor orthopoxvirus, probably related to CPV, encoded all four genes and that specific viruses lost some of them as they specialized in a particular host species. For example, VaV and EV, which are highly species-specific, have retained only one of these genes, suggesting that VaV CrmB and EV CrmD may be more efficient in the human or mouse host, respectively. Interestingly, truncated or mutated genes encoding the other vTNFRs, which are unable to encode active proteins, can still be found in some of the orthopoxvirus species expressing a single active vTNFR, suggesting that the loss of some vTNFR activities has occurred during the recent evolutionary history of these viruses.

Comparative studies on the presence of the four vTNFRs in strains of orthopoxviruses have been performed. In the case of VV, the distribution and activity of vTNFRs varies widely with the virus strain (Alcami *et al.*, 1999). By contrast, genetic and functional analysis of the vTNFRs encoded by EV showed the expression of the same vTNFR in all EV isolates tested (Ribas *et al.*, 2003).

One of the major structural differences among these proteins is the presence of a C-terminal extension in CrmB and CrmD which is unrelated to host TNFRs. It has been shown that the N-terminal CRDs of MV M-T2 and EV CrmD are sufficient for TNF binding (Saraiva, 2002; Schreiber *et al.*, 1996), and thus it is likely that the C-terminal region binds other ligands of the immune system with an important role in antiviral defence.

Evidence accumulated over the years clearly demonstrates a critical role for TNF in the immune response, and this is underscored by the identification of many anti-TNF mechanisms encoded by viruses. However, the experimental evidence demonstrating a critical role of vTNFRs *in vivo* is limited. So far, a MV mutant in the M-T2 gene was found to be attenuated in a rabbit model (Upton *et al.*, 1991), but the evidence that the orthopoxvirus vTNFRs (CrmB, CrmC, CrmD and CrmE) contribute to virus virulence is limited to models that required high virus doses to cause disease: a mild attenuating phenotype in a VV USSR mutant lacking the *crmE* gene and an increased virulence of VV WR recombinants expressing CrmB, CrmC or CrmE (Reading *et al.*, 2002). Better models of poxvirus pathogenesis are needed to address the role of orthopoxvirus vTNFRs in virus pathogenesis.

Anti-TNF therapy, based on either monoclonal antibodies or soluble TNFRs, is available in the clinic as an efficient way of blocking adverse inflammatory responses. A better knowledge of the strategies used by viruses to modulate TNF activity during infection may uncover new anti-TNF strategies that could be applied to the clinic. The use of soluble versions of TNFRs by viruses confirms an adequate strategy used in the clinic. A better understanding of the function of the different vTNFRs *in vivo* may help us to improve even further the potency of soluble TNFRs in blocking adverse inflammatory responses in a number of human disease conditions.

The investigations described here on virus strategies to modulate the activity of TNF illustrate how a better understanding of the interaction of viruses with the host immune system may provide information relevant to virus pathogenesis and suggest new strategies of immune modulation.

ACKNOWLEDGEMENTS

The work in the author's laboratory is funded by grants from the Wellcome Trust, European Union and Spanish Ministry of Education and Science. M. B. Ruiz-Argüello and A. Alejo are recipients of Ramón y Cajal and Comunidad de Madrid Fellowships, respectively.

REFERENCES

Afonso, C. L., Tulman, E. R., Lu, Z., Zsak, L., Osorio, F. A., Balinsky, C., Kutish, G. F. & Rock, D. L. (2002). The genome of swinepox virus. *J Virol* **76**, 783–790.

Afonso, C. L., Delhon, G., Tulman, E. R., Lu, Z., Zsak, A., Becerra, V. M., Zsak, L., Kutish, G. F. & Rock, D. L. (2005). Genome of deerpox virus. *J Virol* **79**, 966–977.

Aggarwal, B. B. (2003). Signalling pathways of the TNF superfamily: a double-edged sword. *Nat Rev Immunol* **3**, 745–756.

Aggarwal, B. B., Eessalu, T. E. & Hass, P. E. (1985). Characterization of receptors for human tumour necrosis factor and their regulation by gamma-interferon. *Nature* **318**, 665–667.

Alcami, A. (2003). Viral mimicry of cytokines, chemokines and their receptors. *Nat Rev Immunol* **3**, 36–50.

Alcami, A. & Koszinowski, U. H. (2000). Viral mechanisms of immune evasion. *Immunol Today* **21**, 447–455.

Alcami, A., Khanna, A., Paul, N. L. & Smith, G. L. (1999). Vaccinia virus strains Lister, USSR and Evans express soluble and cell-surface tumour necrosis factor receptors. *J Gen Virol* **80**, 949–959.

Ashkenazi, A., Marsters, S. A., Capon, D. J., Chamow, S. M., Figari, I. S., Pennica, D., Goeddel, D. V., Palladino, M. A. & Smith, D. H. (1991). Protection against endotoxic shock by a tumor necrosis factor receptor immunoadhesin. *Proc Natl Acad Sci U S A* **88**, 10535–10539.

Banner, D. W., D'Arcy, A., Janes, W., Gentz, R., Schoenfeld, H. J., Broger, C., Loetscher, H. & Lesslauer, W. (1993). Crystal structure of the soluble human 55 kd TNF receptor-human TNF beta complex: implications for TNF receptor activation. *Cell* **73**, 431–445.

Benedict, C. A., Butrovich, K. D., Lurain, N. S., Corbeil, J., Rooney, I., Schneider, P., Tschopp, J. & Ware, C. F. (1999). Cutting edge: a novel viral TNF receptor superfamily member in virulent strains of human cytomegalovirus. *J Immunol* **162**, 6967–6970.

Benedict, C. A., Norris, P. S. & Ware, C. F. (2002). To kill or be killed: viral evasion of apoptosis. *Nat Immunol* **3**, 1013–1018.

Bertin, J., Armstrong, R. C., Ottilie, S. & 10 other authors (1997). Death effector domain-containing herpesvirus and poxvirus proteins inhibit both Fas- and TNFR1-induced apoptosis. *Proc Natl Acad Sci U S A* **94**, 1172–1176.

Beutler, B. (1992). *Tumor Necrosis Factor: the Molecules and their Emerging Role in Medicine*. New York: Raven Press.

Brunetti, C. R., Amano, H., Ueda, Y., Qin, J., Miyamura, T., Suzuki, T., Li, X., Barrett, J. W. & McFadden, G. (2003a). Complete genomic sequence and comparative analysis of the tumorigenic poxvirus Yaba monkey tumor virus. *J Virol* **77**, 13335–13347.

Brunetti, C. R., Paulose-Murphy, M., Singh, R., Qin, J., Barrett, J. W., Tardivel, A., Schneider, P., Essani, K. & McFadden, G. (2003b). A secreted high-affinity inhibitor of human TNF from Tanapox virus. *Proc Natl Acad Sci U S A* **100**, 4831–4836.

Burgert, H. G., Ruzsics, Z., Obermeier, S., Hilgendorf, A., Windheim, M. & Elsing, A. (2002). Subversion of host defense mechanisms by adenoviruses. *Curr Top Microbiol Immunol* **269**, 273–318.

Chan, F. K., Chun, H. J., Zheng, L., Siegel, R. M., Bui, K. L. & Lenardo, M. J. (2000). A domain in TNF receptors that mediates ligand-independent receptor assembly and signaling. *Science* **288**, 2351–2354.

Chen, P. C., DuBois, G. C. & Chen, M. J. (1995). Mapping the domain(s) critical for the binding of human tumor necrosis factor-alpha to its two receptors. *J Biol Chem* **270**, 2874–2878.

Declercq, W., Vandenabeele, P. & Fiers, W. (1995). Dimerization of chimeric erythro-poietin/75 kDa tumour necrosis factor (TNF) receptors transduces TNF signals: necessity for the 75 kDa-TNF receptor transmembrane domain. *Cytokine* **7**, 701–709.

Eck, M. J. & Sprang, S. R. (1989). The structure of tumor necrosis factor-alpha at 2·6 Å resolution. Implications for receptor binding. *J Biol Chem* **264**, 17595–17605.

Farrell, P. J. (1998). Signal transduction from the Epstein-Barr virus LMP-1 transforming protein. *Trends Microbiol* **6**, 175–177; discussion 177–178.

Gubser, C. & Smith, G. L. (2002). The sequence of camelpox virus shows it is most closely related to variola virus, the cause of smallpox. *J Gen Virol* **83**, 855–872.

Hu, F., Smith, C. A. & Pickup, D. J. (1994). Cowpox virus contains two copies of an early gene encoding a soluble secreted form of the type II TNF receptor. *Virology* **204**, 343–356.

Kohno, T., Brewer, M. T., Baker, S. L., Schwartz, P. E., King, M. W., Hale, K. K., Squires, C. H., Thompson, R. C. & Vannice, J. L. (1990). A second tumor necrosis factor receptor gene product can shed a naturally occurring tumor necrosis factor inhibitor. *Proc Natl Acad Sci U S A* **87**, 8331–8335.

Kwon, B., Kim, B. S., Cho, H. R., Park, J. E. & Kwon, B. S. (2003). Involvement of tumor necrosis factor receptor superfamily(TNFRSF) members in the pathogenesis of inflammatory diseases. *Exp Mol Med* **35**, 8–16.

Lee, H. J., Essani, K. & Smith, G. L. (2001). The genome sequence of Yaba-like disease virus, a yatapoxvirus. *Virology* **281**, 170–192.

Lesslauer, W., Tabuchi, H., Gentz, R. & 7 other authors (1991). Recombinant soluble tumor necrosis factor receptor proteins protect mice from lipopolysaccharide-induced lethality. *Eur J Immunol* **21**, 2883–2886.

Locksley, R. M., Killeen, N. & Lenardo, M. J. (2001). The TNF and TNF receptor superfamilies: integrating mammalian biology. *Cell* **104**, 487–501.

Loetscher, H., Pan, Y. C., Lahm, H. W., Gentz, R., Brockhaus, M., Tabuchi, H. & Lesslauer, W. (1990). Molecular cloning and expression of the human 55 kd tumor necrosis factor receptor. *Cell* **61**, 351–359.

Loparev, V. N., Parsons, J. M., Knight, J. C., Panus, J. F., Ray, C. A., Buller, R. M., Pickup, D. J. & Esposito, J. J. (1998). A third distinct tumor necrosis factor receptor of orthopoxviruses. *Proc Natl Acad Sci U S A* **95**, 3786–3791.

Loparev, V. N., Massung, R. F., Esposito, J. J. & Meyer, H. (2001). Detection and differentiation of old world orthopoxviruses: restriction fragment length polymorphism of the *crmB* gene region. *J Clin Microbiol* **39**, 94–100.

Lurain, N. S., Kapell, K. S., Huang, D. D., Short, J. A., Paintsil, J., Winkfield, E., Benedict, C. A., Ware, C. F. & Bremer, J. W. (1999). Human cytomegalovirus UL144 open reading frame: sequence hypervariability in low-passage clinical isolates. *J Virol* **73**, 10040–10050.

Massung, R. F., Esposito, J. J., Liu, L. I. & 12 other authors (1993). Potential virulence determinants in terminal regions of variola smallpox virus genome. *Nature* **366**, 748–751.

McNees, A. L. & Gooding, L. R. (2002). Adenoviral inhibitors of apoptotic cell death. *Virus Res* **88**, 87–101.

Michel, J., Langstein, J., Hofstadter, F. & Schwarz, H. (1998). A soluble form of CD137 (ILA/4-1BB), a member of the TNF receptor family, is released by activated lymphocytes and is detectable in sera of patients with rheumatoid arthritis. *Eur J Immunol* **28**, 290–295.

Moss, B. (2001). *Poxviridae*: the viruses and their replication. In *Fields Virology*, pp. 2849–2883. Edited by D. M. Knipe & P. M. Howley. Philadelphia: Lippincott Williams & Wilkins.

Naismith, J. H., Devine, T. Q., Brandhuber, B. J. & Sprang, S. R. (1995). Crystallographic evidence for dimerization of unliganded tumor necrosis factor receptor. *J Biol Chem* **270**, 13303–13307.

Nophar, Y., Kemper, O., Brakebusch, C., Englemann, H., Zwang, R., Aderka, D., Holtmann, H. & Wallach, D. (1990). Soluble forms of tumor necrosis factor receptors (TNF-Rs). The cDNA for the type I TNF-R, cloned using amino acid sequence data of its soluble form, encodes both the cell surface and a soluble form of the receptor. *EMBO J* **9**, 3269–3278.

Panus, J. F., Smith, C. A., Ray, C. A., Smith, T. D., Patel, D. D. & Pickup, D. J. (2002). Cowpox virus encodes a fifth member of the tumor necrosis factor receptor family: a soluble, secreted CD30 homologue. *Proc Natl Acad Sci U S A* **99**, 8348–8353.

Paulose, M., Bennett, B. L., Manning, A. M. & Essani, K. (1998). Selective inhibition of TNF-alpha induced cell adhesion molecule gene expression by tanapox virus. *Microb Pathog* **25**, 33–41.

Peppel, K., Crawford, D. & Beutler, B. (1991). A tumor necrosis factor (TNF) receptor-IgG heavy chain chimeric protein as a bivalent antagonist of TNF activity. *J Exp Med* **174**, 1483–1489.

Pfeffer, K. (2003). Biological functions of tumor necrosis factor cytokines and their receptors. *Cytokine Growth Factor Rev* **14**, 185–191.

Reading, P. C., Khanna, A. & Smith, G. L. (2002). Vaccinia virus CrmE encodes a soluble and cell surface tumor necrosis factor receptor that contributes to virus virulence. *Virology* **292**, 285–298.

Ribas, G., Rivera, J., Saraiva, M., Campbell, R. D. & Alcami, A. (2003). Genetic variability of immunomodulatory genes in ectromelia virus isolates detected by denaturing high-performance liquid chromatography. *J Virol* **77**, 10139–10146.

Rubins, K. H., Hensley, L. E., Jahrling, P. B. & 7 other authors (2004). The host response to smallpox: analysis of the gene expression program in peripheral blood cells in a nonhuman primate model. *Proc Natl Acad Sci U S A* **101**, 15190–15195.

Saraiva, M. (2002). *Immunomodulation by poxviruses: TNF receptor homologues.* PhD thesis, Cambridge.

Saraiva, M. & Alcami, A. (2001). CrmE, a novel soluble tumor necrosis factor receptor encoded by poxviruses. *J Virol* **75**, 226–233.

Saraiva, M., Smith, P., Fallon, P. G. & Alcami, A. (2002). Inhibition of type 1 cytokine-mediated inflammation by a soluble CD30 homologue encoded by ectromelia (mousepox) virus. *J Exp Med* **196**, 829–839.

Schall, T. J., Lewis, M., Koller, K. J. & 9 other authors (1990). Molecular cloning and expression of a receptor for human tumor necrosis factor. *Cell* **61**, 361–370.

Schreiber, M., Rajarathnam, K. & McFadden, G. (1996). Myxoma virus T2 protein, a tumor necrosis factor (TNF) receptor homolog, is secreted as a monomer and dimer that each bind rabbit TNFα, but the dimer is a more potent TNF inhibitor. *J Biol Chem* **271**, 13333–13341.

Schreiber, M., Sedger, L. & McFadden, G. (1997). Distinct domains of M-T2, the myxoma virus TNF receptor homolog, mediate extracellular TNF binding and intracellular apoptosis inhibition. *J Virol* **71**, 2171–2181.

Seet, B. T., Johnston, J. B., Brunetti, C. R. & 7 other authors (2003). Poxviruses and immune evasion. *Annu Rev Immunol* **21**, 377–423.

Shchelkunov, S. N., Safronov, P. F., Totmenin, A. V., Petrov, N. A., Ryazankina, O. I., Gutorov, V. V. & Kotwal, G. J. (1998). The genomic sequence analysis of the left and right species-specific terminal region of a cowpox virus strain reveals unique sequences and a cluster of intact ORFs for immunomodulatory and host range proteins. *Virology* **243**, 432–460.

Shchelkunov, S. N., Totmenin, A. V., Safronov, P. F. & 12 other authors (2002). Analysis of the monkeypox virus genome. *Virology* **297**, 172–194.

Simonet, W. S., Lacey, D. L., Dunstan, C. R. & 28 other authors (1997). Osteoprotegerin: a novel secreted protein involved in the regulation of bone density. *Cell* **89**, 309–319.

Singh, I., Pahan, K., Khan, M. & Singh, A. K. (1998). Cytokine-mediated induction of ceramide production is redox-sensitive. Implications to proinflammatory cytokine-mediated apoptosis in demyelinating diseases. *J Biol Chem* **273**, 20354–20362.

Smith, C. A., Davis, T., Anderson, D., Solam, L., Beckmann, M. P., Jerzy, R., Dower, S. K., Cosman, D. & Goodwin, R. G. (1990). A receptor for tumor necrosis factor defines an unusual family of cellular and viral proteins. *Science* **248**, 1019–1023.

Smith, C. A., Davis, T., Wignall, J. M., Din, W. S., Farrah, T., Upton, C., McFadden, G. & Goodwin, R. G. (1991). T2 open reading frame from Shope fibroma virus encodes a soluble form of the TNF receptor. *Biochem Biophys Res Commun* **176**, 335–342.

Smith, C. A., Hu, F. Q., Smith, T. D., Richards, C. L., Smolak, P., Goodwin, R. G. & Pickup, D. J. (1996). Cowpox virus genome encodes a second soluble homologue of cellular TNF receptors, distinct from CrmB, that binds TNF but not LT alpha. *Virology* **223**, 132–147.

Song, W. J., Qin, Q. W., Qiu, J., Huang, C. H., Wang, F. & Hew, C. L. (2004). Functional genomics analysis of Singapore grouper iridovirus: complete sequence determination and proteomic analysis. *J Virol* **78**, 12576–12590.

Thome, M. & Tschopp, J. (2001). Regulation of lymphocyte proliferation and death by FLIP. *Nat Rev Immunol* **1**, 50–58.

Tidona, C. A. & Darai, G. (1997). The complete DNA sequence of lymphocystis disease virus. *Virology* **230**, 207–216.

Tortorella, D., Gewurz, B. E., Furman, M. H., Schust, D. J. & Ploegh, H. L. (2000). Viral subversion of the immune system. *Annu Rev Immunol* **18**, 861–926.

Upton, C., DeLange, A. M. & McFadden, G. (1987). Tumorigenic poxviruses: genomic organization and DNA sequence of the telomeric region of the Shope fibroma virus genome. *Virology* **160**, 20–30.

Upton, C., Macen, J. L., Schreiber, M. & McFadden, G. (1991). Myxoma virus expresses a secreted protein with homology to the tumor necrosis factor receptor gene family that contributes to viral virulence. *Virology* **184**, 370–382.

Wallach, D., Bigda, J. & Engelmann, H. (1999a). The tumor necrosis factor (TNF) family and related molecules. In *The Cytokine Network and Immune Functions*, pp. 51–70. Edited by J. Theze. New York: Oxford University Press.

Wallach, D., Varfolomeev, E. E., Malinin, N. L., Goltsev, Y. V., Kovalenko, A. V. & Boldin, M. P. (1999b). Tumor necrosis factor receptor and Fas signaling mechanisms. *Annu Rev Immunol* **17**, 331–367.

Williams, T. (1996). The iridoviruses. *Adv Virus Res* **46**, 345–412.

Wold, W. S., Doronin, K., Toth, K., Kuppuswamy, M., Lichtenstein, D. L. & Tollefson, A. E. (1999). Immune responses to adenoviruses: viral evasion mechanisms and their implications for the clinic. *Curr Opin Immunol* **11**, 380–386.

Evasion of host defence systems by African swine fever virus

Linda K. Dixon

Institute for Animal Health, Pirbright Laboratory, Ash Road, Pirbright, Woking GU24 0NF, UK

INTRODUCTION

African swine fever virus (ASFV) causes an acute haemorrhagic fever with high mortality rates in domestic pigs and asymptomatic, persistent infections in its natural hosts, warthogs, bushpigs and soft ticks of the species *Ornithodoros*. ASFV is the only known DNA arbovirus and the only DNA virus known to cause a haemorrhagic fever. The disease causes major economic losses in affected countries and represents a threat to pig farming worldwide.

ASFV is a large, icosahedral, cytoplasmic DNA virus which shares a similar genome structure and replication strategy to poxviruses. However, the structure of ASF virions differs from that of the poxviruses and ASFV is classified as the only member of a new virus family, the *Asfarviridae*. The virus genome encodes about 150 proteins, including enzymes and factors required for replication and transcription of the virus genome in addition to many other proteins which are not essential for replication in cells, but have a role in facilitating virus survival and transmission in its hosts. Amongst these are proteins which interfere with host defences to facilitate virus persistence. Persistent infections occur in the natural hosts of the virus and in pigs which recover from infection with less virulent isolates.

The virus replicates in cells of the mononuclear phagocytic system and in reticulo-endothelial cells in lymphoid tissues and organs of domestic swine. Widespread cell death caused by apoptosis occurs in both T and B lymphocytes in lymphoid tissues and endothelial cells in arterioles and capillaries. This accounts for the lesions seen in acute

SGM symposium 64: Molecular pathogenesis of virus infections.
Editors P. Digard, A. A. Nash & R. E. Randall. Cambridge University Press. ISBN 0 521 83248 9 ©SGM 2005

disease. Disseminated intravascular coagulation (DIC) develops during acute infections and this may lead to the characteristic haemorrhagic syndrome. Understanding how the function of virus-infected macrophages is modulated is critical for understanding both the mechanisms of virus pathogenesis as well as virus immune evasion mechanisms. This chapter will review current knowledge and focus on recent developments which shed light on the function of virus genes involved in manipulating host cell function.

EPIDEMIOLOGY

ASFV infection has been well established over many years in East and southern Africa in a sylvatic cycle involving transmission between warthogs and the soft tick vector of the virus, *Ornithodoros moubata*. The virus causes inapparent infections in these hosts and in bushpigs, which can remain persistently infected over several years. African swine fever (ASF) was first reported early in the 1900s, when domestic pigs came into contact with infected warthogs. Since then, ASF has been reported in most sub-Saharan African countries. In 1957, the first transcontinental spread of the virus occurred and ASF remained endemic in Spain and Portugal from 1960 until the 1990s. During the 1970s and 1980s, ASF spread to South America and the Caribbean, but outside Africa, ASF is now endemic only in Sardinia. Since the mid-1990s, ASF has caused severe losses in southern Mozambique, Madagascar (1998), in several countries in West Africa [e.g. Ivory Coast (1996), Benin (1997), Togo and Nigeria (1997–1998), Ghana (1999), Burkina Faso (2003)] and it has re-emerged in Kenya after 30 years of absence. In Zambia there have been several outbreaks of ASFV outside the previous endemic area. This reflects the growing pig population in Africa, which provides a cheap source of good quality protein.

Virus can be transmitted from wildlife to pigs by bites from infected ticks. In many African countries, ASF has become established as an enzootic disease in domestic pigs and is maintained in the absence of contact with warthogs. Within pig populations, virus can spread by direct contact between pigs and by bites from infected ticks, which infest pig houses in some areas. Although the original descriptions of ASF were of an acute haemorrhagic fever causing mortality approaching 100 %, less virulent isolates have emerged as disease has circulated in domestic pigs. Pigs which recover from infection can remain persistently infected and excrete virus over long periods thus providing a reservoir for infecting healthy pigs. No vaccine is available so control relies on rapid diagnosis and implementation of quarantine and slaughter policies. The infrastructure to implement these policies is lacking in many countries and ASF continues to pose a threat to pig farming worldwide.

PATHOGENESIS

In pigs infected with virulent and moderately virulent ASFV isolates, onset of viraemia is observed from 3 days post-infection and can rise to a peak of over 10^8 HAD (haemadsorption units)$_{50}$ ml^{-1}. Virulent isolates cause mortality approaching 100 % in pigs, usually between 8 and 12 days post-infection. A proportion of pigs infected with moderately virulent isolates recover from infection and in these recovered pigs virus can persist for periods of up to 6 months (Wilkinson, 1984; Wilkinson *et al.*, 1981). Virus DNA has been detected in peripheral blood leukocytes up to 500 days post-infection (Carrillo *et al.*, 1994). Significantly fewer infected macrophages were present in lymphoid tissues in pigs infected with moderately virulent isolates compared to pigs infected with virulent isolates, although the tissue tropism of the viruses was similar. This suggests that the level of virus replication is a significant factor in the pathogenesis (Oura *et al.*, 1998). In pigs infected with low virulence isolates, only sporadic low viraemia is observed although moderate levels of virus replication are detected in lymphoid tissues. Most pigs infected with these low virulence isolates show no disease signs but some pigs develop a chronic form of disease (Leitao *et al.*, 2001).

In the early stages of infection, virus replication has been observed primarily in macrophages. Only at later stages of disease has infection been reported in a variety of cell types, including endothelial cells, megakaryocytes, platelets, neutrophils and hepatocytes (Carrasco *et al.*, 1992; Fernandez *et al.*, 1992a, b, c; Gomez-Villamandos *et al.*, 1995b, c, 1996; Sierra *et al.*, 1991). This has led to the suggestion that virus infection of macrophages is the primary event leading to the haemorrhagic pathology (del Moral *et al.*, 1999; Gomez-Villamandos *et al.*, 1995a; Oura *et al.*, 1998).

In common with other viral haemorrhagic fevers, ASF is characterized by damage to vascular endothelial cells and this contributes to vascular permeability. In the early stages of disease, this is thought to be caused by factors released from virus-infected macrophages, whereas later in disease direct infection of endothelial cells may occur. *In vitro* infection of endothelial cells by ASFV leads to induction of apoptosis, which could contribute to the destruction of vascular endothelial cells in infected pigs (Vallee *et al.*, 2001). In the later stages of disease, DIC develops as demonstrated by the appearance of fibrin degradation products and the presence of numerous fibrin thrombi in blood (Villeda *et al.*, 1993a, b, 1995). In non-pathogenic conditions, the endothelium maintains a barrier between tissues and blood and it contributes to the control of haemostatic balance by providing a non-coagulant surface. Activation of endothelium by cytokines or pathogenic agents induces a proinflammatory, coagulant surface and is thought to be involved in the development of DIC. Elevated levels of TNF-α and IL-1β are observed in serum from pigs experimentally infected with ASFV (del Moral *et al.*, 1999; Salguero *et al.*, 2002). TNF-α induces vasodilation, an increase in vascular permeability and activation of the vascular endothelium. All of these may tip the balance in favour of pro-coagulation activities favouring the generation of microthrombi. In addition, TNF-α can lead to induction of apoptosis and this may contribute to the destruction of endothelial cells (del Moral *et al.*, 1999).

A striking feature of ASFV infection is the massive apoptosis of lymphocytes which occurs in infiltrate lymphocytes and lymphoid structure in pigs infected with virulent isolates (Carrasco *et al.*, 1996; Gomez-Villamandos *et al.*, 1995a; Oura *et al.*, 1998; Salguero *et al.*, 2004). Neither T nor B lymphocytes are directly infected by the virus. The proximity of apoptotic lymphocytes to virus-infected macrophages and the correlation between the number of virus-infected macrophages and extent of lymphocyte apoptosis suggest that factors produced by infected macrophages are responsible for the induction of lymphocyte apoptosis (Oura *et al.*, 1998).

The observation that chronic infections with less virulent ASFV isolates can lead to hypergammaglobulinaemia led to suggestions that ASFV may be a polyclonal B-cell stimulator (Wardley, 1982). *In vitro* studies showed that this required the presence of ASFV-infected macrophages and could not be induced by supernatants from macrophages. Although virus particles alone did not stimulate proliferation of B cells *in vitro*, they were co-stimulatory with CD154 and IL-4 (Takamatsu *et al.*, 1999). These observations suggested a model whereby B cells are activated as a result of virus infection of macrophages. Due to the dramatic depletion of T cells that occurs by apoptosis induced early in infection, activated B cells are not provided with survival signals from T cells, such as CD154 interaction, and enter apoptosis. The consequently diminished anti-ASFV antibody response would favour virus replication (Takamatsu *et al.*, 1999).

The factors involved in induction of bystander lymphocyte apoptosis have not been fully characterized. However, a role for TNF-α is suggested since apoptosis induced in uninfected lymphocyte cultures by supernatants from ASFV-infected macrophages could be partially abrogated by anti-TNF-α antibodies (del Moral *et al.*, 1999).

In contrast to their stimulatory effect on B cells, ASFV-infected macrophage cultures inhibit the ability of T cells in peripheral blood leukocyte cultures to proliferate in response to mitogens. Similar inhibition of proliferative ability is observed in lymphocytes from infected pigs (Childerstone *et al.*, 1998). A role for the virus-encoded CD2-like protein in this process was suggested *in vitro* from experiments using virus mutants lacking the gene encoding this protein (Borca *et al.*, 1998). A protein released from ASFV-infected macrophages has been demonstrated to have both B-cell mitogenic and immunosuppressive properties in mice. The effects of this 36 kDa protein were shown to be mediated via induction of IL-4 and to a lesser extent IL-10, suggesting that it stimulates a type 2 cytokine response (Arala-Chaves *et al.*, 1988; Ribeiro *et al.*, 1991;

Vilanova *et al.*, 1999). The protein has not been further characterized and may be either a host protein induced by virus infection or a virus-encoded protein.

VIRUS REPLICATION CYCLE

Virus entry

ASFV enters cells by receptor-mediated endocytosis (Alcami *et al.*, 1989a, b). The host receptor(s) required for virus entry have not been defined, although cell surface markers expressed from intermediate stages of monocyte–macrophage differentiation are indicators of cell susceptibility to infection (McCullough *et al.*, 1999). Expression of the scavenger receptor CD163 correlates with permissiveness to infection and recombinant CD163 has been shown to bind to virus particles and to inhibit virus infection, suggesting that it may act as a virus receptor (Saanchez-Torres *et al.*, 2003). Several ASFV proteins have been shown to have roles in the binding and entry of virus into cells. Antibodies against p12, p54 and p72 as well as these recombinant proteins inhibit virus binding to cells and those against p30 inhibit virus internalization (Alcami *et al.*, 1992; Borca *et al.*, 1994a; Gomez-Puertas *et al.*, 1998). Virus entry requires a fusion event between the viral envelope and the limiting membrane of the endosome at low pH (Valdeira *et al.*, 1998). Following entry, virus cores are transported to perinuclear assembly sites using the microtubule network (Alonso *et al.*, 2001; Valdeira *et al.*, 1998).

Replication and transcription

Virus gene expression is independent of host cell RNA polymerase and early gene expression begins in the cytoplasm immediately following virus entry, using virus encoded enzymes and factors packaged in virions. Early genes encode proteins which include enzymes and factors needed for later in the replication cycle. Following the onset of DNA replication in the cytoplasm, a shift in the pattern of gene expression occurs. Late genes encode proteins including virion structural proteins and enzymes and factors such as RNA polymerase and early transcription factors that are packaged into virus particles for use during the next round of infection. Transcription of some early genes continues throughout the replication cycle.

At early stages of infection, replication of subgenomic length DNA fragments is detected in the cell nucleus (Garcia-Beato *et al.*, 1992; Rojo *et al.*, 1999). Although the function of this early nuclear phase of DNA replication is unknown it has been suggested that a nuclear primase may be necessary to initiate replication (Rojo *et al.*, 1999). Replication of full-length genomes occurs in the cytoplasmic factory areas via head-to-head concatemers which are resolved to unit length genomes containing terminal cross-links and inverted terminal repeats (Gonzalez *et al.*, 1986). The mechanism of DNA replication and transcription is similar to that of poxviruses, although there are some differences.

Assembly

The virus particle is complex, multi-layered and contains more than 50 proteins, including those with a structural role as well as enzymes packaged in the virus core for use early in infection.

Virus morphogenesis takes place in perinuclear virus factories that are located adjacent to the microtubule organizing centre. Virus factories resemble aggresomes since they are surrounded by a vimentin cage and increased number of mitochondria (Heath *et al.*, 2001; Rojo *et al.*, 1998). Aggresomes are formed in response to cell stress and function to remove misfolded proteins. Membranes derived from the endoplasmic reticulum are incorporated as internal lipid membranes in virus particles (Andres *et al.*, 1998; Rouiller *et al.*, 1998) and the icosahedral capsid assembles on these membranes (Cobbold & Wileman, 1998). Expression and processing of the p220 polyprotein is required for packaging of the nucleoprotein core and when its expression is suppressed empty virus particles accumulate in factories and can be observed budding through the plasma membrane. Extracellular virus has an additional loose-fitting external lipid envelope probably derived by budding through the plasma membrane (Andres *et al.*, 2002a, b).

Virus is transported to and from sites of assembly on microtubules. The p54 virion protein binds to the LC8 component of the dynein motor complex and may be involved in transport of virus particles to the factory sites (Alonso *et al.*, 2001). Expression of p54 is required to recruit virus membranes into the factories and the interaction of p54 with LC8 dynein is likely to be involved in both this process and in maintaining immature virions in the factory (Rodriguez *et al.*, 2004). The EP120R virion protein is required for virus transport from assembly sites to the plasma membrane (Andres *et al.*, 2001b). The conventional kinesin motor is required for this process (Jouvenet *et al.*, 2004).

VIRUS-ENCODED PROTEINS

The ASFV genome varies between 170 and 190 kbp depending on the virus isolate. The complete genome of a tissue culture adapted isolate from Spain, BA71V, has been determined and about two-thirds of the genome of a field isolate from Malawi (Dixon *et al.*, 1994; Yanez *et al.*, 1995; Yozawa *et al.*, 1994). About 150 major open reading frames are encoded by the BA71V isolate. Variation in genome length between different virus isolates occurs mainly in regions close to the genome termini due to gain or loss of members of five different multigene families encoded by the virus (de la Vega *et al.*, 1990; Yozawa *et al.*, 1994).

Virus-encoded enzymes and structural proteins

It is estimated that about 90 genes encoded by poxviruses are essential for virus replication in cells (Upton *et al.*, 2003). The similarity in replication strategy and virus particle complexity shared by ASFV and poxviruses suggests that ASFV may require a similar number gene complement for replication. The remaining genes are likely to encode proteins with roles in virus/host interactions that are important for survival and transmission in its hosts.

Since ASFV replicates in the cytoplasm, many encoded proteins function in DNA replication and mRNA transcription. To date, 26 ASFV genes have been identified with predicted roles in these processes. The presence of genes encoding components of a DNA repair system may represent an adaptation of the virus to replicate in the oxidizing environment of macrophages, which produce superoxides as part of their antimicrobial defences.

Two enzymes involved in post-translational protein modification (a ubiquitin conjugating enzyme and a serine/threonine protein kinase) are encoded (Baylis *et al.*, 1993; Hingamp *et al.*, 1992). By either phosphorylation, to alter protein function, or ubiquitination, to direct proteins for degradation by the proteasome, these enzymes might regulate the virus replication cycle or modulate host cell function. The virus-encoded *trans*-prenyltransferase is involved in synthesis of isoprenoid compounds and is targeted to membranes in virus factories, suggesting that it may have a role in virus assembly (Alejo *et al.*, 1997, 1999). The virus encodes two proteins involved in redox metabolism, NifS and ERV1 homologues. The ERV1 homologue is required for efficient virion morphogenesis (Lewis *et al.*, 2000).

Virion structural proteins characterized include p72, p30, p12, p17, p22, p54 or j13L, p49, j18L, j5R and EP402R. The products of a 220 kDa protein which is cleaved to give four structural proteins, p150, p37, p14 and p34, and the products of a 62 kDa protein which is cleaved to give two structural proteins, p35 and p15, are also present in virions (Simon-Mateo *et al.*, 1993, 1997). A virus-encoded protease related to the SUMO-1-specific protease family is involved in cleavage of these polyproteins (Andres *et al.*, 2001a). Two DNA-binding proteins, p10 and p14.5 (E120R), are present in virions (Borca *et al.*, 1996; Martinez-Pomares *et al.*, 1997). One virus-encoded protein (B602L) acts as a chaperone for the major capsid protein p72 but is not incorporated into the virus particle (Cobbold *et al.*, 2001).

Multigene families

Five different multigene families, MGF 110, MGF 360, MGF 530/505, MGF 300 and MGF 100, are located in genome regions close to the termini (Almendral *et al.*, 1990;

Gonzalez *et al.*, 1990; Rodriguez *et al.*, 1994; Yanez & Vinuela, 1993; Yozawa *et al.*, 1994). Large length variations between genomes of different isolates occur due to gain or loss of members of these multigene families. Together these represent about 20 % of the coding capacity of the genome and although they are likely to have important roles in virus infection, little is known about their function. Within individual multigene families, conserved domains are present; these are likely to be functionally important. Proteins encoded by some members of MGF 530 and 360 contain a single ankyrin repeat close to the N-terminus, but otherwise the genes share no significant homology with known genes. By comparing infection with wild-type virus with virus deletion mutants from which several copies of MGF 530 and 360 have been deleted, these families have been implicated as macrophage host range determinants and as virulence factors in domestic swine (Zsak *et al.*, 2001). More recently, a role for these proteins has been suggested in inhibition of type I interferon gene transcription and possibly also in inhibiting type I interferon-activated pathways (Afonso *et al.*, 2004). It is not clear whether individual genes in these MGFs have different functions or if multiple genes are required to produce a gene dosage effect. Using deletion mutants, it has been shown that deletion of three MGF 360 members (3HL, IL, LL) can reduce the ability of virus to replicate in *O. moubata* ticks by 100–1000-fold (Burrage *et al.*, 2004). Proteins encoded by members of MGF family 110 contain N-terminal signal sequences and conserved Cys-rich regions. Two of these proteins contain an endoplasmic reticulum retention sequence. These proteins were localized to pre-Golgi compartments. In the absence of other ASFV genes, expression of the XP124L protein had a dramatic effect on the contents of the endoplasmic reticulum that was dependent on the C-terminal sequence KEDL. The data suggest a role for these proteins in preparing the endoplasmic reticulum for its role in virus morphogenesis (Netherton *et al.*, 2004). Other members of the MGF 110 family may be secreted from infected cells but this has not been confirmed.

Interestingly, low-virulence Portuguese ASFV isolates lack multiple copies of MGF 530 and 360, suggesting that these genes may be determinants of virulence in field isolates (Boinas *et al.*, 2004; Duarte, 2000; Leitao *et al.*, 2001).

ASFV proteins involved in modulating host cell function
Proteins which modulate apoptosis. Induction of apoptosis is a host defence response to limit virus replication by inducing cell death before production of progeny virions. Many viruses encode proteins which inhibit apoptosis and prolong cell survival to enable the virus to replicate. The induction of apoptosis in infected cells represents a critical balance between pro- and anti-apoptotic effectors encoded by the host and by the virus. Cell- or virus-specific factors which alter this balance may limit or enhance virus replication and thus play a role in virus pathogenesis. There are many mechanisms

by which host cells sense virus infection and trigger apoptosis. During ASFV infection, apoptosis is triggered in the absence of protein synthesis or virus DNA replication, and this occurs at an early stage in the infection process, possibly during virus uncoating (Carrascosa *et al.*, 2002). The ASFV structural protein p54 binds to the LC8 chain of the microtubule motor protein dynein via a motif similar to that by which the pro-apoptotic Bcl-2 family member binds to LC8. Interaction of p54 with LC8 may displace Bim from dynein, resulting in its translocation to mitochondria and activation of apoptosis. Evidence for this comes from transient expression experiments which showed that p54 expression resulted in effector caspase-3 activation and apoptosis. Interestingly, p54 mutants lacking the 13 aa dynein-binding motif lose the ability to activate caspases and apoptosis, supporting the hypothesis that apoptosis induced by p54 results from its interaction with LC8 (Hernaez *et al.*, 2004). Protein p54 may induce cell death during virus infection either at an early stage when virus enters cells or at late times in infection during virus assembly. ASFV infection induces p53 gene transcription and activates p53-dependent transcription. Amongst the genes activated are those encoding pro-apoptotic proteins such as Bax and p21, which induces cell cycle arrest. These effects may combine with other cellular responses to push cells into apoptosis. Interestingly, mdm2 expression is also induced. Although this normally functions to inactivate p53 in the nucleus, in ASFV-infected cells mdm2 is retained in virus factories (Granja *et al.*, 2004).

Although virus infection triggers the apoptotic pathway early during the replication process, extensive apoptosis in infected cells is observed only at later times during infection. The probable reason is that ASFV encodes three proteins with demonstrated roles in inhibiting apoptosis and these presumably help to delay the onset of apoptosis. Two of these proteins are similar to known apoptosis regulators. The A224L gene encodes a protein with similarity to the IAP family of apoptosis inhibitors. Expression of the A224L protein was shown to substantially inhibit caspase activity and cell death. The A224L protein is expressed late during virus infection and is packaged into virions, thus it may also have a role early during infection following virus entry. In cells infected with an ASFV deletion mutant lacking the A224L gene, caspase-3 activation was substantially increased. The A224L protein was shown to interact with the proteolytic fragment of caspase-3 and inhibits the activity of this protease during ASFV infection (Chacon *et al.*, 1995; Nogal *et al.*, 2001). Expression of the A224L protein was also shown to increase activation of the NF-κB transcription factor, which includes amongst its gene targets the anti-apoptotic cellular IAP genes (Rodriguez *et al.*, 2002). Cellular IAP proteins have been shown to interact with the cytoplasmic tail of the TNF-α receptor to induce NF-κB activation. However, since ASFV encodes at least one protein, A238L, which inhibits NF-κB activation at a stage further down the signal transduction pathway, the importance of this aspect of A224L function in virus infection is not clear.

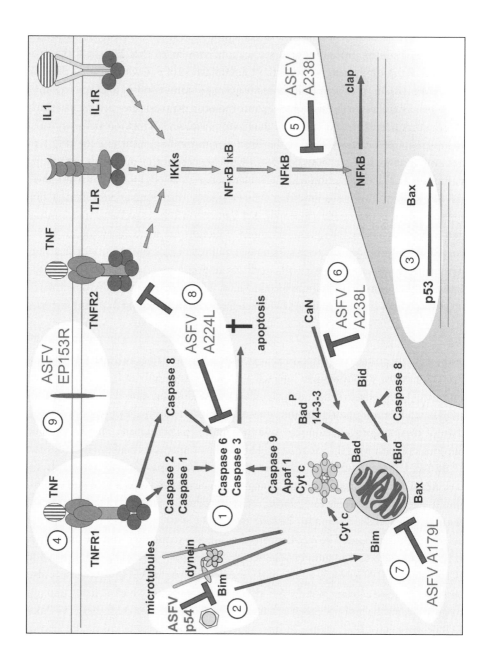

The ASFV Bcl-2 homologue, gene A179L, was shown to protect the human myeloid cell line K562 from apoptosis induced by cycloheximide and actinomycin D or by treatment with cytosine arabinoside. A179L contains a single BH1 domain and mutations in this domain abrogated the anti-apoptotic effect of A179L. The A179L protein is expressed both early and late during virus infection and is thought to be required for virus infection (Afonso *et al.*, 1996; Neilan *et al.*, 1993; Revilla *et al.*, 1997). Activity of the A179L protein was shown to be cell anchorage-dependent, suggesting that it may play different roles in virus-infected circulating macrophages compared to adherent macrophages (Brun *et al.*, 1998). A third ASFV protein, EP153R, has been shown to inhibit apoptosis (Hurtado *et al.*, 2004). This protein is a C-type lectin that is non-essential for virus infectivity and enhances haemadsorption of red blood cells to non-infected cells (Galindo *et al.*, 2000). In cells infected with a virus deletion mutant lacking the EP153R gene, increased levels of caspase-3 activity and of cell death were observed compared to cells infected with wild-type virus. Expression of the EP153R gene resulted in a partial protection of transfected lines from apoptosis induced by virus infection or external stimuli. Although the mechanism by which the EP153R protein acts is not clear, its expression was shown to reduce the transactivating activity of the cellular protein p53 (Hurtado *et al.*, 2004).

The ASFV A238L protein is predicted to have both pro-apoptotic and anti-apoptotic effects. By inhibiting NF-κB activation, A238L is predicted to have an apoptotic effect, since NF-κB is involved in the transcriptional activation of anti-apoptotic cellular proteins including IAP proteins (Wang *et al.*, 1998). However, by inhibiting calcineurin phosphatase activity, A238L is predicted to prevent dephosphorylation and activation of the pro-apoptotic protein Bad and hence inhibit apoptosis induced by Bad (Wang *et al.*, 1999). As yet there has been no clear demonstration of either a pro- or anti-apoptotic effect of A238L.

An outline of modulation of apoptosis in ASFV-infected cells is given in Fig. 1.

Fig. 1. Modulation of apoptosis in ASFV-infected cells. Apoptosis is induced in infected cells by several mechanisms including (1) activation of caspase-3 at an early stage in infection before early gene expression, (2) displacement of pro-apototic protein Bim from microtubules by the ASFV structural protein p54, (3) activation of p53 and p53-dependent gene transcription including the pro-apoptotic protein Bax, (4) interaction of pro-apoptotic molecules such as TNF-α. (5) The ASFV A238L protein is predicted to have a pro-apoptotic role by inhibiting NF-κB-dependent transcription of genes encoding anti-apoptotic proteins such as IAP. (6) A238L is also predicted to have an anti-apoptotic role by inhibiting calcineurin phosphatase activity and hence inhibiting dephosphorylation and activation of the pro-apoptotic protein Bad. (7) The ASFV Bcl-2 homologue A179L acts to inhibit apoptosis. Its mode of action has not been defined. (8) The ASFV IAP homologue, A224L, inhibits apoptosis by binding to caspase-3 and can also activate NF-κB. (9) The ASFV C-type lectin protein, EP153R, also inhibits apoptosis by an undefined mechanism.

ASFV proteins that modulate signalling pathways. Growing evidence suggests that a major mechanism used by ASFV to evade host defence systems is to inhibit activation of host factors involved in immunomodulatory gene transcription. The virus replicates primarily in macrophages, which have a key role in activating both the innate and adaptive immune responses. Hence by modulating macrophage gene transcription, virus infection can have a dramatic effect on these host responses.

The A238L gene encodes a bifunctional protein which both inhibits activation of NF-κB transcription factor and inhibits calcineurin phosphatase and pathways such as activation of NFAT transcription factor that depend on calcineurin (Miskin *et al.*, 1998, 2000; Powell *et al.*, 1996; Revilla *et al.*, 1998). A238L shares similarity, in a region containing ankyrin repeats, with the IκB inhibitor of the host transcription factor NF-κB. The suggestion that A238L acts as a functional homologue of IκB was confirmed by experiments showing that A238L expression inhibits activation of an NF-κB-dependent reporter construct (Powell *et al.*, 1996; Revilla *et al.*, 1998). Members of the IκB protein family, IκBα, IκBβ, IκBε, IκBζ, p105 and p100, bind to the NF-κB transcription factor, retaining it in an inactive state (Baeuerle & Henkel, 1994; Ghosh & Karin, 2002). A variety of stimuli, including TNF-α and IL-1, result in activation of an IκB kinase complex, IKK, which phosphorylates IκB on two serine residues near the N terminus. This leads to poly-ubiquitination of IκB and its degradation by the proteasome. A nuclear localization signal on NF-κB is exposed and results in nuclear translocation of the complex and transcriptional activation of genes containing the appropriate binding sites in their promoter regions. Further levels of regulation are provided by phosphorylation and acetylation of NF-κB. In addition, NF-κB activates transcription of IκB and newly translated IκB enters the nucleus, forming a complex with NF-κB which displaces it from DNA. A nuclear export signal on IκB leads to translocation of the complex to the cytoplasm thus providing a feedback mechanism to control NF-κB-dependent gene transcription. Since IκB does not block the nuclear localization signal on the p50 subunit of NF-κB, a constant shuttling of NF-κB/IκB complexes occurs between nucleus and cytoplasm (Baeuerle & Henkel, 1994; Baldwin, 1996; Ghosh & Karin, 2002; Huxford *et al.*, 1998; Israel, 2000; Jacobs & Harrison, 1998; Naumann & Scheidereit, 1994).

The NF-κB family of transcription factors comprises homodimers or heterodimers of proteins p50, p65, p52, crel and relB. Several lines of evidence suggest that A238L acts to inhibit the p65/p50 heterodimeric form of NF-κB. Firstly, A238L has been shown to co-precipitate with the p65 subunit of NF-κB, suggesting that it may bind directly to p65. Secondly, electrophoretic mobility shift assays suggested that A238L inhibits binding of p65/p50 heterodimers to target DNA sequences. A238L does not displace IκB from NF-κB and is thought to bind to NF-κB following signal-induced degradation of IκB (Revilla *et al.*, 1998; Tait *et al.*, 2000). At earlier times post-infection, most A238L is in the cytoplasm, but at later times, it accumulates in the nucleus (G. Bowick, C. C. Abrams & L. K. Dixon, unpublished results). Expression of A238L does not inhibit nuclear shuttling of NF-κB, suggesting that A238L acts within the nucleus to inhibit NF-κB activity. Recombinant A238L protein added to nuclear extracts inhibited binding of NF-κB to target DNA sequences and displaced NF-κB from preformed complexes with DNA, indicating that A238L could act by preventing NF-κB binding to DNA (Revilla *et al.*, 1998).

The identification of the cellular phosphatase calcineurin (protein phosphatase 2B) as a binding partner of A238L suggested that A238L may have a second function and act like the immunosuppressive drug cyclosporin A to inhibit calcineurin phosphatase activity (Clipstone & Crabtree, 1992; Huai *et al.*, 2002; Jin & Harrison, 2002). This was confirmed, and binding was shown to involve a critical motif close to the C-terminus of A238L that is conserved in NFAT and several other inhibitors of calcineurin (Miskin *et al.*, 1998, 2000). The C-terminal 82 aa domain of A238L acts as a potent inhibitor of calcineurin (C. C. Abrams, S. Najjam & L. K. Dixon, unpublished results). Little is known about how the two functions of A238L are regulated. Two forms of the A238L protein are expressed in infected cells. Since these are expressed from the same cDNA it is assumed that they differ by post-translational modification, but the nature of the modification is not known. The higher molecular mass form of A238L co-precipitates with the p65 subunit of NF-κB (Tait *et al.*, 2000) and this form of the protein accumulates in the nucleus late in infection. A reasonable hypothesis is that one form of the A238L protein forms a complex with NF-κB through the ankyrin repeats in the centre of the protein and the other form binds to calcineurin through its C-terminal domain. Understanding the nature and site of the post-translational modification to A238L is likely to be critical for understanding how the function of the protein is regulated.

NF-κB complexes act as transcriptional transactivators (in some cases suppressors) of a wide range of immunomodulatory genes, including those encoding proinflammatory cytokines and chemokines, anti-apoptotic proteins such as IAP proteins, Bcl-2 and Bcl-x$_L$, acute phase proteins and adhesion molecules (Ghosh *et al.*, 1998). Calcineurin has a wide range of functions, which include activation of transcription factors including NFAT and Elk-1, modulation of the activity of receptors including the inositol-triphosphate IP3 receptor and regulation of apoptosis by activating the pro-apoptotic protein Bad (Clipstone & Crabtree, 1992; Crabtree & Olson, 2002; Graef *et al.*, 2001, 2003; Hamalainen *et al.*, 2002; Lopez-Rodriguez *et al.*, 1999; Macian *et al.*, 2000, 2001; Tian & Karin, 1999; Wang *et al.*, 1999). The activity of NFAT has been best studied in T cells, and although its role in macrophages has not been well defined, NFAT has been shown to be involved in transcription of the IL-12 gene in macrophages (Zhu *et al.*, 2003). Thus A238L is predicted to act as a potent immunomodulatory protein with a broad inhibitory effect on transcription of host genes regulated by NF-κB, NFAT and possibly other factors.

In addition to A238L, ASFV encodes other proteins which have either been shown to inhibit specific host signalling pathways or are predicted to do so. By comparing host gene transcription in macrophages infected with wild-type ASFV or a deletion mutant lacking five members of multigene family 360 and 530 it was shown that these genes are involved in transcriptional activation of type I interferon and possibly also of interferon-activated gene transcription (Afonso *et al.*, 2004). It is not known whether all or a subset of MGF 530 and 360 genes is necessary for these functions and the mode of action of the proteins encoded is unknown.

Knowledge of host proteins that bind to specific ASFV proteins has inferred a role for these proteins in inhibiting host signalling pathways. Thus the ASFV j4R protein binds to the α chain of nascent polypeptide associated complex (α-NAC) (Goatley *et al.*, 2002). α-NAC was first suggested to have a role in translation by preventing non-specific targeting of proteins lacking signal peptides to the secretory pathway (Wiedmann *et al.*, 1994). α-NAC can also act as a transcriptional co-activator potentiating transcription dependent on c-Jun (Moreau *et al.*, 1998; Yotov *et al.*, 1998). By binding to α-NAC, j4R may interfere with the ability of α-NAC to act as a transcriptional co-activator. The transcription factor c-Jun is activated in response to stress and is involved in transcription of many immunomodulatory genes. α-NAC has also been shown to bind to FADD (Fas-associated death domain). FADD is involved in assembly of the death-inducing signalling complex (DISC) at the TNF-α receptor cytoplasmic tail following TNF-α binding and thus in induction of apoptosis. Binding of α-NAC to FADD is suggested to prevent FADD oligomerization and assembly of the DISC in the absence of TNF-α (Stilo *et al.*, 2003). Possibly interaction of j4R with α-NAC in some way modulates assembly of the DISC and induction of apoptosis by TNF-α.

The ASFV ubiquitin conjugating enzyme, UBCv, has been implicated as having a role in modulating host gene transcription since it was shown to bind to a host nuclear protein, SMCy, which contains an ARID DNA-binding domain and is involved in transcription regulation (Bulimo *et al.*, 2000).

The DP71L gene (also named l14L and NLS) encodes a protein with similarity over a conserved C-terminal domain to the herpes simplex virus encoded neurovirulence factor ICP34.5. The ICP34.5 protein is multifunctional and its roles include preventing double-stranded RNA activated shut-off of host protein synthesis. ICP34.5 targets

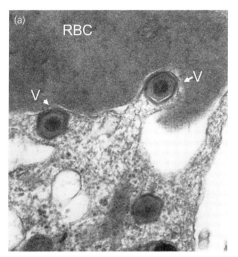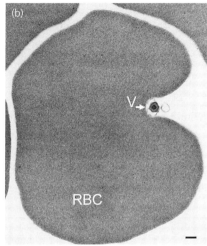

Fig. 2. ASFV-infected macrophages (a) and extracellular virus particles (b) bind to red blood cells. This interaction is mediated by the virus-encoded CD2-like protein. Deletion of the gene encoding this protein delays the onset of viraemia and virus dissemination to tissues. The CD2-like protein is also involved in impairment of lymphocyte activity. EMs were provided by Dr Sharon Brookes, VLA Weybridge, UK.

protein phosphatase 1 to dephosphorylate translation initiation factor eiF2-α, thus returning it to a form active in translation. Although DP71L contains the PP1-binding motif that is present in ICP34.5 and other proteins, the role of DP71L during ASFV infection has not yet been defined.

Proteins involved in cell adhesion

ASFV encodes two transmembrane proteins with similarity to cell adhesion proteins. One shares similarity over the extracellular domain with the T-cell adhesion molecule CD2 (Borca *et al.*, 1994b; Rodriguez *et al.*, 1993). The ligand for host CD2 (CD58 in man and CD49 in rodents) is expressed on a wide range of cells. This interaction stabilizes the contact between T cells and antigen-presenting cells. The cytoplasmic tail of CD2 interacts with an actin-binding adaptor protein, CD2AP, and this is involved in reorganization of the actin cytoskeleton which occurs during formation of the immunological synapse (Dustin *et al.*, 1998; Lee *et al.*, 2003).

The ASFV CD2-like protein (named CD2v or EP402R) is required for the characteristic haemadsorption of red blood cells around virus-infected cells (Fig. 2). The CD2v protein is also incorporated into the external membrane of extracellular virus particles and mediates the binding of virus particles to the surface of red blood cells (Ruiz-Gonzalvo *et al.*, 1996). In pigs infected with virus isolates that cause HAD, most virus is associated with red blood cells (Sierra *et al.*, 1991). When the gene encoding the CD2v protein is deleted from the virus, onset of viraemia and virus dissemination to lymph nodes is delayed. Binding of virus particles and infected cells to red blood cells may help to hide virus and infected cells from components of the immune system hence delaying clearance of virus.

Deletion of the gene encoding the CD2v protein also abrogates the ability of ASFV-infected macrophages to inhibit proliferation of bystander non-infected lymphocytes in response to mitogens, suggesting that the CD2v protein has a role in this process (Borca *et al.*, 1998). The mechanism by which this is mediated has not been elucidated and may involve direct interaction of the extracellular domain of the CD2v protein with lymphocytes. Interestingly, the cytoplasmic tail of the CD2v protein binds to an actin-binding adaptor protein, SH3P7 (also named mabp1 or HIP55) (Kay-Jackson *et al.*, 2004). SH3P7 binds to components of clathrin-coated pits and has been found associated with membrane ruffles and the Golgi network. It has been suggested to have an essential role in endocytosis and in protein transport through the Golgi (Fucini *et al.*, 2002; Kessels *et al.*, 2001; Mise-Omata *et al.*, 2003; Warren *et al.*, 2002) and also to modulate the JNK1 signalling pathway (Ensenat *et al.*, 1999). Interaction of the cytoplasmic tail of the CD2v protein with SH3P7 may modulate protein-trafficking pathways in infected macrophages thus altering the pattern of cell surface or secreted proteins. This could alter co-stimulatory signals provided to bystander lympho-cytes, providing an alternative model for the immunomodulatory effects of CD2v. Within infected cells, most CD2v protein was found in membrane areas that surround virus factories rather than at the cell surface. This provides some support for a model whereby CD2v affects protein trafficking in infected cells (Kay-Jackson *et al.*, 2004).

A second ASFV protein, EP153R, is a type II transmembrane protein containing a C-type lectin domain. This protein is non-essential for virus replication but its expression enhances the haemadsorption of red blood cells around virus-infected cells, suggesting that it has a role in stabilizing the interaction between CD2v and its ligand on red cells. This protein has low sequence similarity with several receptors expressed on the surface of natural killer cells (Galindo *et al.*, 2000). The extracellular ligands for the CD2v and EP153R proteins are not yet known but are likely to be present on other cells as well as red blood cells and contact of these proteins with their ligands on other cells may activate signalling pathways in these cells.

VIRUS GENES WHICH AFFECT VIRULENCE

The targeted deletion of genes from virulent virus isolates has enabled the role of individual genes in virulence of isolates to be assessed. The results of these studies have been reviewed recently (Tulman & Rock, 2001). Deletion of a number of genes impaired the ability of virus to replicate in macrophage cultures *in vitro*. These included two genes encoding enzymes involved in nucleotide metabolism, the thymidine kinase gene and dUTPase gene. The ability of these deletion mutants to replicate in cell lines was not impaired, suggesting that the limiting pool of nucleotides in non-dividing macrophage cultures may restrict replication of virus lacking these enzymes. Deletion of the thymidine kinase gene was shown to reduce virus virulence in pigs (Moore *et al.*, 1998; Oliveros *et al.*, 1999). Deletion of a group of two MGF 530 genes and six MGF 360 genes inhibited growth of virus in macrophages whereas deletion of one to three copies of MGF 360 and MGF 530 gene members had no effect on virus growth in macrophages (Zsak *et al.*, 2001). Deletion of the 9GL gene, which is involved in virion morphogenesis, reduced virus replication in macrophages 100-fold and dramatically reduced virus virulence in pigs (Lewis *et al.*, 2000). This gene is similar to a vaccinia virus gene (E10R) which participates in a cellular redox chain and is also involved in virus assembly. The UK gene is of unknown function and was also demonstrated to encode a virulence factor for pigs (Zsak *et al.*, 1998).

Deletion of the DP71L gene (also named NL, l14L), which shares similarity with the HSV ICP34.5 neurovirulence factor, from the genome of a virulent Spanish ASFV isolate, E70, reduced virulence of the virus for domestic pigs (Zsak *et al.*, 1996), whereas deletion of the gene from the virulent Malawi LIL 20/1 isolate did not reduce virulence. This suggests that the Malawi isolate may encode other genes which can compensate for the loss of l14L (Afonso *et al.*, 1998).

Surprisingly, deletion of several genes encoding host defence evasion proteins, including the A238L gene, the A224L gene (IAP homologue) and the EP153R gene (C-type lectin), did not have any apparent effect on pathogenesis of a virulent isolate in pigs (Neilan *et al.*, 1997a, b, 1999). The most likely explanation is that the virus encodes other genes with similar functions which can compensate for the loss of these genes. Other large DNA viruses, including both pox and herpesviruses, encode several proteins with complementary functions in evasion of host defences (Alcami & Koszinowski, 2000). As discussed in the previous section, deletion of the CD2v gene delayed the onset of viraemia and of virus dissemination but did not reduce the mortality rate of a virulent isolate in pigs (Borca *et al.*, 1998).

HOST TRANSCRIPTIONAL RESPONSES TO ASFV INFECTION

Virus infection of cells induces a cellular response which includes a reprogramming of host gene transcription to activate genes, including those involved in host defence systems. In addition, viruses have mechanisms to manipulate the host transcriptional response in order to facilitate virus survival and replication. Evidence, discussed in the previous sections, indicates that ASFV evades host defence systems by targeting signalling pathways in infected macrophages to interfere with host immunomodulatory gene transcription.

Analysis of gene transcription in ASFV-infected macrophages, either by measuring transcript levels of selected target genes or more globally using a cDNA microarray containing 3000 genes, has provided evidence to support this hypothesis. Two reports showed that ASFV infection of macrophages *in vitro* reduces transcript levels, at late times post-infection, for proinflammatory cytokines, including TNF-α, IL-8 and IFN-α (del Moral *et al.*, 1999; Powell *et al.*, 1996). In one of these studies, an early increase in TNF-α transcript levels was detected as well as secretion of TNF-α (del Moral *et al.*, 1999). More recently, a porcine cDNA microarray has been used to monitor changes in macrophage gene expression following infection with an ASFV virulent isolate (F. Zhang, P. Hopwood, C. Abrams, A. Downing, F. Murray, R. Talbot, A. Archibald, S. Lowden & L. Dixon, unpublished results). This study showed an early increase in expression level of about 100 genes following infection but expression of most of these returned to the base level observed in mock-infected cells at later times post-infection. These genes included proinflammatory cytokines (TNF-α, IL-6 and IFN-β), chemokines, cell surface receptors, adhesion molecules and vesicle sorting proteins. These results are consistent with an early activation of host gene transcription induced by virus infection which is suppressed by virus-encoded proteins, such as A238L and MGF 360/530 proteins. These proteins are expressed following virus entry and onset of early gene expression. A smaller number of genes were increased in expression level later during infection and these may include some encoding proteins that are required by the virus for replication.

A comparison of cytokine gene expression using real-time PCR showed increased transcript levels for IL-6, TNF-α, IL-12 and IL-15 in macrophages infected with a low virulence isolate compared to a high virulence isolate. This suggests that virulence of ASFV isolates may depend on their capacity to regulate the expression of macrophage-derived cytokines relevant for the development of host protective responses (Gil *et al.*, 2003).

In complementary studies, a proteomic approach has been used to monitor host proteins that are altered following ASFV infection (Alfonso *et al.*, 2004). This identified increased levels of host proteins, including those induced by stress such as heat-shock proteins, thioredoxin as well as superoxide dismutase, and apoliprotein E and A. Increased levels of proteins secreted from infected cells included a soluble form of a β-galactoside lectin. These proteins play important roles in regulating apoptosis in the immune system.

CONCLUDING COMMENTS

The remarkable adaptation of ASFV to replicate and persist in its wildlife hosts warthogs and the soft tick vector of the virus, O. *moubata*, shows that the virus has effective mechanisms to evade host defences in both mammalian and arthropod hosts. In its mammalian hosts, the main target cells for infection *in vivo* are macrophages. Macrophages play a key role in activating and orchestrating both the initial innate and the later adaptive immune responses of the host, and thus by modulating macrophage function the virus can dramatically affect both arms of the host response to infection. ASFV and poxviruses are cytoplasmically replicating viruses and are the only DNA viruses that do not require the host RNA polymerase to transcribe their genes. One advantage is that the virus can interfere with host gene transcription without affecting virus gene transcription. Evidence suggests that this is a major mechanism used by ASFV to evade host defence systems. Considerable progress has been made in identifying ASFV immune evasion genes but there are without doubt many remaining to be characterized. The use of genomic and proteomic technologies is beginning to make an impact in improving our understanding of the effects of virus infection and of individual virus proteins on host cell function. Similarly, targeted deletion of genes from the virus genome is helping to dissect the role of individual genes in virus pathogenesis and host cell tropism. The similarity in pathogenesis between ASF and other viral haemorrhagic fevers makes ASFV an interesting and safe, in terms of risk to humans, model for the acutely fatal human haemorrhagic fevers. Finally, progress in understanding the molecular details of ASFV interaction with its hosts will help in the development of effective disease control strategies.

ACKNOWLEDGEMENTS

To the colleagues for helpful discussions; to the BBSRC, EU and DEFRA for funding.

REFERENCES

Afonso, C. L., Neilan, J. G., Kutish, G. F. & Rock, D. L. (1996). An African swine fever virus Bcl-2 homolog, 5-HL, suppresses apoptotic cell death. *J Virol* **70**, 4858–4863.

Afonso, C. L., Zsak, L., Carrillo, C., Borca, M. V. & Rock, D. L. (1998). African swine fever virus NL gene is not required for virus virulence. *J Gen Virol* **79**, 2543–2547.

Afonso, C. L., Piccone, M. E., Zaffuto, K. M. & 8 other authors (2004). African swine fever virus multigene family 360 and 530 genes affect host interferon response. *J Virol* **78**, 1858–1864.

Alcami, A. & Koszinowski, U. H. (2000). Viral mechanisms of immune evasion. *Trends Microbiol* **8**, 410–418.

Alcami, A., Carrascosa, A. L. & Vinuela, E. (1989a). The entry of African swine fever virus into Vero cells. *Virology* **171**, 68–75.

Alcami, A., Carrascosa, A. L. & Vinuela, E. (1989b). Saturable binding sites mediate the entry of African swine fever virus into Vero cells. *Virology* **168**, 393–398.

Alcami, A., Angulo, A., Lopezotin, C., Munoz, M., Freije, J. M. P., Carrascosa, A. L. & Vinuela, E. (1992). Amino acid sequence and structural properties of protein p12, an African swine fever virus attachment protein. *J Virol* **66**, 3860–3868.

Alejo, A., Yanez, R. J., Rodriguez, J. M., Vinuela, E. & Salas, M. L. (1997). African swine fever virus trans-prenyltransferase. *J Biol Chem* **272**, 9417–9423.

Alejo, A., Andres, G., Vinuela, E. & Salas, M. L. (1999). The African swine fever virus prenyltransferase is an integral membrane trans-geranylgeranyl-diphosphate synthase. *J Biol Chem* **274**, 18033–18039.

Alfonso, P., Rivera, J., Hernaez, B., Alonso, C. & Escribano, J. M. (2004). Identification of cellular proteins modified in response to African swine fever virus infection by proteomics. *Proteomics* **4**, 2037–2046.

Almendral, J. M., Almazan, F., Blasco, R. & Vinuela, E. (1990). Multigene families in African swine fever virus: family 110. *J Virol* **64**, 2064–2072.

Alonso, C., Miskin, J., Hernaez, B., Fernandez-Zapatero, P., Soto, L., Canto, C., Rodriguez-Crespo, I., Dixon, L. & Escribano, J. M. (2001). African swine fever virus protein p54 interacts with the microtubular motor complex through direct binding to light-chain dynein. *J Virol* **75**, 9819–9827.

Andres, G., Garcia-Escudero, R., Simon-Mateo, C. & Vinuela, E. (1998). African swine fever virus is enveloped by a two-membraned collapsed cisterna derived from the endoplasmic reticulum. *J Virol* **72**, 8988–9001.

Andres, G., Alejo, A., Simon-Mateo, C. & Salas, M. L. (2001a). African swine fever virus protease, a new viral member of the SUMO-1-specific protease family. *J Biol Chem* **276**, 780–787.

Andres, G., Garcia-Escudero, R., Vinuela, E., Salas, M. L. & Rodriguez, J. M. (2001b). African swine fever virus structural protein pE120R is essential for virus transport from assembly sites to plasma membrane but not for infectivity. *J Virol* **75**, 6758–6768.

Andres, G., Alejo, A., Salas, J. & Salas, M. L. (2002a). African swine fever virus polyproteins pp220 and pp62 assemble into the core shell. *J Virol* **76**, 12473–12482.

Andres, G., Garcia-Escudero, R., Salas, M. L. & Rodriguez, J. M. (2002b). Repression of African swine fever virus polyprotein pp220-encoding gene leads to the assembly of icosahedral core-less particles. *J Virol* **76**, 2654–2666.

Arala-Chaves, M. P., Ribeiro, A. D., Vilanova, M., Porto, M. T., Santarem, M. G. & Lima, M. (1988). Correlation between B-cell mitogenicity and immunosuppressor effects of a protein released by porcine monocytes infected with African swine fever virus. *Am J Vet Res* **49**, 1955–1961.

Baeuerle, P. A. & Henkel, T. (1994). Function and activation of NF-kappa B in the immune system. *Annu Rev Immunol* **12**, 141–179.

Baldwin, A. S. (1996). The NF-kappa B and I kappa B proteins: new discoveries and insights. *Annu Rev Immunol* **14**, 649–683.

Baylis, S. A., Banham, A. H., Vydelingum, S., Dixon, L. K. & Smith, G. L. (1993). African swine fever virus encodes a serine protein kinase which is packaged into virions. *J Virol* **67**, 4549–4556.

Boinas, F. S., Hutchings, G. H., Dixon, L. K. & Wilkinson, P. J. (2004). Characterization of pathogenic and non-pathogenic African swine fever virus isolates from *Ornithodoros erraticus* inhabiting pig premises in Portugal. *J Gen Virol* **85**, 2177–2187.

Borca, M. V., Irusta, P., Carrillo, C., Afonso, C. L., Burrage, T. & Rock, D. L. (1994a). African swine fever virus structural protein p72 contains a conformational neutralizing epitope. *Virology* **201**, 413–418.

Borca, M. V., Kutish, G. F., Afonso, C. L., Irusta, P., Carrillo, C., Brun, A., Sussman, M. & Rock, D. L. (1994b). An African swine fever virus gene with similarity to the T-lymphocyte surface-antigen CD2 mediates hemadsorption. *Virology* **199**, 463–468.

Borca, M. V., Irusta, P. M., Kutish, G. F., Carrillo, C., Afonso, C. L., Burrage, T., Neilan, J. G. & Rock, D. L. (1996). A structural DNA binding protein of African swine fever virus with similarity to bacterial histone-like proteins. *Arch Virol* **141**, 301–313.

Borca, M. V., Carrillo, C., Zsak, L., Laegreid, W. W., Kutish, G. F., Neilan, J. G., Burrage, T. G. & Rock, D. L. (1998). Deletion of a CD2-like gene, 8-DR, from African swine fever virus affects viral infection in domestic swine. *J Virol* **72**, 2881–2889.

Brun, A., Rodriguez, F., Escribano, J. M. & Alonso, C. (1998). Functionality and cell anchorage dependence of the African swine fever virus gene A179L, a viral bcl-2 homolog, in insect cells. *J Virol* **72**, 10227–10233.

Bulimo, W. D., Miskin, J. E. & Dixon, L. K. (2000). An ARID family protein binds to the African swine fever virus encoded ubiquitin conjugating enzyme, UBCv1. *FEBS Lett* **471**, 17–22.

Burrage, T. G., Lu, Z., Neilan, J. G., Rock, D. L. & Zsak, L. (2004). African swine fever virus multigene family 360 genes affect virus replication and generalization of infection in *Ornithodoros porcinus* ticks. *J Virol* **78**, 2445–2453.

Carrasco, L., Fernandez, A., Villamandos, J. C. G., Mozos, E., Mendez, A. & Jover, A. (1992). Kupffer cells and PIMs in acute experimental African swine fever. *Histol Histopathol* **7**, 421–425.

Carrasco, L., de Lara, F. C., Martin de las Mulas, J., Gomez-Villamandos, J. C., Perez, J., Wilkinson, P. J. & Sierra, M. A. (1996). Apoptosis in lymph nodes in acute African swine fever. *J Comp Pathol* **115**, 415–428.

Carrascosa, A. L., Bustos, M. J., Nogal, M. L., de Buitrago, G. G. & Revilla, Y. (2002). Apoptosis induced in an early step of African swine fever virus entry into Vero cells does not require virus replication. *Virology* **294**, 372–382.

Carrillo, C., Borca, M. V., Afonso, C. L., Onisk, D. V. & Rock, D. L. (1994). Long-term persistent infection of swine monocytes/macrophages with African swine fever virus. *J Virol* **68**, 580–583.

Chacon, M. R., Almazan, F., Nogal, M. L., Vinuela, E. & Rodriguez, J. F. (1995). The African swine fever virus IAP homolog is a late structural polypeptide. *Virology* **214**, 670–674.

Childerstone, A., Takamatsu, H., Yang, H., Denyer, M. & Parkhouse, R. M. E. (1998). Modulation of T cell and monocyte function in the spleen following infection of pigs with African swine fever virus. *Vet Immunol Immunopathol* **62**, 281–296.

Clipstone, N. A. & Crabtree, G. R. (1992). Identification of calcineurin as a key signalling enzyme in T-lymphocyte activation. *Nature* **357**, 695–697.

Cobbold, C. & Wileman, T. (1998). The major structural protein of African swine fever virus, p73, is packaged into large structures, indicative of viral capsid or matrix precursors, on the endoplasmic reticulum. *J Virol* **72**, 5215–5223.

Cobbold, C., Windsor, M. & Wileman, T. (2001). A virally encoded chaperone specialized for folding of the major capsid protein of African swine fever virus. *J Virol* **75**, 7221–7229.

Crabtree, G. R. & Olson, E. N. (2002). NFAT signaling: choreographing the social lives of cells. *Cell* **109**, S67–S79.

de la Vega, I., Vinuela, E. & Blasco, R. (1990). Genetic variation and multigene families in African swine fever virus. *Virology* **179**, 234–246.

del Moral, M. G., Ortuno, E., Fernandez-Zapatero, P., Alonso, F., Alonso, C., Ezquerra, A. & Dominguez, J. (1999). African swine fever virus infection induces tumor necrosis factor alpha production: implications in pathogenesis. *J Virol* **73**, 2173–2180.

Dixon, L. K., Twigg, S. R. F., Baylis, S. A., Vydelingum, S., Bristow, C., Hammond, J. M. & Smith, G. L. (1994). Nucleotide sequence of a 55 kbp region from the right end of the genome of a pathogenic African swine fever virus isolate (Malawi LIL20/1). *J Gen Virol* **75**, 1655–1684.

Duarte, M. M. D. (2000). *Bases Moleculares da Virulence e Hemadsorbcao nos Isolados Nacionas Lisboa 60 e Lisboa 68 do Virus Da Peste Suina Africana.* PhD Thesis, Universidade Nova de Lisboa.

Dustin, M. L., Olszowy, M. W., Holdorf, A. D. & 8 other authors (1998). A novel adaptor protein orchestrates receptor patterning and cytoskeletal polarity in T-cell contacts. *Cell* **94**, 667–677.

Ensenat, D., Yao, Z. B., Wang, X. H. S., Kori, R., Zhou, G. S., Lee, S. C. & Tan, T. H. (1999). A novel Src homology 3 domain-containing adaptor protein, HIP-55, that interacts with hematopoietic progenitor kinase 1. *J Biol Chem* **274**, 33945–33950.

Fernandez, A., Perez, J., Carrasco, L., Bautista, M. J., Sanchez-Vizcaino, J. M. & Sierra, M. A. (1992a). Distribution of ASFV antigens in pig tissues experimentally infected with two different Spanish virus isolates. *Zentralbl Veterinarmed B* **39**, 393–402.

Fernandez, A., Perez, J., Carrasco, L., Sierra, M. A., Sanchez-Vizcaino, M. & Jover, A. (1992b). Detection of African swine fever viral antigens in paraffin-embedded tissues by use of immunohistologic methods and polyclonal antibodies. *Am J Vet Res* **53**, 1462–1467.

Fernandez, A., Perez, J., Martin de las Mulas, J., Carrasco, L., Dominguez, J. & Sierra, M. A. (1992c). Localization of African swine fever viral antigen, swine IgM, IgG and C1q in lung and liver tissues of experimentally infected pigs. *J Comp Pathol* **107**, 81–90.

Fucini, R. V., Chen, J. L., Sharma, C., Kessels, M. M. & Stamnes, M. (2002). Golgi vesicle proteins are linked to the assembly of an actin complex defined by mAbp1. *Mol Biol Cell* **13**, 621–631.

Galindo, I., Almazan, F., Bustos, M. J., Vinuela, E. & Carrascosa, A. L. (2000). African swine fever virus EP153R open reading frame encodes a glycoprotein involved in the hemadsorption of infected cells. *Virology* **266**, 340–351.

Garcia-Beato, R., Salas, M. L., Vinuela, E. & Salas, J. (1992). Role of the host cell nucleus in the replication of African swine fever virus DNA. *Virology* **188**, 637–649.

Ghosh, S. & Karin, M. (2002). Missing pieces in the NF-kappa B puzzle. *Cell* **109**, S81–S96.

Ghosh, S., May, M. J. & Kopp, E. B. (1998). NF-kappa B and Rel proteins: evolutionarily conserved mediators of immune responses. *Annu Rev Immunol* **16**, 225–260.

Gil, S., Spagnuolo-Weaver, M., Canals, A., Sepulveda, N., Oliveira, J., Aleixo, A., Allan, G., Leitao, A. & Martins, C. L. V. (2003). Expression at mRNA level of cytokines and A238L gene in porcine blood-derived macrophages infected in vitro with African swine fever virus (ASFV) isolates of different virulence. *Arch Virol* **148**, 2077–2097.

Goatley, L. C., Twigg, S. R. F., Miskin, J. E., Monaghan, P., St-Arnaud, R., Smith, G. L. & Dixon, L. K. (2002). The African swine fever virus protein j4R binds to the alpha chain of nascent polypeptide-associated complex. *J Virol* **76**, 9991–9999.

Gomez-Puertas, P., Rodriguez, F., Oviedo, J. M., Brun, A., Alonso, C. & Escribano, J. M. (1998). The African swine fever virus proteins p54 and p30 are involved in two distinct steps of virus attachment and both contribute to the antibody-mediated protective immune response. *Virology* **243**, 461–471.

Gomez-Villamandos, J. C., Hervas, J., Mendez, A., Carrasco, L., Martin de las Mulas, J., Villeda, C. J., Wilkinson, P. J. & Sierra, M. A. (1995a). Experimental African swine fever: apoptosis of lymphocytes and virus replication in other cells. *J Gen Virol* **76**, 2399–2405.

Gomez-Villamandos, J. C., Hervas, J., Mendez, A., Carrasco, L., Villeda, C. J., Wilkinson, P. J. & Sierra, M. A. (1995b). Pathological changes in the renal interstitial capillaries of pigs inoculated with two different strains of African swine fever virus. *J Comp Pathol* **112**, 283–298.

Gomez-Villamandos, J. C., Hervas, J., Mendez, A., Carrasco, L., Villeda, C. J., Wilkinson, P. J. & Sierra, M. A. (1995c). Ultrastructural study of the renal tubular system in acute experimental African swine fever: virus replication in glomerular mesangial cells and in the collecting ducts. *Arch Virol* **140**, 581–589.

Gomez-Villamandos, J. C., Bautista, M. J., Hervas, J., Carrasco, L., de Lara, F. C. M., Perez, J., Wilkinson, P. J. & Sierra, M. A. (1996). Subcellular changes in platelets in acute and subacute African swine fever. *J Comp Pathol* **115**, 327–341.

Gonzalez, A., Talavera, A., Almendral, J. M. & Vinuela, E. (1986). Hairpin loop structure of African swine fever virus DNA. *Nucleic Acids Res* **14**, 6835–6844.

Gonzalez, A., Calvo, V., Almazan, F., Almendral, J. M., Ramirez, J. C., de la Vega, I., Blasco, R. & Vinuela, E. (1990). Multigene families in African swine fever virus: family 360. *J Virol* **64**, 2073–2081.

Graef, I. A., Chen, F., Chen, L., Kuo, A. & Crabtree, G. R. (2001). Signals transduced by Ca^{2+}/calcineurin and NFATc3/c4 pattern the developing vasculature. *Cell* **105**, 863–875.

Graef, I. A., Wang, F., Charron, F., Chen, L., Neilson, J., Tessier-Lavigne, M. & Crabtree, G. R. (2003). Neurotrophins and netrins require calcineurin/NFAT signaling to stimulate outgrowth of embryonic axons. *Cell* **113**, 657–670.

Granja, A. G., Nogal, M. L., Hurtado, C., Salas, J., Salas, M. L., Carrascosa, A. L. & Revilla, Y. (2004). Modulation of p53 cellular function and cell death by African swine fever virus. *J Virol* **78**, 7165–7174.

Hamalainen, M., Lahti, A. & Moilanen, E. (2002). Calcineurin inhibitors, cyclosporin A and tacrolimus inhibit expression of inducible nitric oxide synthase in colon epithelial and macrophage cell lines. *Eur J Pharmacol* **448**, 239–244.

Heath, C. M., Windsor, M. & Wileman, T. (2001). Aggresomes resemble sites specialized for virus assembly. *J Cell Biol* **153**, 449–455.

Hernaez, B., Diaz-Gil, G., Garcia-Gallo, M., Quetglas, J. I., Rodriguez-Crespo, I., Dixon, L., Escribano, J. M. & Alonso, C. (2004). The African swine fever virus dynein-binding protein p54 induces infected cell apoptosis. *FEBS Lett* **569**, 224–228.

Hingamp, P. M., Arnold, J. E., Mayer, R. J. & Dixon, L. K. (1992). A ubiquitin conjugating enzyme encoded by African swine fever virus. *EMBO J* **11**, 361–366.

Huai, Q., Kim, H. Y., Liu, Y. D., Zhao, Y. D., Mondragon, A., Liu, J. O. & Ke, H. M. (2002). Crystal structure of calcineurin-cyclophilin-cyclosporin shows common but distinct recognition of immunophilin-drug complexes. *Proc Natl Acad Sci U S A* **99**, 12037–12042.

Hurtado, C., Granja, A. G., Bustos, M. J., Nogal, M. L., de Buitrago, G. G., de Yebenes, V. G., Salas, M. L., Revilla, Y. & Carrascosa, A. L. (2004). The C-type lectin homologue gene (EP153R) of African swine fever virus inhibits apoptosis both in virus infection and in heterologous expression. *Virology* **326**, 160–170.

Huxford, T., Huang, D. B., Malek, S. & Ghosh, G. (1998). The crystal structure of the

I kappa B alpha/NF-kappa B complex reveals mechanisms of NF-kappa B inactivation. *Cell* **95**, 759–770.

Israel, A. (2000). The IKK complex: an integrator of all signals that activate NF-kappa B? *Trends Cell Biol* **10**, 129–133.

Jacobs, M. D. & Harrison, S. C. (1998). Structure of an I kappa B alpha/NF-kappa B complex. *Cell* **95**, 749–758.

Jin, L. & Harrison, S. C. (2002). Crystal structure of human calcineurin complexed with cyclosporin A and human cyclophilin. *Proc Natl Acad Sci U S A* **99**, 13522–13526.

Jouvenet, N., Monaghan, P., Way, M. & Wileman, T. (2004). Transport of African swine fever virus from assembly sites to the plasma membrane is dependent on micro-tubules and conventional kinesin. *J Virol* **78**, 7990–8001.

Kay-Jackson, P. C., Goatley, L. C., Cox, L., Miskin, J. E., Parkhouse, R. M. E., Wienands, J. & Dixon, L. K. (2004). The CD2v protein of African swine fever virus interacts with the actin-binding adaptor protein SH3P7. *J Gen Virol* **85**, 119–130.

Kessels, M. M., Engqvist-Goldstein, A. E. Y., Drubin, D. G. & Qualmann, B. (2001). Mammalian Abp1, a signal-responsive F-actin-binding protein, links the actin cytoskeleton to endocytosis via the GTPase dynamin. *J Cell Biol* **153**, 351–366.

Lee, K. H., Dinner, A. R., Tu, C. & 12 other authors (2003). The immunological synapse balances T cell receptor signaling and degradation. *Science* **302**, 1218–1222.

Leitao, A., Cartaxeiro, C., Coelho, R., Cruz, B., Parkhouse, R. M. E., Portugal, F. C., Vigario, J. D. & Martins, C. L. V. (2001). The non-haemadsorbing African swine fever virus isolate ASFV/NH/P68 provides a model for defining the protective anti-virus immune response. *J Gen Virol* **82**, 513–523.

Lewis, T., Zsak, L., Burrage, T. G., Lu, Z., Kutish, G. F., Neilan, J. G. & Rock, D. L. (2000). An African swine fever virus ERV1-ALR homologue, 9GL, affects virion maturation and viral growth in macrophages and viral virulence in swine. *J Virol* **74**, 1275–1285.

Lopez-Rodriguez, C., Aramburu, J., Rakeman, A. S. & Rao, A. (1999). NFAT5, a constitutively nuclear NFAT protein that does not cooperate with Fos and Jun. *Proc Natl Acad Sci U S A* **96**, 7214–7219.

Macian, F., Garcia-Rodriguez, C. & Rao, A. J. N. (2000). Gene expression elicited by NFAT in the presence or absence of cooperative recruitment of Fos and Jun. *EMBO J* **19**, 4783–4795.

Macian, F., Lopez-Rodriguez, C. & Rao, A. (2001). Partners in transcription: NFAT and AP-1. *Oncogene* **20**, 2476–2489.

Martinez-Pomares, L., Simon-Mateo, C., Lopez-Otin, C. & Vinuela, E. (1997). Characterization of the African swine fever virus structural protein p14.5: a DNA binding protein. *Virology* **229**, 201–211.

McCullough, K. C., Basta, S., Knotig, S., Gerber, H., Schaffner, R., Kim, Y. B. & Saalmuller, A. (1999). Intermediate stages in monocyte-macrophage differentiation modulate phenotype and susceptibility to virus infection. *Immunology* **98**, 203–212.

Mise-Omata, S., Montagne, B., Deckert, M., Wienands, J. & Acuto, O. (2003). Mammalian actin binding protein 1 is essential for endocytosis but not lamellipodia formation: functional analysis by RNA interference. *Biochem Biophys Res Commun* **301**, 704–710.

Miskin, J. E., Abrams, C. C., Goatley, L. C. & Dixon, L. K. (1998). A viral mechanism for inhibition of the cellular phosphatase calcineurin. *Science* **281**, 562–565.

Miskin, J. E., Abrams, C. C. & Dixon, L. K. (2000). African swine fever virus protein A238L interacts with the cellular phosphatase calcineurin via a binding domain similar to that of NFAT. *J Virol* **74**, 9412–9420.

Moore, D. M., Zsak, L., Neilan, J. G., Lu, Z. & Rock, D. L. (1998). The African swine fever virus thymidine kinase gene is required for efficient replication in swine macrophages and for virulence in swine. *J Virol* **72**, 10310–10315.

Moreau, A., Yotov, W. V., Glorieux, F. H. & St-Arnaud, R. (1998). Bone-specific expression of the alpha chain of the nascent polypeptide-associated complex, a coactivator potentiating c-Jun-mediated transcription. *Mol Cell Biol* **18**, 1312–1321.

Naumann, M. & Scheidereit, C. (1994). Activation of NF-kappa B in vivo is regulated by multiple phosphorylations. *EMBO J* **13**, 4597–4607.

Neilan, J. G., Lu, Z., Afonso, C. L., Kutish, G. F., Sussman, M. D. & Rock, D. L. (1993). An African swine fever virus gene with similarity to the protooncogene bcl-2 and the Epstein-Barr virus gene BHRF1. *J Virol* **67**, 4391–4394.

Neilan, J. G., Lu, Z., Kutish, G. F., Zsak, L., Burrage, T. G., Borca, M. V., Carrillo, C. & Rock, D. L. (1997a). A BIR motif containing gene of African swine fever virus, 4CL, is nonessential for growth in vitro and viral virulence. *Virology* **230**, 252–264.

Neilan, J. G., Lu, Z., Kutish, G. F., Zsak, L., Lewis, T. L. & Rock, D. L. (1997b). A conserved African swine fever virus I kappa B homolog, 5EL, is nonessential for growth in vitro and virulence in domestic swine. *Virology* **235**, 377–385.

Neilan, J. G., Borca, M. V., Lu, Z., Kutish, G. F., Kleiboeker, S. B., Carrillo, C., Zsak, L. & Rock, D. L. (1999). An African swine fever virus ORF with similarity to C-type lectins is non-essential for growth in swine macrophages *in vitro* and for virus virulence in domestic swine. *J Gen Virol* **80**, 2693–2697.

Netherton, C., Rouiller, I. & Wileman, T. (2004). The subcellular distribution of multigene family 110 proteins of African swine fever virus is determined by differences in C-terminal KDEL endoplasmic reticulum retention motifs. *J Virol* **78**, 3710–3721.

Nogal, M. L., de Buitrago, G. G., Rodriguez, C., Cubelos, B., Carrascosa, A. L., Salas, M. L. & Revilla, Y. (2001). African swine fever virus IAP homologue inhibits caspase activation and promotes cell survival in mammalian cells. *J Virol* **75**, 2535–2543.

Oliveros, M., Garcia-Escudero, R., Alejo, A., Vinuela, E., Salas, M. L. & Salas, J. (1999). African swine fever virus dUTPase is a highly specific enzyme required for efficient replication in swine macrophages. *J Virol* **73**, 8934–8943.

Oura, C. A. L., Powell, P. P. & Parkhouse, R. M. E. (1998). African swine fever: a disease characterized by apoptosis. *J Gen Virol* **79**, 1427–1438.

Powell, P. P., Dixon, L. K. & Parkhouse, R. M. E. (1996). An I kappa B homolog encoded by African swine fever virus provides a novel mechanism for downregulation of proinflammatory cytokine responses in host macrophages. *J Virol* **70**, 8527–8533.

Revilla, Y., Cebrian, A., Baixeras, E., Martinez, C., Vinuela, E. & Salas, M. L. (1997). Inhibition of apoptosis by the African swine fever virus bcl-2 homologue: role of the BH1 domain. *Virology* **228**, 400–404.

Revilla, Y., Callejo, M., Rodriguez, J. M., Culebras, E., Nogal, M. L., Salas, M. L., Vinuela, E. & Fresno, M. (1998). Inhibition of nuclear factor kappa B activation by a virus-encoded I kappa B-like protein. *J Biol Chem* **273**, 5405–5411.

Ribeiro, A. D., Arala-Chaves, M. P., Vilanova, M., Porto, M. T. & Coutinho, A. (1991). Role of B and T lymphocytes in the specific immunosuppression induced by a protein released by porcine monocytes infected with African swine fever virus. *Int Immunol* **3**, 165–174.

Rodriguez, C. I., Nogal, M. L., Carrascosa, A. L., Salas, M. L., Fresno, M. & Revilla, Y. (2002). African swine fever virus IAP-like protein induces the activation of nuclear factor kappa B. *J Virol* **76**, 3936–3942.

Rodriguez, J. M., Yanez, R. J., Almazan, F., Vinuela, E. & Rodriguez, J. F. (1993). African swine fever virus encodes a CD2 homolog responsible for the adhesion of erythrocytes to infected cells. *J Virol* **67**, 5312–5320.

Rodriguez, J. M., Yanez, R. J., Pan, R., Rodriguez, J. F., Salas, M. L. & Vinuela, E. (1994). Multigene families in African swine fever virus: family 505. *J Virol* **68**, 2746–2751.

Rodriguez, J. M., Garcia-Escudero, R., Salas, M. L. & Andres, G. (2004). African swine fever virus structural protein p54 is essential for the recruitment of envelope precursors to assembly sites. *J Virol* **78**, 4299–4313.

Rojo, G., Chamorro, M., Salas, M. L., Vinuela, E., Cuezva, J. M. & Salas, J. (1998). Migration of mitochondria to viral assembly sites in African swine fever virus-infected cells. *J Virol* **72**, 7583–7588.

Rojo, G., Garcia-Beato, R., Vinuela, E., Salas, M. L. & Salas, J. (1999). Replication of African swine fever virus DNA in infected cells. *Virology* **257**, 524–536.

Rouiller, I., Brookes, S. M., Hyatt, A. D., Windsor, M. & Wileman, T. (1998). African swine fever virus is wrapped by the endoplasmic reticulum. *J Virol* **72**, 2373–2387.

Ruiz-Gonzalvo, F., Rodriguez, F. & Escribano, J. M. (1996). Functional and immunological properties of the baculovirus-expressed hemagglutinin of African swine fever virus. *Virology* **218**, 285–289.

Saanchez-Torres, C., Gomez-Puertas, P., Gomez-del-Moral, M., Alonso, F., Escribano, J. M., Ezquerra, A. & Dominguez, J. (2003). Expression of porcine CD163 on monocytes/macrophages correlates with permissiveness to African swine fever infection. *Arch Virol* **148**, 2307–2323.

Salguero, F. J., Ruiz-Villamor, E., Bautista, M. J., Sanchez-Cordon, P. J., Carrasco, L. & Gomez-Villamandos, J. C. (2002). Changes in macrophages in spleen and lymph nodes during acute African swine fever: expression of cytokines. *Vet Immunol Immunopathol* **90**, 11–22.

Salguero, F. J., Sanchez-Cordon, P. J., Sierra, M. A., Jover, A., Nunez, A. & Gomez-Villamandos, J. C. (2004). Apoptosis of thymocytes in experimental African swine fever virus infection. *Histol Histopathol* **19**, 77–84.

Sierra, M. A., Gomez-Villamandos, J. C., Carrasco, L., Fernandez, A., Mozos, E. & Jover, A. (1991). In vivo study of hemadsorption in African swine fever virus infected cells. *Vet Pathol* **28**, 178–181.

Simon-Mateo, C., Andres, G. & Vinuela, E. (1993). Polyprotein processing in African swine fever virus: a novel gene expression strategy for a DNA virus. *EMBO J* **12**, 2977–2987.

Simon-Mateo, C., Andres, G., Almazan, F. & Vinuela, E. (1997). Proteolytic processing in African swine fever virus: evidence for a new structural polyprotein, pp62. *J Virol* **71**, 5799–5804.

Stilo, R., Liguoro, D., di Jeso, B., Leonardi, A. & Vito, P. (2003). The α-chain of the nascent polypeptide-associated complex binds to and regulates FADD function. *Biochem Biophys Res Commun* **303**, 1034–1041.

Tait, S. W. G., Reid, E. B., Greaves, D. R., Wileman, T. E. & Powell, P. P. (2000). Mechanism of inactivation of NF-kappa B by a viral homologue of I kappa B alpha. Signal-induced release of I kappa B alpha results in binding of the viral homologue to NF-kappa B. *J Biol Chem* **275**, 34656–34664.

Takamatsu, H., Denyer, M. S., Oura, C., Childerstone, A., Andersen, J. K., Pullen, L. & Parkhouse, R. M. E. (1999). African swine fever virus: a B cell-mitogenic virus *in vivo* and *in vitro*. *J Gen Virol* **80**, 1453–1461.

Tian, J. M. & Karin, M. (1999). Stimulation of Elk1 transcriptional activity by mitogen-activated protein kinases is negatively regulated by protein phosphatase 2B (calcineurin). *J Biol Chem* **274**, 15173–15180.

Tulman, E. R. & Rock, D. L. (2001). Novel virulence and host range genes of African swine fever virus. *Curr Opin Microbiol* **4**, 456–461.

Upton, C., Slack, S., Hunter, A. L., Ehlers, A. & Roper, R. L. (2003). Poxvirus orthologous clusters: toward defining the minimum essential poxvirus genome. *J Virol* **77**, 7590–7600.

Valdeira, M. L., Bernardes, C., Cruz, B. & Geraldes, A. (1998). Entry of African swine fever virus into Vero cells and uncoating. *Vet Microbiol* **60**, 131–140.

Vallee, I., Tait, S. W. G. & Powell, P. P. (2001). African swine fever virus infection of porcine aortic endothelial cells leads to inhibition of inflammatory responses, activation of the thrombotic state, and apoptosis. *J Virol* **75**, 10372–10382.

Vilanova, M., Ferreira, P., Ribeiro, A. & Arala-Chaves, M. (1999). The biological effects induced in mice by p36, a proteinaceous factor of virulence produced by African swine fever virus, are mediated by interleukin-4 and also to a lesser extent by interleukin-10. *Immunology* **96**, 389–395.

Villeda, C. J., Williams, S. M., Wilkinson, P. J. & Vinuela, E. (1993a). Consumption coagulopathy associated with shock in acute African swine fever. *Arch Virol* **133**, 467–475.

Villeda, C. J., Williams, S. M., Wilkinson, P. J. & Vinuela, E. (1993b). Haemostatic abnormalities in African swine fever: a comparison of two virus strains of different virulence (Dominican Republic '78 and Malta '78). *Arch Virol* **130**, 71–83.

Villeda, C. J., Gomez-Villamandos, J. C., Williams, S. M., Hervas, J., Wilkinson, P. J. & Vinuela, E. (1995). The role of fibrinolysis in the pathogenesis of the haemorrhagic syndrome produced by virulent isolates of African swine fever virus. *Thromb Haemost* **73**, 112–117.

Wang, C. Y., Mayo, M. W., Korneluk, R. G., Goeddel, D. V. & Baldwin, A. S. (1998). NF-kappa B antiapoptosis: induction of TRAF1 and TRAF2 and c-IAP1 and c-IAP2 to suppress caspase-8 activation. *Science* **281**, 1680–1683.

Wang, H. G., Pathan, N., Ethell, I. M. & 7 other authors (1999). Ca2+-induced apoptosis through calcineurin dephosphorylation of BAD. *Science* **284**, 339–343.

Wardley, R. C. (1982). Effect of African swine fever on lymphocyte mitogenesis. *Immunology* **46**, 215–220.

Warren, D. T., Andrews, P. D., Gourlay, C. W. & Ayscough, K. R. (2002). Sla1p couples the yeast endocytic machinery to proteins regulating actin dynamics. *J Cell Sci* **115**, 1703–1715.

Wiedmann, B., Sakai, H., Davis, T. A. & Wiedmann, M. (1994). A protein complex required for signal-sequence-specific sorting and translocation. *Nature* **370**, 434–440.

Wilkinson, P. J. (1984). The persistence of African swine fever in Africa and the Mediterranean. *Prev Vet Med* **2**, 71–82.

Wilkinson, P. J., Wardley, R. C. & Williams, S. M. (1981). African swine fever virus (Malta/78) in pigs. *J Comp Pathol* **91**, 277–284.

Yanez, R. J. & Vinuela, E. (1993). African swine fever virus encodes a DNA ligase. *Virology* **193**, 531–536.

Yanez, R. J., Rodriguez, J. M., Nogal, M. L., Yuste, L., Enriquez, C., Rodriguez, J. F. & Vinuela, E. (1995). Analysis of the complete nucleotide sequence of African swine fever virus. *Virology* **208**, 249–278.

Yotov, W. V., Moreau, A. & St-Arnaud, R. (1998). The alpha chain of the nascent polypeptide-associated complex functions as a transcriptional coactivator. *Mol Cell Biol* **18**, 1303–1311.

Yozawa, T., Kutish, G. F., Afonso, C. L., Lu, Z. & Rock, D. L. (1994). Two novel multigene families, 530 and 300, in the terminal variable regions of African swine fever virus genome. *Virology* **202**, 997–1002.

Zhu, C., Rao, K., Xiong, H. B., Gagnidze, K., Li, F. L., Horvath, C. & Plevy, S. (2003). Activation of the murine interleukin-12 p40 promoter by functional interactions between NFAT and ICSBP. *J Biol Chem* **278**, 39372–39382.

Zsak, L., Lu, Z., Kutish, G. F., Neilan, J. G. & Rock, D. L. (1996). An African swine fever virus virulence-associated gene NL-S with similarity to the herpes simplex virus ICP34.5 gene. *J Virol* **70**, 8865–8871.

Zsak, L., Caler, E., Lu, Z., Kutish, G. F., Neiland, J. G. & Rock, D. L. (1998). A nonessential African swine fever virus gene UK is a significant virulence determinant in domestic swine. *J Virol* **72**, 1028–1035.

Zsak, L., Lu, Z., Burrage, T. G., Neilan, J. G., Kutish, G. F., Moore, D. M. & Rock, D. L. (2001). African swine fever virus multigene family 360 and 530 genes are novel macrophage host range determinants. *J Virol* **75**, 3066–3076.

Murid herpesvirus 4 as a model for gammaherpesvirus pathogenesis

James P. Stewart, David Hughes, Louise Roaden and
Bahram Ebrahimi

Centre for Comparative Infectious Diseases, University of Liverpool, Duncan Building, Daulby
Street, Liverpool L69 3GA, UK

INTRODUCTION

Members of the gamma subfamily of the herpesviruses (γ-herpesviruses) are extremely well represented throughout nature and include viruses that cause a multitude of diseases in humans and animals. Currently, the γ-herpesvirus subfamily is made up of two major groups (or genera): $\gamma1$ and $\gamma2$. $\gamma1$-Herpesviruses (or lymphocryptoviruses) so far have only been detected in primates, and Epstein–Barr virus (EBV) is the representative human $\gamma1$-herpesvirus that naturally infects around 90 % of the world's population (Rickinson & Kieff, 2001). EBV is largely asymptomatic and is usually only clinically obvious upon presentation with infectious mononucleosis (IM). However, in a minority of cases, EBV is associated with the development of Burkitt's lymphoma, AIDS, transplant-associated B-cell lymphoma and other cancers such as nasopharyngeal carcinoma. The $\gamma2$-herpesviruses (or rhadinoviruses), however, appear to be more successful as these viruses are found in many hosts. The human $\gamma2$-herpesvirus, Kaposi's sarcoma-associated herpesvirus or human herpesvirus 8 (KSHV or HHV-8, respectively), is associated with three neoplastic disorders: Kaposi's sarcoma; primary effusion lymphoma; and multicentric Castleman's disease (Schulz, 1998). In animals, particularly in sheep and wildebeest, the malignant catarrhal fever (MCF)-associated herpesviruses are of great interest. MCF is a fatal disease of ruminants and has considerable economic consequences for the farming industry. Alcelaphine herpesvirus 1 (AlHV-1) is spread to susceptible animals (causing African or wildebeest-associated MCF) from wildebeest, which are the asymptomatic hosts for the virus. Similarly, the aetiological agent of sheep-associated MCF in susceptible animals, ovine herpesvirus 2 (OvHV-2), is asymptomatic in its natural host (Coulter et al., 2001).

SGM symposium 64: Molecular pathogenesis of virus infections.
Editors P. Digard, A. A. Nash & R. E. Randall. Cambridge University Press. ISBN 0 521 83248 9 ©SGM 2005

The γ-herpesviruses are therefore important because of the diseases they cause in humans and animals. There are, however, a number of problems in studying these viruses in the context of experimental infection, largely owing to their rather narrow host specificity (Rickinson & Kieff, 2001), or an inability to isolate these viruses as is the case for OvHV-2. These consequently make *in vivo* studies difficult, and highlight the need for amenable animal model systems.

EBV infection of cotton top tamarin, a New World non-human primate, has proven to be a useful tool for studying the oncogenic potential of EBV (Miller *et al.*, 1977; Shope *et al.*, 1973) and possible therapeutic strategies (Mackett *et al.*, 1996). However, the immunobiology of a New World primate such as cotton top tamarin may not accurately model that of humans in response to EBV infection. There are simian homologues of EBV, such as rhesus lymphocryptovirus (Rhesus LCV), that naturally infect rhesus macaques (Old World non-human primate), which can provide a suitable animal model system for EBV pathogenesis (Wang *et al.*, 2001). It was recognized that unlike infecting rhesus macaques with EBV, which results in poor experimental infection, infection of these primates with Rhesus LCV mimicked human EBV infection more effectively (Moghaddam *et al.*, 1997).

A recent study also investigated KSHV infection in simian immunodeficiency virus (SIV)-positive and SIV-negative macaques (Renne *et al.*, 2004). This group was unable to detect any KSHV-specific antibodies, significant levels of mRNA or any sign of the symptoms relating to KSHV pathogenesis (such as Kaposi's sarcoma or lymphoproliferative disease), and therefore concluded that this model system was of little use for investigating KSHV pathogenesis. A more promising model for AIDS-related KSHV disease is experimental infection of SIV-infected macaques with a strain of rhesus rhadinovirus (RRV 17577, a simian homologue of KSHV). These animals developed B-cell hyperplasia reminiscent of lymphoproliferative disorders found in AIDS patients co-infected with KSHV (Wong *et al.*, 1999).

Apart from the numerous problems already mentioned, primate studies are costly and can only be carried out on a small scale. Therefore, infection of laboratory mice with the γ2-herpesvirus murid herpesvirus 4 (MuHV-4, also known as MHV-68 or γHV68) was developed as a suitable model system. This model was pioneered by Tony Nash, Stacey Efstathiou and others at the University of Cambridge in the 1980s. MuHV-4 is closely related to EBV and KSHV (Efstathiou *et al.*, 1990a, b; Virgin *et al.*, 1997) and is capable of infecting both inbred (including transgenic and 'knockout') and out-bred strains of mice (Blaskovic *et al.*, 1980; Macrae *et al.*, 2001; Nash *et al.*, 2001; Rajcani *et al.*, 1985; Stewart, 1999; Sunil-Chandra *et al.*, 1992a) as well as a number of cell lines *in vitro* (Dutia *et al.*, 1999b; Sunil-Chandra *et al.*, 1993; Usherwood *et al.*, 1996b). More recently, bacterial artificial chromosome clones of the virus have been developed (Adler *et al.*, 2000; Jia *et al.*, 2004), enabling the functions of virus sequences to be analysed by specific mutation of loci. For these reasons, experimental infection with MuHV-4 has proven to be a powerful and amenable model system for investigating γ-herpesvirus pathobiology.

The aim of this chapter is to describe how, using the MuHV-4 system, manipulation of the genetics of the virus and the host has enabled mechanisms underlying the pathogenesis of the γ-herpesviruses to be better understood.

NATURAL HISTORY OF MuHV-4

MuHV-4 is a natural pathogen of free-living murid rodents (Blasdell *et al.*, 2003; Blaskovic *et al.*, 1980). MuHV-4 (originally called MHV-68) was isolated, along with four similar herpesviruses (MHV-60, MHV-72, MHV-76 and MHV-78), during a study into small animal viruses in Slovakia in 1976 (Blaskovic *et al.*, 1980). Since then, three more related herpesviruses (MHV-Sumava, MHV-4556 and MHV-5682) have been isolated by the same group (Mistrikova *et al.*, 2000). Molecular analyses (Macakova *et al.*, 2003; Macrae *et al.*, 2001) have shown that at least MHV-76 and MHV-72 (and almost certainly the other isolates) are related strains of MuHV-4. Given the number of strains now discovered, we have chosen to adopt the International Committee on Taxonomy of Viruses nomenclature (MuHV-4) for this virus.

A notable advantage of MuHV-4 is its ability to replicate to high titres in tissue culture and form plaques (Ciampor *et al.*, 1982; Sunil-Chandra *et al.*, 1992a; Svobodova *et al.*, 1982). Initially, MuHV-4 was classified as an α-herpesvirus (Svobodova *et al.*, 1982). However, subsequent pathological features, such as severe pneumonia and systemic viraemia (Blaskovic *et al.*, 1984; Rajcani *et al.*, 1985), and sequence analysis (Efstathiou *et al.*, 1990a, b) have led to the reclassification of MuHV-4 as a γ-herpesvirus.

MuHV-4 strain 68, along with strains 60 and 72, were originally isolated from bank voles (*Clethrionomys glareolus*) and strains 76 and 78 were isolated from the yellow-necked mouse (*Apodemus flavicollis*) (Blaskovic *et al.*, 1980). Serological studies had been reported which determined the prevalence of MuHV-4-like viruses in wild rodents in Slovakia (Kozuch *et al.*, 1993; Mistrikova & Blaskovic, 1985). These studies, however, failed to define the prevalence of MuHV-4 infection within a particular species of rodent. A more recent epidemiological survey by Blasdell *et al.* (2003) on the presence of the virus in the UK revealed that *Apodemus* spp. (in this case *Apodemus sylvaticus*, the wood mouse) is in fact the natural host with the bank vole being occasionally infected (Blasdell *et al.*, 2003). Although called mice, wood and yellow-necked mice belong to a separate genus from the free-living relative of laboratory mice, *Mus*

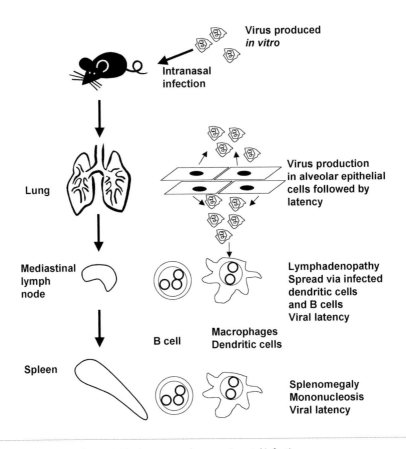

Fig. 1. Pathogenesis of MuHV-4 in the mouse after experimental infection.

musculus/domesticus. Further serological and molecular studies by the Liverpool group have revealed that MuHV-4 does not naturally infect free-living *M. musculus*. In future studies, we intend to compare the experimental infection of the model host (laboratory mice) with that of its natural hosts (wood mice and bank voles).

EXPERIMENTAL INFECTION WITH MuHV-4 *IN VIVO*

The natural infection of the host by γ-herpesviruses is modelled by experimental infection of laboratory mice with MuHV-4. A schematic diagram of the course of MuHV-4 infection in the mouse is shown in Fig. 1. The natural route of MuHV-4 infection is not known. However, a likely route of infection is the respiratory tract (Nash *et al.*, 2001). Recent evidence suggests that neonatal transmission may also occur via the mammary gland (Raslova *et al.*, 2001). The majority of research groups working with MuHV-4 prefer the respiratory route for introducing the virus into the host (Macrae *et al.*, 2001; Stewart *et al.*, 1998; Sunil-Chandra *et al.*, 1992a, 1994). However, others have opted for the intraperitoneal route (Jacoby *et al.*, 2002; Weck *et al.*, 1996, 1999), which may be equally valid and is certainly useful when investigating replication-deficient viruses. However, differing routes of infection are not equivalent and care therefore needs to be taken when interpreting results (Jacoby *et al.*, 2002; van Berkel *et al.*, 2002).

Infection of laboratory mice with MuHV-4 via the respiratory route results in an acute, productive replication in lung epithelial cells and also occasionally within alveolar macrophages (Sunil-Chandra *et al.*, 1992a). There is a reactive inflammatory response to infection characterized by a severe increase in cellularity with evidence of necrosis and mononuclear and lymphocyte infiltration. Dissecting the inflammatory response by analysing the bronchial alveolar lavage, it was shown that the macrophage response peaks around day 3, followed by infiltration of CD8$^+$ T cells, which peaks by day 7. The acute productive infection is cleared around day 10–14 post-infection (p.i.) mainly by virus-specific CD8$^+$ T cells (see later).

Like all herpesviruses, MuHV-4 persists for the lifetime of the infected host and the virus becomes latent in selected cell types and at distinct anatomical locations. Latency involves genome maintenance in the absence of virus particle production. So, after the clearance of acute infection in the lung, MuHV-4 becomes latent in epithelial cells and B lymphocytes at this site (Flano *et al.*, 2003; Stewart *et al.*, 1998). During the first 5 days of infection, MuHV-4 also spreads to the spleen, where it also becomes latent. Like EBV and KSHV, MuHV-4 latency is mainly seen within B lymphocytes, but also in macrophages and dendritic cells (Flano *et al.*, 2000; Sunil-Chandra *et al.*, 1992b; Usherwood *et al.*, 1996c; Weck *et al.*, 1999).

Spread to the spleen and the establishment of latency is associated with a marked increase in cell numbers in the spleen (splenic mononucleosis) (Usherwood *et al.*, 1996a) and a subsequent mononucleosis in the bloodstream that is reminiscent of IM (glandular fever) caused by primary infection of humans by EBV (Tripp *et al.*, 1997). Splenomegaly and splenic mononucleosis, which peak at day 14 p.i., are driven by CD4$^+$ T cells (Ehtisham *et al.*, 1993; Usherwood *et al.*, 1996a) and are dependent on MuHV-4-infected B cells in the spleen (Usherwood *et al.*, 1996c; Weck *et al.*, 1996). At the same time as the mononucleosis in the spleen, there is a sharp rise in the number of latently infected B cells. This increase in latently infected B cells triggers a CD8$^+$ T-cell response specific to virus proteins expressed during latency. Virus-specific T cells halt the rise in latently infected cells but do not result in clearance and a constant baseline level of latently infected B cells is maintained for the lifetime of the mouse (Ehtisham *et al.*, 1993; Tibbetts *et al.*, 2002; Weck *et al.*, 1996). CD8$^+$ T cells along with antibody (Kim *et al.*, 2002; Stewart *et al.*, 1998) are important in the long-term control of persistent infection (Cardin *et al.*, 1996; Stewart *et al.*, 1998; Weck *et al.*, 1996).

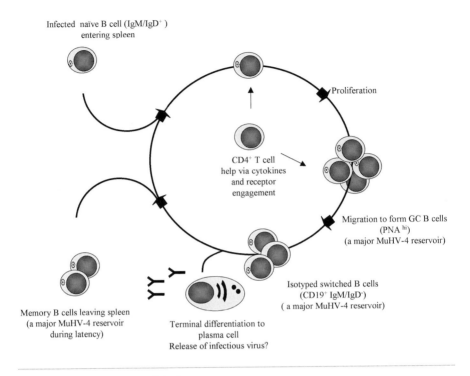

Infected naïve B cell (IgM/IgD⁺)
entering spleen

Proliferation

CD4⁺ T cell
help via cytokines
and receptor
engagement

Migration to form GC B cells
(PNA ʰⁱ)
(a major MuHV-4 reservoir)

Isotyped switched B cells
(CD19⁺ IgM/IgD⁻)
(a major MuHV-4 reservoir)

Memory B cells leaving spleen
(a major MuHV-4 reservoir
during latency)

Terminal differentiation to
plasma cell
Release of infectious virus?

Fig. 2. MuHV-4 life cycle in B cells in the mouse.

The infection of B lymphocytes is a key part of MuHV-4 persistence. Early in infection, many B-cell subsets, including naïve B cells, harbour MuHV-4, but late in infection (2 months or more) the virus is found almost exclusively in germinal centre (GC) B cells (Flano *et al.*, 2000, 2003; Willer & Speck, 2003). GC B cells can give rise to memory B cells, which are long-lived, or to antibody producing plasma cells. Memory B cells can circulate to the bone marrow, spleen and other lymphoid tissues where they may go through terminal differentiation. Indeed, bone marrow is a major site of MuHV-4 latency. Therefore, by preferentially residing in GC B cells, MuHV-4, in a similar fashion to EBV and KSHV, has evolved mechanisms to infect long-lived, re-circulating B cells. These properties help long-term MuHV-4 maintenance as well as spread of infected cells to other organs (Fig. 2).

The survival of circulating B cells requires signals from T cells as well as signals through B-cell receptor. In EBV infection, a number of viral proteins (LMP-1 and LMP-2A) have been shown to provide T-cell-independent survival signals to EBV-infected memory B cells (Caldwell *et al.*, 1998). However, MuHV-4 does not encode homologues of LMP-1 or LMP-2A. This would suggest that either these survival signals are provided by as-yet-

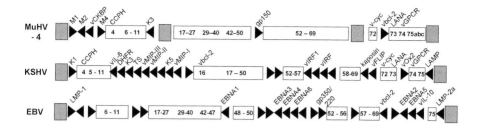

Fig. 3. Organization of the MuHV-4 genome as compared to that of KSHV (Russo *et al.*, 1996) and EBV (Baer *et al.*, 1984). The open boxes show homologous herpesvirus core gene blocks. These are genes involved in DNA replication and virus structure. Terminal repeat (TR) elements are shown by hashed bars. Gammaherpesvirus-specific genes and functions are indicated as triangles. Functions are abbreviated as follows: vCKBP, viral chemokine-binding protein; CCPH, complement control protein homologue; vIL-6, viral interleukin-6; DHFR, dihydrofolate reductase; TS, thymidylate synthase; vMIP, viral macrophage inflammatory factor; vIRF, viral interferon regulatory protein; vFLIP, viral FLICE inhibitor protein; LANA, latent nuclear antigen; vOx2, viral Ox2; vGPCR, viral G-protein coupled receptor; LAMP, latency-associated membrane protein; EBNA, Epstein–Barr virus nuclear antigen; LMP, latent membrane protein; vIL-10, viral interleukin-10; v-cyc, viral cyclin D; vbcl-2, viral Bcl-2.

unidentified MuHV-4 protein(s) or MuHV-4 uses other novel mechanisms mediated by proteins of cellular origin.

THE MuHV-4 GENOME

The MuHV-4 genome contains 118 kbp of unique double-stranded DNA flanked by variable numbers of 1·23 kbp terminal repeat regions (Efstathiou *et al.*, 1990a, b). The genome has been sequenced by two laboratories and contains approximately 73 protein-coding open reading frames (ORFs), the majority of which are collinear and homologous to those of other gammaherpesviruses (Nash *et al.*, 2001; Virgin *et al.*, 1997). Fig. 3 shows a diagrammatic representation of the MuHV-4 genome and the genomes of KSHV and EBV for comparison. All three viruses contain four conserved gene blocks comprising core herpesvirus genes (open boxes). These are numbered (ORF4 through ORF75) according to homologous genes in the prototypic herpesvirus saimiri (Albrecht *et al.*, 1992). Outside these gene blocks lie genes that are largely unique to each individual virus and encode functions that are important for virus pathogenesis in the host. In MuHV-4, these are numbered M1 through M11. There are also eight viral tRNA-like genes that are positioned at the left end of the unique portion of the genome. Some gene homologues (complement control protein homologue, viral cyclin, latency-associated nuclear antigen, viral Bcl-2 and viral G-protein coupled receptor) are shared between MuHV-4 and KSHV and this has allowed direct comparison of functions in the MuHV-4 model *in vivo* (Fowler *et al.*, 2003; Gangappa *et al.*, 2002b; Moorman *et al.*, 2003a, b; van Dyk *et al.*, 2000). The functions of some of the so-called 'unique' genes, however, still remain cryptic.

MOLECULAR DETERMINANTS OF MuHV-4 LYTIC AND LATENT INFECTIONS

As identified above, latency is a fundamental aspect of γ-herpesvirus biology, allowing persistence in the host. The occasional reactivation of virus particle production is equally as important to enable spread to naïve hosts. Also, since B cells undergo sporadic activation and cell division, occasional reactivation of virion production may be needed to maintain a pool of latently infected cells within the host. Therefore, understanding the molecular basis of latency and the switch to virus production will help future therapeutic interventions and vaccine design. The use of virus genetics in the MuHV-4 model has provided valuable insights as to which components of virus and host-derived components may be responsible for the maintenance of latency and the switch between lytic and latent stages of virus infection.

A number of MuHV-4 genes have been associated with virus latency. These are the eight viral tRNA-like sequences, M2, M3, ORF12 (K3), v-cyclin (ORF72), ORF73, ORF65 (previously M9), vBcl-2 (previously M11) and G-protein coupled receptor (vGPCR) (Bowden *et al.*, 1997; Fowler *et al.*, 2003; Husain *et al.*, 1999; Moorman *et al.*, 2003b; Simas *et al.*, 1999; Virgin *et al.*, 1999).

M2 was the first MuHV-4 gene to be associated with virus latency (Husain *et al.*, 1999). M2 encodes a 30 kDa plasma-membrane-associated protein and is predominantly expressed in B cells (Macrae *et al.*, 2003). However, the precise role of M2 protein in MuHV-4 persistence is not entirely clear. Data obtained from infection with a natural deletion mutant of MuHV-4 (strain 76) and *in vitro* generated M2 mutants of MuHV-4 suggested that long-term virus persistence can be established in the absence of M2 (Clambey *et al.*, 2002; Jacoby *et al.*, 2002; Macrae *et al.*, 2001, 2003; Simas *et al.*, 2004). It has, however, been postulated that the M2 gene product may be required for efficient persistence in splenic follicles (Simas *et al.*, 2004). Recent work has also suggested that M2 may inhibit the cellular interferon response to MuHV-4 (Liang *et al.*, 2004).

Similarly, genetic mutants of other candidate latency-associated genes (M3, K3, v-cyclin, vGPCR and vBcl-2) have shown that none is essential for the establishment of virus latency in the mouse model although the efficiency of establishment and/or reactivation is affected by the deletion of these genes (Bridgeman *et al.*, 2001; Gangappa *et al.*, 2002b; Moorman *et al.*, 2003a; Stevenson *et al.*, 2002; van Berkel *et al.*, 2002; van Dyk *et al.*, 2000). Hence MuHV-4 persistence, albeit in a much reduced form, can take place in the absence of a number of viral genes associated with latency.

In contrast to other latency-associated genes, the MuHV-4 ORF73 is essential for establishing latency *in vivo* (Fowler *et al.*, 2003; Moorman *et al.*, 2003b). These

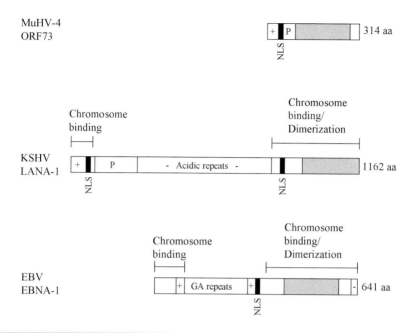

Fig. 4. MuHV-4 ORF73 and related genes in KSHV and EBV. Acidic (–) and basic (+) areas are shown, with nuclear localization (NLS) motifs in black and sections of repetitive sequence indicated. The grey boxes represent regions of conserved secondary structures. However, the amino acid sequence conservation between these related genes is low.

mutational analyses have shown that whereas ORF73 was dispensable for the initial productive phase of infection in the lung, latent infection was severely compromised *in vivo*. These studies also raised the question of whether the deficit in establishing virus persistence with ORF73 mutants was due to the failure of the virus to either (1) establish latency or (2) reactivate from latency in co-culture assays (or plaque assays) *in vitro*. The approach taken by Fowler and colleagues has shown that this deficit is more likely because ORF73 mutant virus cannot establish latency in lymphoid tissues rather than failure to reactivate when cultured with permissive cells *in vitro* (Fowler *et al.*, 2003).

The ORF73 protein is conserved amongst γ-herpesviruses. The direct sequence homologue of MuHV-4 ORF73 in KSHV is the latency-associated nuclear antigen (LANA-1). Although not a direct sequence homologue, the EBNA-1 protein of EBV shares structural and functional conservation with MuHV-4 ORF73 and LANA-1 (Fig. 4). KSHV LANA-1 is a multi-functional protein (reviewed by Viejo-Borbolla & Schulz, 2003). LANA-1 is critical for the persistence of the viral genome as an episome during latency, analogous to EBV EBNA-1. To perform this function, LANA-1 tethers the viral episome to host chromatin by binding to chromatin via its N-terminal region and to the terminal repeat element of the KSHV genome through its C-terminal region. LANA-1 can also act as a modulator of cellular and viral transcription. By binding to p53, LANA-1 inhibits p53-mediated apoptosis and induces E2F-dependent promoters by binding to the retinoblastoma protein. It also interacts through its C-terminus with several members of the fsh family of BET proteins which are involved in transcriptional regulation. MuHV-4 ORF73 is considerably smaller than LANA-1 but does show structural homology and sequence homology, particularly with the C-terminal region of LANA (Fig. 4). It is not currently known how much functional conservation there is between MuHV-4 ORF73 and LANA-1. Thus while it seems likely, based on sequence comparisons, that the MuHV-4 ORF73 will have episomal maintenance and transcriptional modulation functions, this awaits empirical confirmation.

Like the establishment of latency, one protein has been found to be critical for the reactivation of MuHV-4 from latency, the replication and transcription activator (Rta), which is encoded by ORF50 and is the main transcriptional transactivation factor during the productive phase of infection (Virgin *et al.*, 1997). Like ORF73, this key protein is conserved throughout γ-herpesviruses with direct sequence homologues in KSHV and EBV (Fig. 5). MuHV-4 Rta was found to be functionally analogous to EBV and KSHV Rta in that it was able to induce directly lytic replication of viral DNA and was essential for reactivation of MuHV-4 from latently infected B lymphocytes (Wu *et al.*, 2000). Significantly, it was also demonstrated that overexpression of Rta was detrimental to the establishment of latency *in vivo* (May *et al.*, 2004; Moorman *et al.*, 2004; Rickabaugh *et al.*, 2004). Taken together, these important observations showed that Rta was an important molecular switch which controls the lytic and latent phases of MuHV-4 infection.

Further experiments suggest that there may be a crucial interplay between Rta and ORF73 and that the balance between these two proteins is important for determining whether the virus undergoes productive replication or becomes latent. Thus over-expression of Rta in a recombinant virus resulted in the failure of MuHV-4 to establish efficient latency *in vivo* (Boname *et al.*, 2004; Rickabaugh *et al.*, 2004). In another γ-herpesvirus, herpesvirus saimiri, ORF73 could inhibit lytic virus replication by directly binding to the Rta promoter (Schafer *et al.*, 2003). Therefore, a picture is emerging whereby the protein encoded by ORF73 influences latency by: (1) repression of lytic cycle genes by inhibiting Rta; and (2) tethering the viral episome to the chromatin to allow division of virus genome alongside daughter cells during mitosis. In contrast, induction of Rta expression can break latency as well as being essential for productive replication (Fig. 6). Further experiments on cellular (and viral) factors important for the activity of Rta and ORF73 gene promoters will therefore be required to help our understanding of the latency/productive cycle switch.

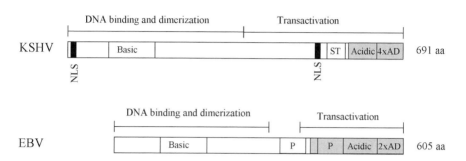

Fig. 5. MuHV-4 Rta and related genes in KSHV and EBV. The putative nuclear localization signal (NLS), serine/threonine-rich region (ST) and poly-proline motifs (P) are indicated. The basic regions are boxed and acidic areas are shaded grey. EBV and KSHV Rta both contain a set of conserved repeats consisting of acidic residues regularly dispersed with bulky hydrophobic amino acids in their transactivation domains (AD).

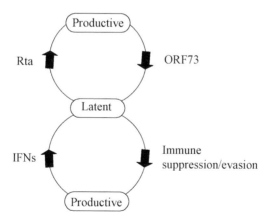

Fig. 6. Molecular switches controlling MuHV-4 productive and latent infections. ORF73 is essential for latency and may repress Rta expression. Induction of Rta expression reactivates virus production from latency and is essential for productive infection. In turn, host responses such as interferons (IFNs) inhibit and clear productive infection but immune suppression can result in reactivation.

ANTIVIRAL IMMUNE RESPONSES TO MuHV-4 INFECTION

Both T- and B-lymphocyte-derived immune responses have been shown to play a crucial role in controlling both lytic and persistent phases of MuHV-4 infection by allowing immune recognition and subsequent elimination of virus-infected cells.

The primary role of B cells during viral infection is the production of neutralizing antibodies against the invading pathogen. MuHV-4-specific IgM antibodies are evident within the first week after infection, leading to a prominent class-switched antibody response by the second week post MuHV-4 infection. The latter is dominated by IgG2a and IgG2b isotypes (Sangster *et al.*, 2000; Stevenson & Doherty, 1998). A notable feature in these experiments was the delayed B-cell-derived responses to the lytic infection, the reasons for which are not entirely clear.

Studies in mice which lack mature B cells and hence cannot make antibodies (μMT mice) have demonstrated that B-cell responses are important in controlling long-term MuHV-4 infection *in vivo* (Stewart *et al.*, 1998; Weck *et al.*, 1996). It has also been shown that passive transfer of immune serum to mice infected with MuHV-4 significantly inhibited virus reactivation (Gangappa *et al.*, 2002a). Currently, there is no evidence that antibody responses are directed against latently infected cells. Rather, antibody responses may control the reactivation of MuHV-4 from latently infected cells. It is plausible that low levels of reactivation may contribute to the latently infected pool of B cells (Flano *et al.*, 2003). This may explain the presence of activated CD4[+] and CD8[+] T cells specific for viral lytic antigens throughout the MuHV-4 infection (Flano *et al.*, 2001; Stevenson *et al.*, 1999). There is, however, no direct evidence to support significant episodic reactivation of MuHV-4 *in vivo*.

The precise role of CD4[+] T cell mediated antiviral responses in MuHV-4 infection is less clear. It has been shown that CD4[+] T cells can transiently control the MuHV-4 lytic infection via cytokines such as IFN-γ and by facilitating antibody responses by B cells (Christensen & Doherty, 1999; Sparks-Thissen *et al.*, 2004). A limited number of MuHV-4 epitopes recognized by CD4[+] T cells have been identified (Flano *et al.*, 2001). CD4[+] T cells can potentially remove virus-infected cells by functioning as cytotoxic T cells (CTLs). For example, EBV EBNA-1-specific CD4[+] T cells could cause direct lysis of Burkitt's lymphoma cells (Paludan *et al.*, 2002). Currently, there is no evidence of such activity by CD4[+] T cells in MuHV-4 infection. Nevertheless, CD4[+] T cells are necessary for the efficient B-cell responses by providing growth factors such as cytokines and by engaging receptors such as CD40 on B cells. This is because CD40–CD40L interactions are vital in maturation, proliferation and survival of antigen primed B cells. Indeed, lack of CD40–CD40L interactions was shown to result in short-lived CD40[+] B cells, which in turn resulted in reduced memory B cells and hence reduced virus load during the persistent phase of MuHV-4 infection *in vivo* (Kim *et al.*, 2003). Taken together, these observations suggest that the major roles of CD4[+] T cells in MuHV-4 infection are (1) efficient activation and expansion of CD8[+] T and B cells and (2) maintenance of long-term B-cell-mediated antibody responses.

CD8[+] T lymphocytes, which upon receiving appropriate signals can become CTLs, are crucial in controlling most virus infections. CD8[+] T cells play an important role in the early (lytic) phase of MuHV-4 infection. It was shown that CTLs, independent of CD4[+] T and B cells, could efficiently clear the MuHV-4 lytic infection *in vivo* (Cardin *et al.*, 1996; Christensen & Doherty, 1999; Ehtisham *et al.*, 1993; Stevenson & Doherty, 1998). This is because CTLs recognize a number of MHC class I restricted epitopes within both lytic (e.g. single-stranded DNA-binding protein) as well as latent viral antigens (e.g. M2) (Obar *et al.*, 2004; Stevenson *et al.*, 1999). Despite these strong CTL responses, MuHV-4 can still establish and maintain latency *in vivo*. This is probably due to specific viral immune evasion strategies. Indeed, MuHV-4 encodes a protein which can interfere with MHC class I mediated immune responses. This viral protein, K3, is a homologue of KSHV K3 (Ishido *et al.*, 2000; Virgin *et al.*, 1997). MuHV-4 K3 was shown to down-regulate the surface expression of MHC class I. Using a K3-deficient virus, it was also demonstrated that K3 does not play a significant role during the lytic phase of infection, but may help maintain viral load during the latent phase (Stevenson *et al.*, 2000, 2002).

Another immune evasion mechanism employed by MuHV-4 is the chemokine-binding protein (vCKBP) encoded by the *M3* gene (Parry *et al.*, 2000; van Berkel *et al.*, 1999, 2000). Rapid trafficking of immune cells to the site of infection constitutes an important part of host immunity to viral infections. This trafficking is regulated by a large number of chemoattractant proteins or chemokines. Studies using MuHV-4 mutants lacking functional vCKBP have shown that the presence of this decoy receptor may provide an immune evasion strategy by (1) reducing the trafficking of inflammatory cells, e.g. CD8[+] T cells, to the site of infection and (2) preventing/reducing the emigration of virus antigen-loaded dendritic cells to distinct locations within secondary lymphoid tissues where efficient priming/activation of immune cells takes place (Bridgeman *et al.*, 2001; Jensen *et al.*, 2003; van Berkel *et al.*, 2002).

There is therefore a host–virus balance which enables MuHV-4 to persist. MuHV-4 induces strong antiviral immune responses which serve to limit viral infection. However, the virus in turn expresses immune evasion proteins which enable the virus to establish latency and persist in the face of host immunity.

CONTROL OF MuHV-4 INFECTION BY TYPE I INTERFERONS

In addition to specific immunity, cell-independent host-derived antiviral responses also play a vital role in controlling the initial phase of MuHV-4 infection in the naïve host. Infection of mice lacking type I interferon (IFN) receptor or interferon regulatory protein-1 (IRF-1), unlike immunocompetent mice, with MuHV-4 resulted in severe disease because of uncontrolled lytic virus load (Dutia *et al.*, 1999a). Therefore, type I

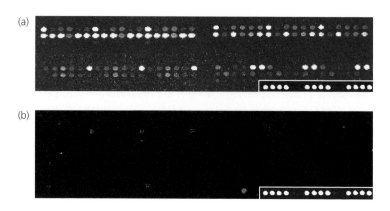

Fig. 7. MuHV-4 productive infection causes loss of cellular mRNA. DNA microarrays were used to profile the impact of MuHV-4 infection on cellular transcripts before (a) and after (b) infection. Note loss of transcripts in (b) after MuHV-4 infection. The insets are control microbial RNA. From Ebrahimi *et al.* (2003).

IFNs play a direct role in controlling the lytic phase of MuHV-4 infection and more than likely prevent reactivation from latency. However, the molecular bases of IFN-mediated responses are far from clear.

Using MuHV-4-specific DNA microarrays (or DNA chips), it was possible to detect all classes of MuHV-4 transcripts (immediate early, early, and late) even in the presence of plaque inhibitory levels of type I IFNs (L. Roaden & B. Ebrahimi, unpublished data). This observation suggested that at least viral attachment, uncoating or entry of viral DNA into the cell nucleus was not inhibited by IFN treatment. However, using more sensitive assays such as PCR, IFN treatment did cause a significant reduction in mRNA expression of Rta, ORF73, as well as glycoproteins gL and gp150 (L. Roaden & B. Ebrahimi, unpublished data). Therefore, direct down-regulation of viral genes by type I IFNs may, at least in part, interfere with MuHV-4 lytic infection.

Viruses in general and γ-herpesviruses specifically have evolved a number of strategies to evade host-derived antiviral responses. For example, KSHV encodes cellular homologues of IRF. Moreover, vIRF-1 and -3 have been shown to inhibit antiviral responses mediated by IFNs (Burysek *et al.*, 1999; Zimring *et al.*, 1998). In addition, KSHV also encodes a homologue of cellular IL-6 (vIL-6), which contains two putative IFN-stimulated response element (ISRE) sequences. It is believed that vIL-6 antagonizes the IFN action by disrupting the phosphorylation of Tyk2, hence decoupling the Jak/STAT pathway (Chatterjee *et al.*, 2002). The Jak/STAT pathway is crucial in IFN signalling. Based on sequence analyses, MuHV-4 does not encode homologues of IL-6 or IRFs. However, recent observations that the M2 protein can interfere with STAT-1/2 signalling pathways may provide a clue as to how MuHV-4 escapes IFN-mediated antiviral responses *in vivo* (Liang *et al*., 2004).

Another mechanism by which MuHV-4 may subvert antiviral responses is interference with cellular transcription machinery. It was shown previously that MuHV-4 caused a significant loss of cellular transcripts after infection (Ebrahimi *et al*., 2003) (Fig. 7). More recent observations suggest that type I IFN can diminish MuHV-4-mediated degradation of cellular mRNA (L. Roaden & B. Ebrahimi, unpublished data). Therefore, type I IFNs can limit MuHV-4 infection in two ways: (1) direct suppression of viral gene expression; and (2) inhibition of virus-mediated loss of cellular transcripts.

CONCLUSION

MuHV-4 infection of mice is an important model system for the study of the pathogenesis of an important group of pathogens, the γ-herpesviruses. As shown in this chapter, its real power is the ability to be able to translate *in vitro* observations on individual γ-herpesvirus functions into their role in the host. The functions of many γ-herpesvirus genes and non-coding regions have still to be elucidated and intelligent genetic alterations of MuHV-4 will be important in determining their roles in viral pathogenesis and persistence. In addition, there are currently no effective vaccines or therapies against γ-herpesvirus-mediated disease, and therefore MuHV-4 will have an important role in formulating successful future therapeutic interventions.

ACKNOWLEDGEMENTS

The authors wish to acknowledge the Biotechnology and Biological Sciences Research Council, the National Institutes of Health (USA) and the Royal Society for support. J. P. S. is a Royal Society University Research Fellow.

REFERENCES

Adler, H., Messerle, M., Wagner, M. & Koszinowski, U. H. (2000). Cloning and mutagenesis of the murine gammaherpesvirus 68 genome as an infectious bacterial artificial chromosome. *J Virol* **74**, 6964–6974.

Albrecht, J. C., Nicholas, J., Biller, D. & 8 other authors (1992). Primary structure of the herpesvirus saimiri genome. *J Virol* **66**, 5047–5058.

Baer, R., Bankier, A. T., Biggin, M. D. & 9 other authors (1984). DNA sequence and expression of the B95-8 Epstein-Barr virus genome. *Nature* **310**, 207–211.

Blasdell, K., McCracken, C., Morris, A., Nash, A. A., Begon, M., Bennett, M. & Stewart, J. P. (2003). The wood mouse is a natural host for *Murid herpesvirus 4*. *J Gen Virol* **84**, 111–113.

Blaskovic, D., Stancekova, M., Svobodova, J. & Mistrikova, J. (1980). Isolation of five strains of herpesviruses from two species of free living small rodents. *Acta Virol* **24**, 468.

Blaskovic, D., Stanekova, D. & Rajcani, J. (1984). Experimental pathogenesis of murine herpesvirus in newborn mice. *Acta Virol* **28**, 225–231.

Boname, J. M., Coleman, H. M., May, J. S. & Stevenson, P. G. (2004). Protection against wild-type murine gammaherpesvirus-68 latency by a latency-deficient mutant. *J Gen Virol* **85**, 131–135.

Bowden, R. J., Simas, J. P., Davis, A. J. & Efstathiou, S. (1997). Murine gammaherpesvirus 68 encodes tRNA-like sequences which are expressed during latency. *J Gen Virol* **78**, 1675–1687.

Bridgeman, A., Stevenson, P. G., Simas, J. P. & Efstathiou, S. (2001). A secreted chemokine binding protein encoded by murine gammaherpesvirus-68 is necessary for the establishment of a normal latent load. *J Exp Med* **194**, 301–312.

Burysek, L., Yeow, W. S., Lubyova, B., Kellum, M., Schafer, S. L., Huang, Y. Q. & Pitha, P. M. (1999). Functional analysis of human herpesvirus 8-encoded viral interferon regulatory factor 1 and its association with cellular interferon regulatory factors and p300. *J Virol* **73**, 7334–7342.

Caldwell, R. G., Wilson, J. B., Anderson, S. J. & Longnecker, R. (1998). Epstein-Barr virus LMP2A drives B cell development and survival in the absence of normal B cell receptor signals. *Immunity* **9**, 405–411.

Cardin, R. D., Brooks, J. W., Sarawar, S. R. & Doherty, P. C. (1996). Progressive loss of CD8+ T cell-mediated control of a gamma-herpesvirus in the absence of CD4+ T cells. *J Exp Med* **184**, 863–871.

Chatterjee, M., Osborne, J., Bestetti, G., Chang, Y. & Moore, P. S. (2002). Viral IL-6-induced cell proliferation and immune evasion of interferon activity. *Science* **298**, 1432–1435.

Christensen, J. P. & Doherty, P. C. (1999). Quantitative analysis of the acute and long-term CD4(+) T-cell response to a persistent gammaherpesvirus. *J Virol* **73**, 4279–4283.

Ciampor, F., Stancekova, M., Svobodova, J. & Mistrikova, J. (1982). Electron microscopy of rabbit embryo fibroblasts infected with herpesvirus isolates from *Clethrionomys glareolus* and *Apodemus flavicollis*. *Acta Virol* **25**, 101–107.

Clambey, E. T., Virgin, H. W., IV & Speck, S. H. (2002). Characterization of a spontaneous 9·5-kilobase-deletion mutant of murine gammaherpesvirus 68 reveals tissue-specific genetic requirements for latency. *J Virol* **76**, 6532–6544.

Coulter, L. J., Wright, H. & Reid, H. W. (2001). Molecular genomic characterization of the viruses of malignant catarrhal fever. *J Comp Pathol* **124**, 2–19.

Dutia, B. M., Allen, D. J., Dyson, H. & Nash, A. A. (1999a). Type I interferons and IRF-1 play a critical role in the control of a gammaherpesvirus infection. *Virology* **261**, 173–179.

Dutia, B. M., Stewart, J. P., Clayton, R. A., Dyson, H. & Nash, A. A. (1999b). Kinetic and phenotypic changes in murine lymphocytes infected with murine gamma-herpesvirus-68 *in vitro*. *J Gen Virol* **80**, 2729–2736.

Ebrahimi, B., Dutia, B. M., Roberts, K. L., Garcia-Ramirez, J. J., Dickinson, P., Stewart, J. P., Ghazal, P., Roy, D. J. & Nash, A. A. (2003). Transcriptome profile of murine gammaherpesvirus-68 lytic infection. *J Gen Virol* **84**, 99–109.

Efstathiou, S., Ho, Y. M., Hall, S., Styles, C. J., Scott, S. D. & Gompels, U. A. (1990a). Murine herpesvirus 68 is genetically related to the gammaherpesviruses Epstein-Barr virus and herpesvirus saimiri. *J Gen Virol* **71**, 1365–1372.

Efstathiou, S., Ho, Y. M. & Minson, A. C. (1990b). Cloning and molecular characterization of the murine herpesvirus 68 genome. *J Gen Virol* **71**, 1355–1364.

Ehtisham, S., Sunil-Chandra, N. P. & Nash, A. A. (1993). Pathogenesis of murine gammaherpesvirus infection in mice deficient in CD4 and CD8 T cells. *J Virol* **67**, 5247–5252.

Flano, E., Husain, S. M., Sample, J. T., Woodland, D. L. & Blackman, M. A. (2000). Latent murine gamma-herpesvirus infection is established in activated B cells, dendritic cells, and macrophages. *J Immunol* **165**, 1074–1081.

Flano, E., Woodland, D. L., Blackman, M. A. & Doherty, P. C. (2001). Analysis of virus-specific CD4(+) T cells during long-term gammaherpesvirus infection. *J Virol* **75**, 7744–7748.

Flano, E., Kim, I. J., Moore, J., Woodland, D. L. & Blackman, M. A. (2003). Differential gamma-herpesvirus distribution in distinct anatomical locations and cell subsets during persistent infection in mice. *J Immunol* **170**, 3828–3834.

Fowler, P., Marques, S., Simas, J. P. & Efstathiou, S. (2003). ORF73 of murine herpesvirus-68 is critical for the establishment and maintenance of latency. *J Gen Virol* **84**, 3405–3416.

Gangappa, S., Kapadia, S. B., Speck, S. H. & Virgin, H. W., IV (2002a). Antibody to a lytic cycle viral protein decreases gammaherpesvirus latency in B-cell-deficient mice. *J Virol* **76**, 11460–11468.

Gangappa, S., van Dyk, L. F., Jewett, T. J., Speck, S. H. & Virgin, H. W., IV (2002b). Identification of the in vivo role of a viral bcl-2. *J Exp Med* **195**, 931–940.

Husain, S. M., Usherwood, E. J., Dyson, H., Coleclough, C., Coppola, M. A., Woodland, D. L., Blackman, M. A., Stewart, J. P. & Sample, J. T. (1999). Murine gamma-herpesvirus M2 gene is latency-associated and its protein a target for CD8(+) T lymphocytes. *Proc Natl Acad Sci U S A* **96**, 7508–7513.

Ishido, S., Wang, C., Lee, B. S., Cohen, G. B. & Jung, J. U. (2000). Downregulation of major histocompatibility complex class I molecules by Kaposi's sarcoma-associated herpesvirus K3 and K5 proteins. *J Virol* **74**, 5300–5309.

Jacoby, M. A., Virgin, H. W., IV & Speck, S. H. (2002). Disruption of the M2 gene of murine gammaherpesvirus 68 alters splenic latency following intranasal, but not intraperitoneal, inoculation. *J Virol* **76**, 1790–1801.

Jensen, K. K., Chen, S. C., Hipkin, R. W., Wiekowski, M. T., Schwarz, M. A., Chou, C. C., Simas, J. P., Alcami, A. & Lira, S. A. (2003). Disruption of CCL21-induced chemotaxis in vitro and in vivo by M3, a chemokine-binding protein encoded by murine gammaherpesvirus 68. *J Virol* **77**, 624–630.

Jia, Q., Wu, T. T., Liao, H. I., Chernishof, V. & Sun, R. (2004). Murine gammaherpesvirus 68 open reading frame 31 is required for viral replication. *J Virol* **78**, 6610–6620.

Kim, I. J., Flano, E., Woodland, D. L. & Blackman, M. A. (2002). Antibody-mediated control of persistent gamma-herpesvirus infection. *J Immunol* **168**, 3958–3964.

Kim, I. J., Flano, E., Woodland, D. L., Lund, F. E., Randall, T. D. & Blackman, M. A. (2003). Maintenance of long term gamma-herpesvirus B cell latency is dependent on CD40-mediated development of memory B cells. *J Immunol* **171**, 886–892.

Kozuch, O., Reichel, M., Lesso, J., Remenova, A., Labuda, M., Lysy, J. & Mistrikova, J. (1993). Further isolation of murine herpesviruses from small mammals in southwestern Slovakia. *Acta Virol* **37**, 101–105.

Liang, X., Shin, Y. C., Means, R. E. & Jung, J. U. (2004). Inhibition of interferon-mediated antiviral activity by murine gammaherpesvirus 68 latency-associated M2 protein. *J Virol* **78**, 12416–12427.

Macakova, K., Matis, J., Rezuchova, I., Kudela, O., Raslova, H. & Kudelova, M. (2003). Murine gammaherpesvirus (MHV) M7 gene encoding glycoprotein 150 (gp150): difference in the sequence between 72 and 68 strains. *Virus Genes* **26**, 89–95.

Mackett, M., Cox, C., Pepper, S. D., Lees, J. F., Naylor, B. A., Wedderburn, N. & Arrand, J. R. (1996). Immunisation of common marmosets with vaccinia virus expressing Epstein-Barr virus (EBV) gp340 and challenge with EBV. *J Med Virol* **50**, 263–271.

Macrae, A. I., Dutia, B. M., Milligan, S., Brownstein, D. G., Allen, D. J., Mistrikova, J., Davison, A. J., Nash, A. A. & Stewart, J. P. (2001). Analysis of a novel strain of murine gammaherpesvirus reveals a genomic locus important for acute pathogenesis. *J Virol* **75**, 5315–5327.

Macrae, A. I., Usherwood, E. J., Husain, S. M. & 7 other authors (2003). Murid herpesvirus 4 strain 68 M2 protein is a B-cell-associated antigen important for latency but not lymphocytosis. *J Virol* **77**, 9700–9709.

May, J. S., Coleman, H. M., Smillie, B., Efstathiou, S. & Stevenson, P. G. (2004). Forced lytic replication impairs host colonization by a latency-deficient mutant of murine gammaherpesvirus-68. *J Gen Virol* **85**, 137–146.

Miller, G., Shope, T., Coope, D., Waters, L., Pagano, J., Bornkamn, G. & Henle, W. (1977). Lymphoma in cotton-top marmosets after inoculation with Epstein-Barr virus: tumor incidence, histologic spectrum antibody responses, demonstration of viral DNA, and characterization of viruses. *J Exp Med* **145**, 948–967.

Mistrikova, J. & Blaskovic, D. (1985). Ecology of the murine alphaherpesvirus and its isolation from lungs of rodents in cell culture. *Acta Virol* **29**, 312–317.

Mistrikova, J. R. H., Mrmusova, M. & Kudelova, M. (2000). A murine gammaherpesvirus. *Acta Virol* **44**, 211–226.

Moghaddam, A., Rosenzweig, M., Lee-Parritz, D., Annis, B., Johnson, R. P. & Wang, F. (1997). An animal model for acute and persistent Epstein-Barr virus infection. *Science* **276**, 2030–2033.

Moorman, N. J., Virgin, H. W., IV & Speck, S. H. (2003a). Disruption of the gene encoding the gammaHV68 v-GPCR leads to decreased efficiency of reactivation from latency. *Virology* **307**, 179–190.

Moorman, N. J., Willer, D. O. & Speck, S. H. (2003b). The gammaherpesvirus 68 latency-associated nuclear antigen homolog is critical for the establishment of splenic latency. *J Virol* **77**, 10295–10303.

Moorman, N. J., Lin, C. Y. & Speck, S. H. (2004). Identification of candidate gammaherpesvirus 68 genes required for virus replication by signature-tagged transposon mutagenesis. *J Virol* **78**, 10282–10290.

Nash, A. A., Dutia, B. M., Stewart, J. P. & Davison, A. J. (2001). Natural history of murine gamma-herpesvirus infection. *Philos Trans R Soc Lond B Biol Sci* **356**, 569–579.

Obar, J. J., Crist, S. G., Gondek, D. C. & Usherwood, E. J. (2004). Different functional capacities of latent and lytic antigen-specific CD8 T cells in murine gammaherpesvirus infection. *J Immunol* **172**, 1213–1219.

Paludan, C., Bickham, K., Nikiforow, S., Tsang, M. L., Goodman, K., Hanekom, W. A., Fonteneau, J. F., Stevanovic, S. & Munz, C. (2002). Epstein-Barr nuclear antigen 1-specific CD4(+) Th1 cells kill Burkitt's lymphoma cells. *J Immunol* **169**, 1593–1603.

Parry, C. M., Simas, J. P., Smith, V. P., Stewart, C. A., Minson, A. C., Efstathiou, S. & Alcami, A. (2000). A broad spectrum secreted chemokine binding protein encoded by a herpesvirus. *J Exp Med* **191**, 573–578.

Rajcani, J., Blaskovic, D., Svobodova, J., Ciampor, F., Huckova, D. & Stanekova, D. (1985). Pathogenesis of acute and persistent murine herpesvirus infection in mice. *Acta Virol* **29**, 51–60.

Raslova, H., Berebbi, M., Rajcani, J., Sarasin, A., Matis, J. & Kudelova, M. (2001). Susceptibility of mouse mammary glands to murine gammaherpesvirus 72 (MHV-72) infection: evidence of MHV-72 transmission via breast milk. *Microb Pathog* **31**, 47–58.

Renne, R., Dittmer, D., Kedes, D., Schmidt, K., Desrosiers, R. C., Luciw, P. A. & Ganem, D. (2004). Experimental transmission of Kaposi's sarcoma-associated herpesvirus (KSHV/HHV-8) to SIV-positive and SIV-negative rhesus macaques. *J Med Primatol* **33**, 1–9.

Rickabaugh, T. M., Brown, H. J., Martinez-Guzman, D., Wu, T. T., Tong, L., Yu, F., Cole, S. & Sun, R. (2004). Generation of a latency-deficient gammaherpesvirus that is protective against secondary infection. *J Virol* **78**, 9215–9223.

Rickinson, A. B. & Kieff, E. (2001). Epstein-Barr virus. In *Fields Virology*, pp. 2575–2628. Edited by D. M. Knipe & P. M. Howley. New York: Lippincott Williams & Wilkins.

Russo, J. J., Bohenzky, R. A., Chien, M. C. & 8 other authors (1996). Nucleotide sequence of the Kaposi sarcoma-associated herpesvirus (HHV8). *Proc Natl Acad Sci U S A* **93**, 14862–14867.

Sangster, M. Y., Topham, D. J., D'Costa, S., Cardin, R. D., Marion, T. N., Myers, L. K. & Doherty, P. C. (2000). Analysis of the virus-specific and nonspecific B cell response to a persistent B-lymphotropic gammaherpesvirus. *J Immunol* **164**, 1820–1828.

Schafer, A., Lengenfelder, D., Grillhosl, C., Wieser, C., Fleckenstein, B. & Ensser, A. (2003). The latency-associated nuclear antigen homolog of herpesvirus saimiri inhibits lytic virus replication. *J Virol* **77**, 5911–5925.

Schulz, T. F. (1998). Kaposi's sarcoma-associated herpesvirus (human herpesvirus-8). *J Gen Virol* **79**, 1573–1591.

Shope, T., Dechairo, D. & Miller, G. (1973). Malignant lymphoma in cottontop marmosets after inoculation with Epstein-Barr virus. *Proc Natl Acad Sci U S A* **70**, 2487–2491.

Simas, J. P., Swann, D., Bowden, R. & Efstathiou, S. (1999). Analysis of murine gammaherpesvirus-68 transcription during lytic and latent infection. *J Gen Virol* **80**, 75–82.

Simas, J. P., Marques, S., Bridgeman, A., Efstathiou, S. & Adler, H. (2004). The *M2* gene product of murine gammaherpesvirus 68 is required for efficient colonization of splenic follicles but is not necessary for expansion of latently infected germinal centre B cells. *J Gen Virol* **85**, 2789–2797.

Sparks-Thissen, R. L., Braaten, D. C., Kreher, S., Speck, S. H. & Virgin, H. W., IV (2004). An optimized CD4 T-cell response can control productive and latent gamma-herpesvirus infection. *J Virol* **78**, 6827–6835.

Stevenson, P. G. & Doherty, P. C. (1998). Kinetic analysis of the specific host response to a murine gammaherpesvirus. *J Virol* **72**, 943–949.

Stevenson, P. G., Belz, G. T., Altman, J. D. & Doherty, P. C. (1999). Changing patterns of dominance in the CD8+ T cell response during acute and persistent murine gamma-herpesvirus infection. *Eur J Immunol* **29**, 1059–1067.

Stevenson, P. G., Efstathiou, S., Doherty, P. C. & Lehner, P. J. (2000). Inhibition of MHC class I-restricted antigen presentation by gamma 2-herpesviruses. *Proc Natl Acad Sci U S A* **97**, 8455–8460.

Stevenson, P. G., May, J. S., Smith, X. G., Marques, S., Adler, H., Koszinowski, U. H., Simas, J. P. & Efstathiou, S. (2002). K3-mediated evasion of CD8(+) T cells aids amplification of a latent gamma-herpesvirus. *Nat Immunol* **3**, 733–740.

Stewart, J. P. (1999). Of mice and men: murine gammaherpesvirus 68 as a model. *EBV Rep* **6**, 31–35.

Stewart, J. P., Usherwood, E. J., Ross, A., Dyson, H. & Nash, T. (1998). Lung epithelial cells are a major site of murine gammaherpesvirus persistence. *J Exp Med* **187**, 1941–1951.

Sunil-Chandra, N. P., Efstathiou, S., Arno, J. & Nash, A. A. (1992a). Virological and pathological features of mice infected with murine gamma-herpesvirus 68. *J Gen Virol* **73**, 2347–2356.

Sunil-Chandra, N. P., Efstathiou, S. & Nash, A. A. (1992b). Murine gammaherpesvirus 68 establishes a latent infection in mouse B lymphocytes *in vivo*. *J Gen Virol* **73**, 3275–3279.

Sunil-Chandra, N. P., Efstathiou, S. & Nash, A. A. (1993). Interactions of murine gammaherpesvirus 68 with B and T cell lines. *Virology* **193**, 825–833.

Sunil-Chandra, N. P., Arno, J., Fazakerley, J. & Nash, A. A. (1994). Lymphoproliferative disease in mice infected with murine gammaherpesvirus 68. *Am J Pathol* **145**, 818–826.

Svobodova, J., Blaskovic, D. & Mistrikova, J. (1982). Growth characteristics of herpesviruses isolated from free living small rodents. *Acta Virol* **26**, 256–263.

Tibbetts, S. A., van Dyk, L. F., Speck, S. H. & Virgin, H. W., IV (2002). Immune control of the number and reactivation phenotype of cells latently infected with a gammaherpesvirus. *J Virol* **76**, 7125–7132.

Tripp, R. A., Hamilton-Easton, A. M., Cardin, R. D., Nguyen, P., Behm, F. G., Woodland, D. L., Doherty, P. C. & Blackman, M. A. (1997). Pathogenesis of an infectious mononucleosis-like disease induced by a murine gamma-herpesvirus: role for a viral superantigen? *J Exp Med* **185**, 1641–1650.

Usherwood, E. J., Ross, A. J., Allen, D. J. & Nash, A. A. (1996a). Murine gamma-herpesvirus-induced splenomegaly: a critical role for CD4 T cells. *J Gen Virol* **77**, 627–630.

Usherwood, E. J., Stewart, J. P. & Nash, A. A. (1996b). Characterization of tumor cell lines derived from murine gammaherpesvirus-68-infected mice. *J Virol* **70**, 6516–6518.

Usherwood, E. J., Stewart, J. P., Robertson, K., Allen, D. J. & Nash, A. A. (1996c). Absence of splenic latency in murine gammaherpesvirus 68-infected B cell-deficient mice. *J Gen Virol* **77**, 2819–2825.

van Berkel, V., Preiter, K., Virgin, H. W., IV & Speck, S. H. (1999). Identification and initial characterization of the murine gammaherpesvirus 68 gene M3, encoding an abundantly secreted protein. *J Virol* **73**, 4524–4529.

van Berkel, V., Barrett, J., Tiffany, H. L., Fremont, D. H., Murphy, P. M., McFadden, G., Speck, S. H. & Virgin, H. I. (2000). Identification of a gammaherpesvirus selective chemokine binding protein that inhibits chemokine action. *J Virol* **74**, 6741–6747.

van Berkel, V., Levine, B., Kapadia, S. B., Goldman, J. E., Speck, S. H. & Virgin, H. W., IV (2002). Critical role for a high-affinity chemokine-binding protein in gamma-herpesvirus-induced lethal meningitis. *J Clin Invest* **109**, 905–914.

van Dyk, L. F., Virgin, H. W. & Speck, S. H. (2000). The murine gammaherpesvirus 68 v-cyclin is a critical regulator of reactivation from latency. *J Virol* **74**, 7451–7461.

Viejo-Borbolla, A. & Schulz, T. F. (2003). Kaposi's sarcoma-associated herpesvirus (KSHV/HHV8): key aspects of epidemiology and pathogenesis. *AIDS Rev* **5**, 222–229.

Virgin, H. W., Latreille, P., Wamsley, P., Hallsworth, K., Weck, K. E., Dal Canto, A. J. & Speck, S. H. (1997). Complete sequence and genomic analysis of murine gamma-herpesvirus 68. *J Virol* **71**, 5894–5904.

Virgin, H. W., IV, Presti, R. M., Li, X. Y., Liu, C. & Speck, S. H. (1999). Three distinct regions of the murine gammaherpesvirus 68 genome are transcriptionally active in latently infected mice. *J Virol* **73**, 2321–2332.

Wang, F., Rivailler, P., Rao, P. & Cho, Y. (2001). Simian homologues of Epstein-Barr virus. *Philos Trans R Soc Lond B Biol Sci* **356**, 489–497.

Weck, K. E., Barkon, M. L., Yoo, L. I., Speck, S. H. & Virgin, H. W. (1996). Mature B cells are required for acute splenic infection, but not for establishment of latency, by murine gammaherpesvirus 68. *J Virol* **70**, 6775–6780.

Weck, K. E., Kim, S. S., Virgin, H. W. & Speck, S. H. (1999). Macrophages are the major reservoir of latent murine gammaherpesvirus 68 in peritoneal cells. *J Virol* **73**, 3273–3283.

Willer, D. O. & Speck, S. H. (2003). Long-term latent murine Gammaherpesvirus 68 infection is preferentially found within the surface immunoglobulin D-negative subset of splenic B cells in vivo. *J Virol* **77**, 8310–8321.

Wong, S. W., Bergquam, E. P., Swanson, R. M., Lee, F. W., Shiigi, S. M., Avery, N. A., Fanton, J. W. & Axthelm, M. K. (1999). Induction of B cell hyperplasia in simian immunodeficiency virus-infected rhesus macaques with the simian homologue of Kaposi's sarcoma-associated herpesvirus. *J Exp Med* **190**, 827–840.

Wu, T. T., Usherwood, E. J., Stewart, J. P., Nash, A. A. & Sun, R. (2000). Rta of murine gammaherpesvirus 68 reactivates the complete lytic cycle from latency. *J Virol* **74**, 3659–3667.

Zimring, J. C., Goodbourn, S. & Offermann, M. K. (1998). Human herpesvirus 8 encodes an interferon regulatory factor (IRF) homolog that represses IRF-1-mediated transcription. *J Virol* **72**, 701–707.

INDEX

References to tables/figures are shown in italics